# Whole Earth

○

## ALSO BY JOHN MARKOFF

*Machines of Loving Grace*

*What the Dormouse Said*

*Takedown*
(with Tsutomu Shimomura)

*Cyberpunk*
(with Katie Hafner)

*The High Cost of High Tech*
(with Lenny Siegel)

# Whole Earth

## The Many Lives of Stewart Brand

### JOHN MARKOFF

PENGUIN PRESS
NEW YORK
2022

PENGUIN PRESS

An imprint of Penguin Random House LLC

penguinrandomhouse.com

LIBRARY OF CONGRESS CATALOGING-IN-PUBLICATION DATA

Names: Markoff, John, author.

Title: Whole Earth : the many lives of Stewart Brand / John Markoff.

Description: New York : Penguin Press, [2022] | Includes bibliographical references and index.

Identifiers: LCCN 2021039442 (print) | LCCN 2021039443 (ebook) |
ISBN 9780735223943 (hardcover) | ISBN 9780735223950 (ebook)

Subjects: LCSH: Brand, Stewart. | Technologists—United States—Biography. |
Appropriate technology—United States—History. | Whole Earth catalog (Menlo Park, Calif.) |
Counterculture—United States—History. | Technology—California, Northern—History. |
Futurologists—United States—Biography. | Technological innovations—Social aspects—United States. |
Journalism, Technical—United States—History. | California, Northern—Biography.

Classification: LCC T40.B64 M37 2022 (print) | LCC T40.B64 (ebook) |
DDC 609.2 [B]—dc23/eng/20211208

LC record available at https://lccn.loc.gov/2021039442

LC ebook record available at https://lccn.loc.gov/2021039443

Printed in the United States of America
1st Printing

Book design by Daniel Lagin

*In memory of Douglas Engelbart, whose ideas,
like many things, Stewart Brand discovered early on*

# Contents

*"Throw a Brand in the river and they will float upstream."*

—Brand family wisdom

# Prologue

A SOLITARY FIGURE HUDDLED ON A SAN FRANCISCO ROOFTOP gazing over the city's skyline. On a wintry afternoon, sitting above his North Beach apartment, he felt as if he were vibrating. He had taken half a tab of LSD, then wrapped himself in a blanket and gone up intent on spending the afternoon doing nothing in particular.

Shaking from the chill and the chemical reaction, the man on the roof saw the world as if through a fish-eye lens and he realized that the tall buildings in the financial district weren't perfectly parallel: they diverged slightly. In his mind's eye he rose above San Francisco, and the planet suddenly became as a glorious globe.

Seeing Earth from space would transform the way we view our planet and ourselves, he realized. He spent the rest of the afternoon obsessing over how it would be possible to capture a photograph of the entire planet to bring home the point that all of humanity shared a single home. After a few hours, he headed downstairs, a question now in mind: "Why haven't we seen a photograph of the whole Earth yet?"[1]

TODAY STEWART BRAND IS LARGELY KNOWN AS THE CREATOR OF the *Whole Earth Catalog*, a compendium of tools, books, and other intriguing ephemera that became a bible for a generation of young Americans during the 1960s and 1970s. That, however, is only part of his legacy. Since then, in both small and large ways, he has reformed the way we look at the world. A quixotic intellectual troubadour, he has prosecuted a series of discrete visions united only by a potent sense of curiosity and a provocative optimism.

Earlier, he played an instrumental role in the emergence of the 1960s American counterculture, which exploded upon the American scene in many ways and still resonates in the political and cultural battles taking place today. He was in the thick of the events that led to the modern environmental movement during the 1970s. At various times he has dabbled in journalism, photography, activism, multimedia, social media, and business consulting.

He has also been a provocateur. A member of Ken Kesey's band of Merry Pranksters, he helped light the spark that led to the Summer of Love and the San Francisco music scene. Among the first to catch wind of the significance of cyberspace and personal computing, in the 1980s he helped create an early online virtual community.

Although he was largely an observer of the technical community that created Silicon Valley, his various ideas and crusades around the *Whole Earth Catalog*, which he created in the fall of 1968, foreshadow and resonate with the techno-utopian culture that the Valley spawned. He went on to rethink modern architecture from a biological perspective and later publicly broke with the environmental movement over nuclear power and GMO food. Both as a young man and more recently, he first figuratively and then literally set out to "play God," initially by making the claim that humans had the power of gods and then during the past decade by creating an organization to save and restore endangered species with modern biotechnology. He has done it all from a single base, Northern California. The arc of his life helped to create a very particular California sensibility, a state of mind that has gone on to spread throughout the entire world.

O

IN FEBRUARY OF 1966 A SQUANDERED AFTERNOON IN THE COM-
pany of a hallucinatory drug was not a completely unusual way to
spend a day. Yet Brand was not a beatnik, nor would he become a hippie. He was far too ambitious to fit in comfortably with his peers. As
often as not he has found a way to go against the grain. He has floated
upstream.

When he wandered onto his rooftop, he was a lanky twenty-eight-
year-old with already thinning blond hair, an aquiline nose, and a long
face often painted with a toothy grin. People who met him then described him as socially awkward, overly earnest, and cerebral. Six years
earlier, as a college senior, he'd discovered San Francisco's North Beach
and immediately fallen in love with its libertine world of artists and Beat
poets. An aspiring photojournalist, he was living off a small family inheritance that allowed the freedom to dive deeply into San Francisco's
bohemia.

Years later, when the two met at the Santa Fe Institute, mathematician and economist W. Brian Arthur asked Brand what he did. His response was "I find things and I found things." Many of the things that
Brand has attempted in the past six decades would fail, but some have
succeeded spectacularly. He has retained an uncanny ability to stumble
into trends early, and sometimes to create them. One of the most important roles he has played, first for the counterculture generation and
more recently for millennials, is as a model for how to live one's life.

Steve Jobs fastened onto the Brand approach in recounting his own
life in his influential 2005 Stanford commencement address. He found
inspiration in the closing page of Brand's *Whole Earth Epilog*. It captured
the sensibility that emerged on the western edge of the continent during
the sixties. The back cover of the 1974 edition of the publication was a
photograph of the kind of open road a hitchhiker might find while waiting in the morning sun on a California highway. Adventure and unlimited possibilities await the traveler. At the top of the page is a photo of

sunrise as seen from space, and in between are the words "Stay Hungry. Stay Foolish." Brand intended the photo as a way to describe living a life open to serendipity. As Jobs put it, "I have always wished that for myself."

Telling the story of Stewart Brand's life poses a puzzle, for he isn't someone who can be neatly categorized. Perhaps it is so difficult to put him in a box because he has such an uncanny knack for seeing the world from outside the box. After playing a major role in the creation of the environmental movement, decades later he would push back on the antinuclear movement, defining himself as an "eco-pragmatist." He did so because he was opposed not to environmentalism, only to environmentalists, whom he viewed as being antiscience ideologues.

Today he sees himself as a conservationist rather than a preservationist. Humans are very much a part of the natural world, he argues. They have coevolved with it and they are duty bound to care for it. His thinking has not so much changed as evolved. While he is labeled as an early techno-optimist, during the past two decades it has been the existential threat of climate change that has framed his thinking and actions.

In editing the *Whole Earth Catalog*, he embraced and celebrated amateurs. Brand turned to the world of cybernetics to reject hierarchy in preference to the idea of a networked world in which the center of control constantly moves to wherever is most "relevant and useful." It was a worldview that both resonated and broke with the New Left, for Brand rejected traditional politics and focused instead on what he called direct power—a focus on tools and skills for the individual—emerging from his early libertarian sympathies.

A self-described member of an American "silent generation," which came of age in the 1950s, Brand has acted more in the mold of the "action intellectuals," identified by Theodore H. White in a 1967 *Life* article that described a "Kennedy type" of educated anticommunist intellectual who operated beyond the boundaries of the university. However, unlike White's characters, who were largely inside-the-Beltway creatures of the Kennedy administration, Brand would become the consummate outsider, a coast away.

He has reinvented himself frequently enough that it is tempting to describe him as a Zelig, for decades turning up seemingly everywhere as new things were developing. Indeed, Brand's thinking has evolved in many ways—from anti- to pronuclear, from environmentalism to conservationism, and from libertarianism to something closer to traditional liberalism. But the Zelig reference is the wrong way to describe him, for there has been a consistent through line that has connected his various campaigns, crusades, and inquiries over more than six decades.

In particular, Brand has been constant in his commitment to science, which he refers to as the only "true news"; in his commitment to bottom-up democracy (with a small *d*); and in his relentless curiosity. He has displayed an eerie knack for showing up first at the onset of some social movement or technological inflection point and then moving on just when everyone else catches up. He has retained a remarkable aversion to orthodoxy, which has on occasion made him the target for criticism from the right, and sometimes from the left.

Brand first made a cameo in the opening pages of Tom Wolfe's *The Electric Kool-Aid Acid Test*. Since then almost two dozen books have sought to document Brand's contribution to politics and culture. In each instance some aspect of his life has been used to make a particular point, whether the point related to Native American politics, environmentalism, innovation, architecture, computer networks, social media, or personal computing.

Recently Brand has been heralded as the first internet Utopian. Indeed, in several contemporary books he has been called out as the person responsible for the libertarian ideals that grew out of the creation of the internet. The American view of the internet—and by extension Silicon Valley—performed a 180-degree reversal in the wake of the 2016 presidential election. Facebook, Google, and Twitter went from being the nation's darlings to being castigated as enablers of Russian subversion of American democracy. It is striking that both Franklin Foer in *World Without Mind: The Existential Threat of Big Tech* (2017) and Jonathan Taplin in *Move Fast and Break Things: How Facebook, Google, and Amazon Cornered Culture and Undermined Democracy* (2017) open their jeremiads with profiles of Brand as a harbinger of an emergent cyberculture that originally promised digital freedom and instead has brought a post-Orwellian state of soft control and concentration of power. (It should be noted that Foer and Taplin differ on Brand's contribution. Taplin, accurately I believe, sees Brand's original utopianism as being subverted by a later generation of internet capitalists.)

None of these critiques actually do justice to the arc of Brand's life. His restlessness has been his greatest strength, though it has taken its toll. His older brother Mike Brand was forever pained that his younger brother chose to leave family behind and not look back. "Stewart has always wanted to be at the cutting edge," he reflected, "and he is trapped there."

That has left the younger Brand isolated in some ways. The rest of the world, however, has benefited from his compulsion to stay at the edge. Throughout his life Brand has taken pride in creating new institutions. The impact of his thinking and his work has been to create what Robert Horvitz, an artist and designer who served as the art editor for the *CoEvolution Quarterly*, Brand's follow-on to the *Catalog*, describes as a "Whole Earth Culture."[2] The most profound aspect of Brand's legacy can be summarized in a single symbol of the whole Earth. In the late 1960s the image symbolized a new sense of optimism and an end to fear: it replaced the mushroom cloud as the defining symbol of an era. Even today it is unique among all the political and commercial symbols that define modern life. All of the others divide us from one another. It would await Stewart Brand and a photograph on the cover of the *Whole Earth Catalog* to conjure the significance of a unified planetary culture; that image has been his and our polestar ever since.

# Chapter 1

# Shoppenagon

PRESERVED IN A LARGE PHOTOGRAPH, HE WAS A MYTHIC PRES-
ence in the living room of the cabin where Stewart Brand spent
his childhood summers. David Shoppenagon was a Chippewa Indian and a popular hunting and fishing guide for the wealthier residents
of frontier Saginaw, Michigan, when in 1875 he brought Stewart Brand's
great-great-grandfather, former New York congressman Lorenzo Burrows, and his family to Higgins Lake. The lake would soon become the
summer gathering spot for four closely knit Saginaw families—the Bar-
nards, Brands, Burrowses, and Morleys.

In a world that was collapsing for his people, Shoppenagon was an
anomaly. Most of his compatriots had either been defeated in battle,
herded onto reservations, sickened, or starved, but he had managed to
coexist with an encroaching white civilization.

Shoppenagon lived more than a century after the period in which
James Fenimore Cooper's historical novel *The Last of the Mohicans* was
set, but in many ways, he fit the story line well. A member of a dying
race, he was an anachronism who offered the newcomers a window into
the world their civilization was obliterating. A journalist who had been
invited to tour timber country by the railroads described him as wearing a fur turban topped by a crown of eagle plumes. From his belt hung
deer hooves, eagle claws, shells. A band of leather over his shoulder was
covered with a rattlesnake skin.[1]

To the white families who encountered him, Shoppenagon represented the "good Indian."[2] He became a fixture at the summer hunting
camps, where he sold handcrafted items, including paddles, blankets,
and baskets made by his relatives. He gave visitors Chippewa nicknames, and on occasions when he visited the home of Brand's great-
grandfather, Lorenzo's son George, he slept in front of the fireplace on
the living room floor.[3]

During the summer the Saginaw families would use the lumber trains
to escape to the lake from the hot and mosquito-infested city. Established
in 1875, the first campsite, known as Lakeside, remained a seasonal encampment for several years. At first they stayed in tents, which were later
supplanted by small shacks and then eventually summer homes. Soon
other small communities were built adjoining Lakeside, including Cottage Grove, where the Brand family, part owners of Saginaw's flour mill,
summered. (The product of a marriage between the two camps, Brand
would summer at Lakeside.) Brand's grandfather Ralph Chase Morley,
heir to the Saginaw-based Morley Brothers wholesalers, ran what by
1892 had become the nation's second-largest hardware business, as well
as one of the largest horse and buggy retailers in the country. The business prospered and expanded through World War I, as the Morley
Brothers were major suppliers to the US Army. Soon the Morleys opened
a bank as well. They became part of a prosperous, upwardly mobile
business class that built Saginaw during the great Michigan timber
boom. As the frontier outpost became a thriving city, they prospered,
giving their families freedom during holidays to escape to the countryside. And so each summer the Saginaw families would come with an en-
tourage of cooks and nannies to the lake.

The family's role in the horse and carriage business might have
blinded them to changing transportation technology. Ralph Morley is
said to have turned down a neighbor, Henry Ford, who offered him the
opportunity to invest in his new motorcar company. Sometime later, another Morley, AJ, who had been sent to the West Coast to extend the
family timber interests, turned down a similar offer to invest in a new
airplane company named Boeing.[4]

Attracted by the Michigan logging boom, George Burrows arrived in Saginaw and set up a private bank in 1862, the year before the Morley brothers arrived. He would become one of Saginaw's city fathers. He was treasurer of the company that built the city's first streetcar line; he served as chief of the Saginaw Volunteer Fire Department and in that capacity oversaw the construction of the city's Water Works. He became a Saginaw investor in mining and forest land, and an early venture capitalist as well, providing a loan that helped establish the Eastman Kodak company.

The Morleys were able to navigate the logging boom and the following bust, growing their hardware supply business statewide beyond logging and creating a department store as well. Yet logging remained in the blood, and the families followed it west. Members of the Morley and Barnard families went to Washington and Oregon and created what would become the Western Logging Company. Decades later they sold it to Georgia-Pacific in a deal that some family members perceived as a giveaway.

The immigrant West Coast Brands could have been the Stampers, the tough-as-nails Oregon logging family imagined in Ken Kesey's novel *Sometimes a Great Notion*. Stewart Brand's oldest brother, Mike, spent three summers in the woods in the family logging business in the Northwest and at one time hoped that he might someday be able to take over the logging operation. A decade later Stewart would follow in Mike's footsteps.

Michigan's Lower Peninsula had been completely logged by the end of the nineteenth century, and the Indians had almost entirely vanished as well.* However, for the young man who summered each year on the shores of Higgins Lake from the late 1930s into the 1950s, both towered in his imagination. Years later the photograph of David Shoppenagon on the summer cottage wall remained an influence on Brand when he reinterpreted America's view of Indians and the role they played in caring for the land on which they lived. It would be a message that he would

---

* There is an enduring controversy over whether to call America's indigenous people American Indians or Native Americans. In keeping with Stewart Brand's original usage, I have used the former identifier.

bring to an emerging American counterculture in the sixties, and it would have a powerful resonance with a budding modern environmental movement as well.

O

IT WOULD BE A MODEST CONTINUING FAMILY INHERITANCE FROM these various commercial ventures that would give a young Stewart Brand, then fresh from the army in 1962, the ability to avoid immediately getting a job and the freedom to pursue a variety of potentially sketchy notions throughout the sixties.

The Lakeside community was rustic in the best sense of the word. Even as late as World War II there was only one radio (it belonged to Brand's grandfather Ralph) and one phone in the camp. It was a perfect summer refuge for a boy whose connection to the outdoors began early and ran deep. At age seven he would memorize *Outdoor Life*'s Conservation Pledge: "I give my pledge as an American to save and faithfully to defend from waste the natural resources of my country—its air, soil, and minerals, its forests, waters, and wildlife." Even decades later, after he'd run afoul of the modern environmental movement he had helped create, Brand could recite the pledge by heart.

He in effect ran a wild animal refuge each summer in the Michigan woods. Every Easter he would be given two baby ducklings, named Fry and Riley, which he would care for until the family left for the lake—a four-hundred-mile drive—on Memorial Day, at which point the ducks would be released into the local pond around the corner on the banks of the Rock River. Once at the lake, Brand would tame chipmunks, squirrels, and feral cats. It was not unusual to find him with a peanut tied to the line of a fishing pole as bait, carefully training a chipmunk. At one point he had one such critter so comfortable that it would ride around all day in his shirt pocket.

Brand and his siblings played in the forest and swam and sailed on a lake that a century earlier had been described by the *Detroit Free Press* as "the most beautiful of Michigan's 3000 gems."[5] A generation earlier, when their parents had first summered at Higgins, the lake had been so pristine that the vacationers drank from it directly. All of the Brand children lived for summer, when for several months, freed from the bonds of city life, they would become "free-range" kids.

It was like a Nick Adams boyhood taken from the pages of Ernest Hemingway's short story collection *In Our Time*. Indeed, Brand's childhood was in many ways parallel to Hemingway's. Several decades earlier Hemingway had grown up in the suburbs of Chicago and summered at Windemere Cottage on Walloon Lake in northern Michigan, less than a two hours' drive northwest of Higgins Lake. Hemingway spent his first twenty-one summers fishing and hunting at Windemere.

The Brand family lived a day's drive from Higgins Lake in Rockford, Illinois, a heavy machinery and manufacturing center, where Stewart's father was a partner in a small advertising agency. On Memorial Day weekend each summer, the entire family, including his mother, Julia Morley Brand, sister, Mary Clare, and brothers, Mike and Peter, would pack up and make the annual summer road trip to Higgins Lake. On occasion they would loop under Lake Michigan, and more frequently they would take the ferry from Milwaukee to Muskegon. The family would stay at the lake until September, while Stewart's father would return to the city periodically to run his advertising agency.

Brand's group of Lakeside friends nicknamed him Screwy Stewy—not in censure but because they were in awe of his nonstop scheming and offbeat ideas. It was Brand who would persuade them to leap off the woodshed behind the family cabin while holding only an umbrella as a parachute, explore one of the old logging camps that dotted the countryside, or undertake a fifteen-mile forced march to a nearby town in search of postcards.

Yet there was something more. His young friends could see it even before they were teenagers. To be sure, Brand was a natural leader. But what his young friends sensed was not simply leadership but an unceasing font of ideas that would sweep them along in pursuit of each new whim.

And occasionally Stewart Brand would have a great notion.

O

R OCKFORD, ILLINOIS, WAS THE QUINTESSENTIAL MIDWESTERN
  town.

The September 12, 1949, issue of *Life* magazine featured a photo
essay taken by the legendary Margaret Bourke-White. Her subject:
Rockford, which the magazine's editors had decided was the most typical city in America.

For Brand, it was a world circumscribed by the range of his bicycle,
including downtown Rockford, which *Life* celebrated: "Parking meters
line the streets of the shopping district. A fleet of cabs line up at the station to meet the seven daily passenger trains. On Saturdays, farmers
pour in from the surrounding country to shop while their children go
to the movies or roller-skate across the street from C.I.O. headquarters."[6]

The article divided American society into classes from "lower-lower"
to "upper-upper." The magazine chose a factory worker and his wife to
represent the bottom of the pyramid. Their home was in a trailer park
on the outskirts of town, and the magazine noted that their neighbors
included "blacks who were moving in to take unskilled jobs." At the
other end of the scale, it placed an heiress of the founder of the Gunite
Foundries Corporation, who continued to reside in the "old Forbes house"
in a "fine old residential area."

The Brand family fell somewhere in between. Stewart, born December 14, 1938, was the youngest, ten years younger than his sister, Mary
Clare, whom the family called Clare; seven years younger than oldest
brother Mike; and five years younger than his middle brother, Peter. Although Peter was closer in age to Stewart, it would be Mike whom the
youngest Brand would worship and feel a kinship with. Pete went on to
become an electrical engineer and then changed professions and became a physician. Clare married a career army officer. Stewart would
harbor a mild suspicion that he had been an accident.

Trained as a mechanical engineer at MIT, Stewart's father, Arthur
"Bob" Brand, had moved to Rockford after college to work in the ma-

chine tool industry that then dominated the city but gravitated to the advertising business. He would become a partner in a local advertising agency. Without fail he ended his letters with the valediction "73," the mark of a committed ham radio operator.

Not only had Stewart's mother gone to Vassar, but her sister and her mother were graduates as well. The school was legendary for its skeptical academic mantra, "Go to primary sources," an outlook that was repeatedly conveyed to Brand through the maternal side of his family. Julia was also an avid reader, and when the children were growing up, she would inevitably be devouring three or four books simultaneously. It was a habit that a young Brand would inherit; throughout his life, much of his outlook on the world has come from reading. If one of the children developed an interest, their mother would make sure they had all the books imaginable to pursue their subject. If Julia and Arthur Brand thought Stewart was their brightest child, they gave him few clues. He was startled once, after he'd done something particularly inept, when his father, in an exasperated tone of voice, said, "You're supposed to be the smart one!" That was news to him.

At one point so many books had made it into the third-story room where Brand would hide away and read that his father expressed concern that the house might tip over. The family was nominally a member of the Second Congregational Church in town, but after several weeks of Sunday school, Brand complained to his mother that it was screwing up his Sunday reading, so she gave him a pass and that was the end of it.

Brand grew up believing that he was part of upper-class society in a midwestern city that defined the America of the 1950s. Julia was president of the Junior League and the family belonged to both the Rockford Country Club and the University Club. He was educated mostly in private schools; his parents had attended two of the finest colleges in the nation; and his extended family was populated by department store owners, bankers, and timber magnates.

But by opting not to join a family business and settling his family away from Michigan, Brand's father had set out on his own and started fresh. That could be seen most clearly in the Brand family home, on

Harlem Boulevard, a broad avenue divided by a grassy, tree-lined median. The home itself was a solid but modest three-story affair with a steeply pitched roof, a front porch, and a driveway running down the left side of the lot to a garage.

Harlem Boulevard ran parallel to National Avenue, a leafier and quieter six-block stretch that fronted the Rockford River, where the homes were expansive mansions. The Brand home, in contrast, felt crowded inside, with a small backyard that was shaded by a single apple tree.

At the corner of Harlem and Guard Street was the home of Mr. Pennypacker, headmaster of the Keith Country Day School, which Brand attended through the eighth grade. Next door was another equally stately home where Brand's closest childhood friend, John Edson, lived.

Harlem Boulevard ran through what was a solidly middle-class neighborhood. Several miles up the river, however, the family had purchased a choice piece of property they called the Hill. Close to the Rockford Country Club and with the same elevated on-the-river status as the National Avenue mansions, the Hill was an integral part of the Brand family plan—one day they would build their dream home on the river, a beacon to signify completely establishing their clan as part of Rockford's upper class.

The reality, however, was that the Brands were on a budget. At home, neither money nor sex was ever talked about. It was made clear that private educations, while a priority, were expensive. Dinners most nights were eaten at a table in the center of a crowded kitchen. With four kids there was no room to walk around the table, which was squeezed up against the stove. And the Hill would remain a pipe dream, never built.

O

IN THE EARLY 1950S, STEWART'S OLDEST BROTHER, MIKE, WAS GET-ting straight As and playing football at West Rockford High School when his father decided that he had become too much of a big fish in a small pond and put him on a train to Phillips Exeter Academy, the elite New Hampshire preparatory school that has educated generations of the New England establishment. While a star athlete at Exeter, Mike was already looking at other horizons beyond the predictable—which at Exeter meant heading off to the Ivy League. Instead, he applied to Stanford after reading in an article in the *Saturday Evening Post* that the university had rehired its football coach even after it had lost every game the previous season. That struck Mike as remarkable: "That's the place I want to go. The place that realizes that this is a game and you don't have to be training all year."

Mike's decision to attend Stanford would prove momentous for his younger brother, first and foremost in giving Stewart and his parents a reason to head west to visit the campus. On the two-day train trip from Chicago to Oakland, Stewart, along with his parents on his first real adventure, camped out in the top story of the domed observation car of the *California Zephyr* most of the way. As the train came through the Sierra Nevada, he was transfixed by the forests and mountains that stretched off into the distance in every direction. On the final day of the journey, the train made its way out of the mountains and through the Feather River Canyon, leaving him with a deep desire to return and explore the countryside.

It was the next day, on his arrival at the California shore, when he waded into the Pacific Ocean that his fate was sealed. First the mountains and then the ocean—it all grabbed him viscerally. Even as a twelve-year-old he realized he was not long for Illinois.

O

A LMOST A DECADE AFTER MIKE BRAND WAS SENT TO EXETER, Stewart followed in his footsteps. Arriving on campus in the fall of 1954, he immediately found his way to a tall spruce tree with a grand view of the campus. *Rear Window*, the Alfred Hitchcock thriller starring Jimmy Stewart, had appeared that year, and Brand, now enamored with the idea of photojournalism, would use the spruce as a regular perch to survey the Exeter campus, occasionally with camera in hand.

In the weeks after he arrived at Exeter, there was no evidence of a homesick sixteen-year-old boy away from his family for the first time. He was studying math, history, French, English, and chemistry and waiting tables at his dormitory. His first-year roommate was a brilliant teenager named Gordon Park Baker. The son of a New York City attorney and a biochemist, Baker would go on to study math at Harvard and then philosophy at Oxford. The two became close enough that Brand would invite Baker to visit him in his spruce hideaway.

While at Exeter, Brand would be awarded a scholarship that would have paid most of his college tuition if there was a family need, but his father was too proud to accept any aid and so it would remain an honorary award. However, he soon learned that he had a financial safety net to call on. In the fall of 1954, within a week after taking the train to Boston and then on to Exeter, he posted a short airmail letter to his mother: "I hate to write for money but now it's necessary. They charge heavily for books and athletic equipment. Send whatever you think is adequate.—Stew"

*Adequate.* It would establish a pattern that would repeat itself throughout college and even after the army when, fascinated by the Beats, he settled in San Francisco. And it would continue in the 1970s, which he spent trying to reinvent himself. Stewart Brand was wealthy in an "as needed" way. Whether it was several thousand dollars in family stock from his grandfather Ralph Morley or the occasional check for $100 or $500 from home, it meant freedom to be "lazy" and follow his interests wherever they led. Those supplements changed the world.

B RAND'S FATHER HAD DECIDED ON EXETER FOR MIKE AND STEWart because of its well-known Socratic teaching approach, known as the Harkness method, in which a small group of students sit around an oval table with a teacher who, rather than engaging in dry lectures, en-

courages participation. Initially, Brand was overwhelmed by the capability and the intelligence of his fellow students—a sense of inferiority
underscored by the fact that he had dropped back a grade when he arrived, repeating his junior year of high school. His challenge was further complicated by the fact that his brother Mike had been not only a
prize student but also captain of both the football and the basketball
teams. But Exeter taught him how to study and learn independently.
Never an academic star, he compensated with a polymath's intense interest in many subjects. Moreover, he soon learned an even more useful coping mechanism that would become an operating manual for
him throughout his adult life. Brand figured out that the best way to
compete was not to follow the crowd but to instead chart his own iconoclastic path. It was simple—go where the others weren't, and you had
the field to yourself. But it was a strategy that had downsides. At Exeter, and more intensely beyond, Brand often saw attachment as a
functional need, not an emotional one. Again and again he would insist on moving forward to the next great thing—project or person—
without looking back or reminiscing.

When he arrived at Exeter, Stewart Brand was a bright and curious
young man who was a perfect match for the school's Harkness method,
which made it possible for students to build close bonds with their in-
structors. Exeter offered a deeply personal education, giving the students a close set of faculty connections right up to the principal, William
Gurdon Saltonstall, who took a deep interest in his young charges.

Brand's Exeter education not only equipped him with a set of skills
that gave him the capacity to learn independently, it also exposed him
to members of an American elite and allowed him to get close enough
to see their character. During the height of the Joseph McCarthy Red
Scare era, Robert Oppenheimer, just a year after he had been stripped
of his security clearance over suspicions of communist sympathies,
came to spend a week at Exeter and to give a campus lecture on science
and society that Brand described as "one of the most brilliant and sincere speeches I have ever heard."

Soon afterward, Brand's faculty mentor tracked him down and invited him to tea with Oppenheimer. Too intimidated to ask questions, Brand remained captivated by the physicist. He appeared to be a normal, even an innocuous, figure until Brand looked at his eyes. "He has the wisest and friendliest, and most sincere eyes I have ever seen," he wrote in his journal. Above all, he was struck by Oppenheimer's utter humility.

At Exeter, Brand's passion for conservation continued to sharpen, but with an idiosyncratic twist. His roommate had read a *New York Times* profile of a watershed conservation activist in western Massachusetts and planned to visit the man as part of his research for a history paper. Brand decided to accompany him on the visit. They traveled by train to meet with the director of the Upper Hoosac Valley Water Association. His roommate's paper focused on watershed conservation, and Brand wrote a paper profiling the activist as an example of leadership.

At Exeter, he would read *Sweet Thursday* and *Cannery Row*, John Steinbeck's lightly fictionalized accounts of marine biologist Ed Ricketts. In Rockford, Brand had become close to a high school science teacher who would take him bird-watching. That had sparked an early interest in biology. Now it was deepened; Brand was captivated by Steinbeck's portrait of an iconoclastic scientist and philosopher. He decided he wanted to be Ricketts, or someone very much like him. Reading Steinbeck from rural New England only intensified the siren call of California.

O

THE SUMMER AFTER HIS FIRST YEAR AT EXETER, BRAND TACKED a handwritten note that read "Gone Fission" on his bedroom door. He had been lent the family Pontiac station wagon, and the day after he arrived home from the East, he headed to California with Rockford friends John Edson and Hank Taylor. Edson's relatives had a small cabin on a mining claim on the Feather River—the same river Brand had glimpsed on his train ride to California—which seemed a perfect spot to search for gold. The idea of a summer road trip caught their collective imagination, and because they would be passing through much of the West on their way, Edson decided that they could take some time and hunt for uranium as well.

Brand had spent the spring of his first Exeter year meticulously planning the adventure, studying US Geological Survey (USGS) maps for hours as he traced an ideal route that included a detour through Fort Sill, the vast army base near Lawton, Oklahoma, where his brother Mike was serving in an army artillery unit. From there they would swing south to stop and see Carlsbad Caverns, and then on to the Grand Canyon before heading north to Reno and then the Feather River claim. In addition to the fishing and camping gear they brought, he purchased a Geiger counter at Radio Shack and talked his father into lending the trio two field telephones. On Brand's initiative, they also acquired a set of scarlet felt hats. "If all three of us have them," he wrote, "we would have a uniform and we could spot each other easier in the canyon."

Before they were ready to get on the road, there was one final argument. When John and Hank saw that Brand was bringing a .22 rifle on the trip, they were alarmed. Both boys were committed fishermen, but neither of them was a hunter—if anything, they were antigun. Brand overruled them and the rifle went along.

They crossed the Mississippi before midday and stopped that night at a farm outside of Springfield, Missouri. The farmer offered them food and water and a place to sleep outside under the stars. The next night they made it to Fort Sill, where they spent several days with Brand's brother.

When they arrived at Bright Angel Lodge, on the Grand Canyon rim, they read about the trail to the bottom and decided to take a look. Captivated by Higgins Lake Indian lore, Brand had ordered a pair of moccasins before the trip began that he was proudly wearing as the three of them skipped down the trail. Oblivious to the soaring temperature, they hadn't bothered to think about the wisdom of carrying some water.

It was almost ten miles and a 4,400-foot drop in elevation to the Colorado River. After an hour or so, they realized it was getting really hot and they were barely halfway down. The boys turned around and headed back uphill, but now the sun was in full force. The trio struggled to find the shade of a boulder above the trail and huddled under it. Edson and Taylor were on the edge of delirium. Finally, two rangers leading a string of burros passed them. Seeing their distress, one ranger said, "You stupid kids. Here's my canteen. Go ahead and drink it up. I'll get something at the bottom."

Back in the lodge they drank endless cups of water and celebrated with two scoops of strawberry ice cream. Recovered, in the middle of the night they hopped in the car and headed off. The road through the Arizona desert was unbelievably straight—"You could see the main street of a town 20 miles away," Edson wrote. California, here we come.

O

BY THE END OF JUNE, THEY HAD ESTABLISHED THEMSELVES AT the mining claim on the Feather River. There was a cable across the river, and they used it to get some of their luggage to the cabin, fashioned from a train caboose that had rolled down the hill from the railroad bed up above. Getting the rest of the equipment across required a lot of wading back and forth carrying loads from the car.

After several days, they determined a routine. Brand organized a more elaborate cable car system for getting across the river, and they cleaned the cabin and then spent their days alternating between fishing, panning for gold, and exploring. They met an old miner who lived on the other side of the river and had just two remaining teeth. He was very proud, however, because, as he said: "And they meet."

An early sweep of the surroundings with the Geiger counter revealed that they weren't going to strike a nuclear-age bonanza, so they settled on fishing and hunting for gold nuggets. Several days after their arrival Edson and Taylor had panned two small gold nuggets each, and the fishing wasn't much better. Three weeks after they had set up camp, they packed up and headed home, this time through the Donner Pass.

Back at Exeter after his summer adventure, Brand, now an upperclassman, took control of his room in Williams House before his new roommate arrived, quickly claiming the bed that was under the light, the desk by the windows and the radiator, the better bureau, and the larger closet.

His new roommate, Ben Mason, was nice enough, Brand thought, but an "athletic type" who lacked the intellectual qualities of Gordon Baker. He soon discovered that Mason was different in other ways as well. Returning to their dorm room one evening, the two got into a prolonged argument about the purpose of college. Brand mentioned that he was tired of people who decided to go to college just to be collegians, crowding out those who were actually seeking an education. Mason responded that the only reason to go to college was to make friends, ready oneself for a successful career, and have a good time. Indeed, he took pity on people like Brand who did not realize the value of friends, a good time, and money and who wasted their time studying as an end in itself.

Brand didn't acknowledge it, but Mason's dismissal flattered him. While for the most part a B and C student, he had thrown himself in with the campus intellectuals. It wasn't quite clear that they saw Brand as one of their ranks, but if Mason did—well, that was something.

O

THE SUMMER BRAND WAS FOURTEEN A FOREST FIRE BROKE OUT at one end of Higgins Lake and a crew was sent from the state prison in Jackson to control it. Brand took it upon himself to volunteer to help fight the blaze and was accepted as a member of the crew. It wasn't a big fire—midwestern fires were not like the conflagrations in the West. Afterward, however, he was as proud as he had ever been.

As he considered life after Exeter, it occurred to Brand that forestry and firefighting could make for a career out west. He soon stumbled upon the University of Idaho, which offered a PhD in his newly chosen vocation. Coincidentally, his Rockford friend Hank Taylor had similar thoughts, and the two struck upon the idea of studying forestry together in Idaho.

The plan was foiled, however, when Taylor's mother accidentally opened a letter from Brand about the plan. She promptly let his mother know, and letters immediately came from both his father and his brother counseling him that he might be setting his sights too low.

Additionally, his parents enlisted the aid of Darcy Curwen, a legendary Exeter English instructor who had taught Mike and was a friend of the family. Curwen had studied at Harvard and had been offered a one-year appointment in the history department but had come to Exeter to teach instead. Brand was transferred into Curwen's English class, and it was largely his new teacher who convinced him that he should set his sights higher.

"Well, you know, if you really want to get a graduate degree in firefighting, you can do that after you get an undergraduate degree somewhere else," Curwen told him.

The idea hadn't occurred to Brand and he began casting around for better schools—still in the West—adding Reed College, Whitman College, and Stanford University to his list of possibilities.

His brother Mike wrote him a letter suggesting he take a look at Reed but noted in passing that there was talk that the school had communist tendencies. Brand immediately sent an inquiry to Reed noting that "the college had a reputation of being 'pink,'" and that many of its students seemed "a shade odd." An admissions counselor quickly assured Brand that Reed's student body was no "odder" than those attending Oberlin, Swarthmore, Haverford (to name a few colleges with programs similar to Reed's), or "your own eastern private colleges and universities." The college's reply seemed to put Brand's mind at ease, and he tentatively decided to apply. However, after discussing it with Curwen that evening, he changed his mind. The next morning he sent a letter to Stanford applying for admission.

## Chapter 2

# On the Golden Shore

ONE EVENING IN SEPTEMBER OF 1956, STEWART BRAND AR-rived in the Bay Area long after the sun had gone down. Accustomed to the open roads of the West, he suddenly found himself with nowhere to easily pull off. He spent his first night at college in the back of his station wagon in a campus parking lot.

After graduating from Exeter, Brand had returned home to Rockford just long enough to pack. For the second summer in a row, he had loaded the family station wagon and driven west. This time the car was his, a present the previous Christmas.

Freshly minted from two years at an eastern prep school, Brand was something of a square peg when he arrived at Stanford. He didn't know all the dance steps and didn't quite fit in. He was a little too earnest and a little too talkative.

He was also full of ambition. Stewart Brand went to college wanting to make his mark. To do so, he followed an unusual path—in a sea of 1950s conformity, he charted his own course.

Brand's long-standing love of the natural world had deepened at Exeter after reading John Steinbeck's lightly fictionalized accounts of the iconoclastic marine biologist Ed Ricketts, and he had arrived in New Hampshire intending to study biology. But Darcy Curwen woke him up to writing as well, and so early on at Stanford, he began thinking about becoming a journalist.

Stewart's landing at Stanford was softened by his connection to an old acquaintance of his brother Mike's, a man named Dick Raymond, and his wife, Ann Meilstrup. Raymond was an entrepreneur who had befriended Mike Brand and his wife after Mike left the army to work for Tektronix, the upstart electronics firm in Portland, Oregon. Meilstrup, a painter, had been a childhood friend of Mike's wife, Gale. By the time Stewart arrived at Stanford, the Raymond family had moved to the Midpeninsula, the suburban stretch between San Francisco and San Jose, and Dick Raymond had gone to work for the Stanford Research Institute (SRI), a think tank that then was part of the university. Although SRI was a major military contractor, it was actively trying to grow a civilian consulting business and hired people like Raymond to develop new business.

Raymond had attended college at Miami University in Ohio and then obtained a Harvard MBA. He referred to himself as an economic consultant who specialized in land use, recreational economics, and community development, but his entrepreneurial spirit was expansive. (A serious runner, often putting in more than one hundred miles a week, he helped create Shoe Goo to extend the life of his running shoes, and in 1973 he started a company to sell it.) His clients at SRI ranged from the city of Menlo Park to the 1962 Seattle World's Fair—he had convinced the fair organizers that the buildings, including the Space Needle, should be permanent.

Dick and Ann had moved to the tiny neighborhood of Ladera, in Portola Valley. Decades before Silicon Valley would blossom and transform Portola Valley into a woodsy and elegant retreat for technologists and venture capitalists, it was a rustic community in the hills west of Stanford that had a distinct bohemian tinge. Ladera was located on Alpine Road, just a short drive from Perry Lane—described by writer Ed McClanahan as "the lunatic fringe on Stanford's stiff upper lip"[1]—where Ken Kesey and the Merry Pranksters first took root in the early sixties, and around the corner from where the Grateful Dead took up residence in an old mansion in the midsixties.

Prepped by Mike, the Raymonds met Brand the morning after he arrived at Stanford and promptly drove him to Brooks Brothers in San

Francisco to buy clothes for his new college career. They spent the rest of the day exploring the city. Ann sensed a spirit of mischief in Brand, which she shared. Not that much older than the college student who would frequently come to their home for dinner, she would become like an older sister.

Stanford in the 1950s was not yet the elite international institution it would become. The insular campus had little contact with the towns that surrounded it, in marked contrast to the University of California at Berkeley, across the bay, which spilled noisily and publicly out of its official boundaries. By decree, the Stanfords, its founders, mandated that the area around the school should be dry, and liquor was still not sold close to campus in the 1950s.

There were only hints of the school that would become instrumental in the creation of Silicon Valley, but during his freshman year Brand got a first intimation of what later became a key part of the region's lore. In his dorm, students were playing with an illicit technology that preceded a gadget, later perfected by a group of outlaw hobbyists, that would become known as a blue box—a simple electronic device that allowed you to make free phone calls from a pay phone. When Brand was at Stanford, placing a transistor in contact with the telephone handset and grounding it against the hinges of the phone booth produced a dial tone and—presto!—you were able to make a free phone call. Schooled in basic electronics by his ham-radio-operator father, Brand obtained a stack of the necessary transistors from Zack Electronics, a local supply store, and passed them out freely. He was ahead of his time: thirteen years later Steve Jobs and Steve Wozniak would famously meet Captain Crunch (John Draper), an early hacker, who taught them how to build blue boxes, which they then sold in Wozniak's dormitory at Berkeley.

From the start his regular letters home were peppered with optimism and superlatives: "The class this year looks the best ever"; "the top man in the philosophy department"; "the liveliest guy in the classics department"; "experience excellent"; "a damn good man." His first-quarter courses included French, English, Western civilization, biology, and ROTC, the military officer training program that was then on most college campuses. When the first essay test was assigned in Western civ, he watched as his classmates found the experience terrifying. "Hearing these complaints, I suddenly began to appreciate Exeter a great deal," he wrote in his journal. "For two years now I have regularly been having objective-essay tests covering a large area and limited to 50 minutes. Consequently, I was right at home with the first test, and I think I may have done fairly well. I wonder in what else I have a head start on the other guys here."

It took just three days to realize his optimism might have been misplaced. His first essay came back with a C grade, leaving him mortified. "Big Exeter hero: know how to study, how to take essay exams. Hah!" he noted.

He decided that he was too reflective most of the time and needed to find a way to "hurry up" his mind. But competing against his goal of learning to be a better scholar were an increasing number of distractions as he got deeper into his first year of college. For the first time in his life, he was dating (Exeter didn't become coed until 1970), although it was awkward in that particular way that 1950s college life made almost inevitable. Colleges practiced what was known as in loco parentis control of their students, and for Stanford women (far outnumbered by men) it meant they were locked in their dormitories at night. Brand's Pontiac made it possible to drive to San Francisco for a dinner date and a movie or to head up into the redwoods behind campus. At one point he even obtained a portable record player and placed it in the back seat of the car.

Brand was coming of age in a world where the operating assumption was that women were at school to meet someone and marry. For many of the men on campus in the 1950s, female students were distinguishable mostly by qualities exclusive of brainpower. (One of Brand's chemistry professors illustrated the difference between qualitative and quantitative methods by explaining, "Qualitatively you might say a certain girl is shapely; the quantitative terms would be 36–24–36.") That traditional worldview did not immediately leave Brand.

At the end of his first quarter, encouraged by his parents, he went through fraternity rush. Mike had been a member of the Phi Delta Theta fraternity, and as a legacy, Stewart had a high likelihood of being accepted as a pledge. Brand found the whole process odd and hypocritical. But there was something more to it than just that.

"The thing is I really don't want to commit myself anywhere till I have a better idea who I am," he wrote his parents. "I'm changing, I know that much and am enjoying it, but I'm not too certain of direction yet." Ultimately, it was a nonissue. The Phi Delts weren't interested in him either, nor were any of Stanford's eating clubs.

Instead, he found common ground with some outsiders—specifically Stanford's foreign students. Beginning in his freshman year he became increasingly involved in the Institute for International Relations (IIR), putting him in contact with students who were often far less provincial than the typical upper-middle-class Stanford freshman.

His classmates would later remember him as a tall, straitlaced and earnest, even "square" young man with close-cut hair. Though he came across as square, his intensity caught the attention of Vartan Gregorian, an Iranian-born Armenian who had arrived as a twenty-two-year-old freshman the same year as Brand. The two were both deeply involved in the IIR, and although they were not close, Vartanian would remember Brand for his military posture on campus. He could frequently be seen in his dress ROTC uniform, something that would permanently impress the young Armenian, who graduated from Stanford in just two years and then obtained a PhD before eventually heading the New York Public Library, Brown University, and the Carnegie Corporation.

The IIR was Brand's window into the upper echelons of the Stanford administration, and it also gave him a more intimate view of the faculty. For a while he became excited about organizing the students to evaluate the faculty. He was promptly shot down by the dean of students, who told him, "Well, that's admirable initiative. . . . But I'm not sure you realize that the faculty is actually the best judge of who's the best faculty." Brand cringed and backed down. It was something he would regret for years afterward.

Reading was almost always the portal through which Brand veered off in new directions. In the spring quarter, he was thrilled by a powerful Western civ lecture on population, food, and conservation. Things were bleak, but the lecturer speculated that there might yet be hope— South America might have enough unused capacity to support a burgeoning human population. After class, he cornered his professor and was told to read *Our Plundered Planet*. Written by Fairfield Osborn in 1948, it was one of several books that would touch off a Malthusian revival and had influenced Paul Ehrlich, a Stanford biologist who would be Brand's adviser at Stanford during his senior year.

Eventually, Brand would become far more optimistic about technology and the human condition. But he left that class on fire, writing at the end of the day: "All these things grab me right by the ambition. South America, conservation, biology. I came out of class with my eyes on the stars."

In rapid succession, two more events—one later that evening and another the following day—conspired to create a new certainty that he was ready to set off on a path to become a scientist. The campus film that night was *Our Mr. Sun*. Produced the previous year and directed by Frank Capra, it was the first of what became known as the Bell System Science Series. Blending clever animation with authoritative presentations by scientists, the films would captivate American schoolchildren on rainy days for decades to come.

"It was educational, exciting, and—yes—inspiring," Brand wrote afterward. "Maybe I'm doomed to be a scientist. Can't get rid of those damned stars!"

The next day he attended a lecture by Aldous Huxley, the author of *Brave New World*, who argued that in the future, biology would become the world's dominant science. Inspired, Brand decided that the path was inescapable. Huxley was also the author of *The Doors of Perception*, which recounted his experiences with psychedelics (in his case, mescaline) as tools to unlock an otherwise barricaded path to spiritual ascendance. It was a book that became a bible for many who experimented with psychedelics, and one that Brand would initially find valuable—and then ultimately find useless.

In the spring of 1957, he was finally baptized as a true Californian. Lying in the grass outside of his freshman dormitory, he felt the earth shake for half a minute while the sky held still. California was in the midst of a long quiet period without major earthquakes. This one measured just 5.3, but it was strong enough to kill one person and to injure forty.

He had been lying on his back under a sunny sky, talking with friends. It was a moment of an unsettling loss of trust in the stability of the ground that he had always taken for granted. He was struck by the calm sky over a suddenly very uncalm earth. It gave him a new perspective on the Golden State.

O

FROM FAMILY LORE ABOUT LOGGING IN THE NORTHWOODS, TO fighting fires at Higgins Lake, to Hollywood movies that captured the spirit of adventure in the West, Brand had long been looking to prove his character in a way that would be unlikely to be within the bounds of his comfortable and cloistered Stanford education. Although the family had sold Western Logging to Georgia-Pacific, Ed Morley was still running Tree-O, a small "highball" logging firm that was logging near Sweet Home, Oregon, and Brand successfully wrote his way into a job there. *Highball* meant moving at a fast clip, the kind of work that would provide exactly the kind of test that Brand was looking for.

With another freshman, a Stanford gymnast, he showed up in July in Sweet Home, a small logging boom town set along the banks of the Santiam River northeast of Eugene. With a third young logger, they rented a motel room, splitting the thirty-five-dollar-a-month rent (one got the bed, another the cot, and the last a mattress on the floor, though the gymnast friend quit after a couple of weeks, making things a bit less crowded).

Each day would start when the first alarm went off at 4:55 a.m., and they would be in bed and fall asleep each night a little after 9:00 p.m. They were paid $2.25 an hour. Brand was hired as a choker setter—the logger who attached a special cable end to a skyline that would pull the log up to a central area. The job required wrangling a cable through brush on a steep hillside, tying it to the downed tree, and then getting the hell out of the way.

The Tree-O logging operation was conducted on terrain so steep, it might have been illegal because of the danger of causing erosion and landslides, but Brand came away from the summer without serious injury as well as a certainty that someone had to write a *Moby-Dick* about logging. The characters were so extraordinary, so violent, amazing, and wonderful—surely the logging world was at least as interesting as whaling.

But who could do it, and how? "God, I don't know if I can do fiction," Brand wrote. "The only thing I can do is non-fiction. What am I going to do?" The dilemma hung over him for years until, after leaving the army, he read Ken Kesey's *Sometimes a Great Notion*. Kesey nailed it, Brand decided. (Kesey never logged but knew the world well—partly thanks to extensive research in the local bars.)

At the end of August, Brand spent a last day taking photos of the logging operation and then headed to Higgins Lake for a family vacation. On one of his last nights at the lake he was reading in the cabin when he heard his brother Pete outside.

"Stew, come on out! The Northern Lights are back. There's red, green, and purple. Get Mom."

He wrapped himself in a blanket and went out and lay on the boat dock on the lake. The night air was cold, and a three-quarter moon hanging above reflected off the ripples a breeze blew in on the lake. The moon was surrounded by the sheets of light that made up the aurora borealis, which stretched across the entire sky. The sky was completely alive. At one point a single shaft of light stabbed toward an apex in the north and then the lights subsided.

He stayed on the dock for a long time that night. He would later recall it as a night apt for visitations from God, he thought, maybe for murdering kings. At the least, it was a night for wonder and reflection.

O

I N OCTOBER 1956, SOVIET TANKS ROLLED INTO HUNGARY, CRUSH-
ing a revolt against the Communist regime. Brand was studying military history in his ROTC class, being trained as a soldier with the
nuclear standoff between the United States and the USSR as backdrop.
In March he wrote a gloomy three-page entry in his journal speculating
on a surprise Soviet attack on the United States, acknowledging that up
until then he had never considered that America might lose the war and
"become a distant but very important industrially [*sic*] province of the
Union of Soviet Socialist Republics. . . . That I would become part of the
Communist Civilization. That my life would necessarily become small,
a gear with its place on a certain axle of the Communist machine. Perhaps only a tooth on the gear."

If it came to war, he would fight, he decided, but not for the president
or even for democracy but rather for his "individualism and personal
liberty. If I must be a fool, I want the freedom to be foolish in my own
way—not somebody else's. I will fight to avoid becoming a number—to
others and myself."

At Stanford he read both *The Fountainhead* and *Atlas Shrugged,* and
was seduced by Ayn Rand's romantic view of free-market capitalism as
well as her view of businessmen as heroic.

It is tempting to read a great deal of significance into Brand's journal entry. However, in the mid-1950s taking a stance against communism was, to put it mildly, very much in vogue. In high school, like many
Americans, he was a fan of the TV series *I Led Three Lives*, which celebrated "counterspying" on communist infiltrators in America. He was
also part of a generation too young to have ever known the Soviets as
an ally. But Brand was anything but a textbook anticommunist. Growing up in a relatively liberal family, he feared and reviled Joe McCarthy
and his methods. Within a few years, as an army lieutenant, Brand
would have to make a personal decision about whether he actually did

want to fight against communism, not in America but in Asia—and when he was faced with that decision, he chose not to.

Nevertheless, his stated fear of communism, described in his journal soon after he'd read Eric Hoffer's *The True Believer*, was revealing. Hoffer was a German immigrant and a San Francisco longshoreman who had become an acclaimed social philosopher, and Brand was channeling his warnings about the dangers of mass movements and their threat to personal freedom.

In February of 1958, one of Brand's classmates, a crew-cut tennis player named Thomas Cordry III, picked up a high school girl who lived across the street from him in Palo Alto and shot her in the head with a .22 rifle. He then drove her body to the police station, confessed, and turned himself in. He told the police that he had been beset with the "urge to kill" for some time and had planned to rape her and bury her body, before changing his mind.

From the vantage point of six decades, a single senseless killing in a world in which mass murder with military-grade automatic weapons is routine seems almost quaint. In the winter of 1958, however, the murder stunned the Stanford campus and forced Brand to look more closely at his classmates. He sent newspaper clippings describing the killing home to his father with a reflective letter about the shortcomings of his generation. He concluded that Cordry's evil was lurking just below the surface in many of his fellow students and that the only thing keeping such insanity at bay was a binding web of friendships and family. It occurred to him that there was a basic flaw in the American way of life—the nation was moving so fast that it was getting much harder to form meaningful relationships. Even he had felt the pressure to accelerate. Toward the end of his first quarter at Stanford, he had summarized his frustrations in an essay titled "Stewart Brand, Slow Thinker." The world was divided into fast and slow thinkers, he wrote, and the way to make the transition would be through diligence and focus: "By the end of this year, you may expect to hear an extraordinary freshman who has his studying down to a science, who cools his tests with an easygoing competence, who inspires all those who hear him with his brilliant conversation, who is accustomed to making decisions with the speed and accuracy of a well-oiled calculating machine. This will be Stew Brand, fast thinker." Decades later he would still be thinking of the slow/fast dichotomy and its impact on society. It would be a perspective that would become instrumental in the formation of the Long Now Foundation, an organization dedicated to promoting long-term thinking.

Mike Brand had counseled his brother not to focus on selecting individual courses but rather to find the best professors. One of those he recommended was a charismatic professor of religion, Frederic Spiegelberg.

Brand had been more or less agnostic through his late teens, but at Stanford he had become seduced into a loose Christian mysticism by listening to Huxley and several of his professors as well as falling under the spell of Nikos Kazantzakis, the Greek novelist, socialist, and spiritual seeker. However, when Spiegelberg walked into the huge lecture hall early in the quarter and announced in his booming voice, "We study the East, not for escape, but that we may find ourself [*sic*]," Brand was duly smitten.

Spiegelberg, a German Jewish immigrant who'd arrived at Stanford in 1941, would become a significant bridge between Eastern and Western thought in America. In Germany he had studied with both Martin Heidegger and Paul Tillich, modern philosophers who had explored existentialism, phenomenology, and Christianity, and he carried the seeds of their thought with him to America. In 1949 he traveled to India on a Fulbright scholarship, visiting ashrams and meeting notable religious figures, including Ramana Maharshi and Sri Aurobindo. With Alan Watts, he helped establish the American Academy of Asian Studies in San Francisco in 1950, a powerful catalyst that introduced Zen Buddhist and other Eastern mystical ideas into the San Francisco community of Beat poets. The academy became a magnet for a remarkable group of artists during the 1950s: not only were Gary Snyder and Allen Ginsberg regular attendees but other important Beats such as Michael McClure and David Meltzer also fell into its orbit.[2]

During one of his lectures, Spiegelberg shook Brand by casually commenting, "I don't actually read the news. I think the important news shows up with contemporary poets, and so I read the Beat poets and I'm very interested in what Allen Ginsberg has to say, what Michael Mc-Clure has to say, what Gary Snyder has to say." Remarkably enough, all those people were living in North Beach, the San Francisco neighborhood an hour's drive north of the campus. Finding the Beats would eventually open a window into a world that Brand had never encountered in the Midwest. The quirky San Francisco bohemian community lit him up intellectually. Still, there was no direct path to enlightenment. That would require a couple of years and a bit of serendipity.

O

STANFORD HAD BEEN DUBBED "THE FARM" FOR A REASON. MRS. Stanford had designed the school intending to separate the students from the outside world, a sort of prep school insularity, and Brand could feel the inward-looking culture of the university closing in around him—so much so that at times it felt more like prison than palace.

Decades before the western exit from campus led to the roost of Silicon Valley's venture capitalists and the world's most expensive real estate, Sand Hill Road offered the quickest route from Stanford to the Pacific Ocean and the California coast. It was a gateway to the fog-shrouded Santa Cruz Mountains and a way out through La Honda and the ribbon of Highway 1 running along the luminescent coastline down to Big Sur and Slates Hot Springs. This was a California that offered everything Brand hadn't found in midwestern Illinois or even at Higgins Lake. Sand Hill Road became his get-out-of-jail pass.

He first discovered Big Sur at an IIR foreign policy retreat held at Slates Hot Springs (now more commonly known simply as Big Sur Hot Springs). Located a little more than an hour's drive south of Monterey on Highway 1, the springs were owned by the Murphy family, who had purchased the land in 1910 and had originally intended to create a European-style health spa once the highway was completed. The spa, however, wouldn't materialize until the early sixties, when one of the

Murphys' renegade sons was able to persuade the family to transform the hot springs into a New Age retreat.

Brand rhapsodized about his first weekend in a letter home. There were "jungles of redwood canyons," "velvety spring nights," "the magnificent steep Big Sur coast: sand, cliffs, waterfall and buried jade," he wrote to his mother.

Brand's encounters in Big Sur would introduce him to two distinct and peculiarly Northern California communities—Big Sur writers like Henry Miller and John Steinbeck, and a pioneering group of San Francisco photographers exemplified by Ansel Adams and Imogen Cunningham, known as f/64. Several years later, in the early 1960s, he became deeply immersed in both worlds, bathing himself in the Beat culture of North Beach through his friendships with Beats like the poet Gerd Stern, and Jean Varda, the European-born collage artist who had introduced Miller to Big Sur.

Brand was the student organizer for the IIR weekend, and he was attempting to be the unofficial trip photographer as well. On Saturday evening, hanging out in the lodge, the Stanford students were introduced to one of the two sons of the Murphy family. Their father was John Steinbeck's best friend, and their grandfather had delivered Steinbeck. It had even been rumored that Michael and Dennis Murphy (both Stanford graduates) were the models for the two brothers in Steinbeck's autobiographical novel *East of Eden* (something that Dennis would deny).

That evening the Stanford students drank from an "inexhaustible supply of Gallo wine." Brand wanted to photograph Dennis Murphy, who was becoming a national celebrity based on his bestselling first novel, *The Sergeant*, "but he and the wine had weaved a spell I was afraid to break," he wrote afterward. "No matter," he decided. "I got a lot out of the evening without pictures."

Brand's school year ended on a spooky note. In addition to contemplating a career as a journalist, he continued to fantasize about life as a novelist. That spring he wrote a short story about logging using the characters he had met in the Oregon woods the previous summer. The story revolved around an imaginary fatal grisly accident that befell one of the loggers on a small-scale crew like the one he had worked for.

Immediately after finishing the story, he opened a letter from Frankie Brawn, his boss in the logging crew from the previous summer—and the man he had killed off in his story—asking for any photos he had taken of one of the loggers who had worked with Brand that summer. Three weeks earlier—just when he had dreamed up the plot for his story—the logger had been involved in a bad accident and had taken two days to die. Brand couldn't shake the feeling that somehow he had been to blame.

O

IN THE SUMMER OF 1958, BRAND BOARDED THE *AROSA KULM*, A CON-verted troop transport that was one year away from the junkyard, and headed for Europe to spend time with a Swiss family in Lausanne as part of the Experiment in International Living student exchange program, which sent students abroad to live with families and be immersed in a foreign culture. He spent much of the passage feeling seasick and wrote home inquiring if it would be possible to consider coming home by plane—it was hardly any more expensive, he noted.

His group of exchange students reached port in Le Havre, France, and went first to Paris to see the sights of the city. The next day he headed to Lausanne by train.

Much of the summer was taken up with language lessons and getting to know the city. There were also bike trips and climbing adventures in the Swiss countryside. The locals got something out of it too: Brand had brought his Frisbee and helped introduce them to the sport.

Throughout the summer he had a lot of contact with the coeds who were in the program, but like his Stanford dates, the encounters were entirely cerebral. And so in August he returned to Paris with a fellow exchange student and headed to a seedy neighborhood known as Pigalle for consummation of a less noble sort. To steel themselves for the challenge of picking up two prostitutes, they had consumed a large quantity of beer—more than Brand had ever had, and when it came time to lose his virginity, all he could think about was the stale taste in his mouth. There was no excitement and everything was over almost before it began. "No fanfare, nothing romantic," he wrote later in his journal, "just a Paris whore at Pigalle." Afterward he decided that it had been a learning experience, although it was not clear what the lesson had been.

Back at school, he continued in his volunteer role at IIR. That brought him in closer contact with a tall blond woman who was a new regular at foreign student events. Joan Squires was younger than Brand, but years more worldly. The daughter of an American business executive based in India, she was horrified by being abruptly thrust into a largely non-intellectual upper-middle-class American environment at Stanford and spent her first year desperately trying to transfer to Radcliffe. Failing to find an intelligent community interested in more than beer and football, she took refuge with Stanford's foreign students and in doing so met Brand.

They were an unexpected pair. Brand, remarkably straitlaced, still had a Boy Scout quality about him. At the same time, he was intensely curious and clearly trying to escape his midwestern skin. Squires was a revelation. He found in her a mature world traveler who looked him directly in the eye and at the same time was "scary nice." And while she was two years behind him, her Western civilization professor, Peter King, and his wife, Fox, had clued her in to the North Beach cultural scene, and she already knew the world of City Lights Books, the Co-Existence Bagel Shop, the Hungry i nightclub, the Beat poets, and artists like Varda. When Brand first joined her in exploring North Beach, she worried that her friends might think him hopelessly square, the leader of a Boy Scout troop in a hipster world, but he evolved with lightning speed. This, it seemed, was water he could take to.

Before he could immerse himself any deeper, however, Brand still had to finish school and his military commitment. Although both his brother Mike and his brother-in-law were in the army, the Marine Corps sounded tougher. However, the Marine Corps demanded a three-year commitment, and so he decided to enlist in the army after graduating.

ROTC "summer camp" was in Seattle. Brand planned to attend with Jeff Fadiman, a classmate and friend who also intended to join the army. Fadiman, who was a nephew of Clifton Fadiman, the public intellectual and early television personality, was a couple of years older than Brand, but the two became close friends and drinking buddies. Fadiman was a larger-than-life figure who had spent several years abroad. Although he was an anti-Soviet, unlike Brand he considered himself a leftist. Like Brand he was thinking about becoming a writer.

Before joining Fadiman in Seattle, Brand went home to visit his parents and ran into John Edson, then a premed student at Harvard. One evening the two of them sat for a long time near the Edson family garage and discussed their common conviction that humanity, fueled by population explosion and the development of advanced weapons such as intercontinental ballistic missiles, was running out of time. They speculated about whether fascism could reemerge and agreed that some sort of cataclysm was likely inevitable.

Brand—relying on the evolutionary biology he had been studying—said he thought there were three likely outcomes: (1) Humanity would find itself in an "evolutionary cul-de-sac" in which the collective intelligent action that underlay civilization would also ultimately be its undoing; (2) the collapse would occur, but a reduced "post–*Homo sapiens*" species would survive using intelligence to offset Malthus; (3) humanity would evolve quickly enough based on technology—"books"—to offset the coming crisis.

Edson thought there was little possibility of the third path, making it inevitable that the species would die off. At first Brand reluctantly agreed, concluding that destruction was both inevitable and imminent. "Like the weather, we can do nothing about it."

Then he hesitated. It didn't make sense that there was no way out. Pursuing the weather analogy, he said, "When it begins to storm, we go inside."

"But people won't move until they get wet or until they feel the pinch, and it will be just chaos," his friend replied.

That left Brand wondering about personal responsibility. The alternative might be to just dig a hole, go off in the woods, and hide. But that left him feeling unsatisfied. Wasn't that sort of thinking the very cause of the destruction? he wondered. He believed in the priority of independent and individual decision making, but "Fuck you, Jack, I've got mine" was a morally bankrupt philosophy as far as he was concerned.

That still left the question: What could one person do? Together they decided that there was some value in messengers and that they had crucial and unique knowledge about the human condition and a duty to warn the world. The next day Brand wrote: "Suppose, just suppose, I put all this down in a book. If done well, that book would be read by the entire world. They would hate me; I would have destroyed myself. But would it save the world? Or only prolong the agony?" He added that "the judgment, the decision, is a God-like one. Who are John and I to settle the fate of the planet?"

It was the first sign of a complicated call to activism that Brand would exhibit in an on-again, off-again fashion throughout his life: his education in biology colliding with his romanticism. He had been reading Kazantzakis, whose novels included *The Last Temptation of Christ* and *Zorba the Greek*, and flush with mystical and romantic notions about the role of faith and salvation, he now wondered whether the world might not be ready for a new religion, one in which God was directly aided by man to save the species—the ultimate existential act. A decade later Brand's *Whole Earth Catalog* would begin with the premise "We are as gods and might as well get good at it."

The next evening Brand went to see the Gregory Peck movie *Pork Chop Hill*, about a fierce battle for an insignificant military objective. The hill is seized at great cost, only because the US military decides to reinforce the troops when it resolves that it will show tenacity to the Chinese.

When he got home, he sketched some ideas for a book that he would title *The Last Generation*.

"I wanted to act on the feelings and thoughts I had merely talked about before," he wrote in his journal. "The only other act possible to me was to dig a hole. Perhaps I can still do that."

He added, "I do not believe that this book will be any help at all." And then he crossed out his last sentence.

O

IN THE EARLY 1950s TWO PROFESSORS SHARED THE TITLE OF "MOST popular" Stanford faculty. One was Frederic Spiegelberg. The other was Harry Rathbun, who taught a course on business law heavy in its emphasis on personal ethics and values. Like Spiegelberg, Rathbun was a charismatic figure whose classrooms were regularly filled to the point of overflow. Brand's brother Mike and Michael Murphy, who would later cofound the Esalen Institute, had taken classes from both men and come away transformed by the experience. Now, in the winter quarter of his senior year, it would be Stewart's turn. Thinking it might be good preparation for the course, before classes began that fall he attended a two-week retreat in the Santa Cruz Mountains that focused on Rathbun's particular flavor of Christian mysticism.

Rathbun's life had been transformed when he and his wife, Emilia, attended a 1935 wilderness retreat in Canada led by Henry B. Sharman, a wealthy retired Canadian theologian exploring the historical records relating to the teachings of Jesus. After returning to Palo Alto, the Rathbuns began conducting Sharman-style study groups for Stanford students in their home. The sessions were later expanded to include a retreat at a center that was established in the mountains about twenty miles southwest of campus near the then sleepy beach town of Santa Cruz. The Rathbuns were intent on building a new religion that would draw not only from the teachings of Jesus and Sharman but from Quaker values as well (the Rathbuns were members of the Palo Alto Friends Gathering).[3] The sessions were known as the Sequoia Seminars and ultimately, in the 1970s, they would spin off a series of cultlike groups—

including the Creative Initiative Foundation, Beyond War, and Women Building the Earth for the Children's Sake—that attracted a broad, largely upper-middle-class following centered in the emerging Silicon Valley. In many cases people who joined these groups would sell their homes and personal belongings and dedicate their lives completely to these New Age communities.

Among those deeply affected by the Sequoia Seminars was Myron Stolaroff, an executive at Ampex, one of the pioneering technology companies on the Midpeninsula. On a lark in 1954, Stolaroff attended the Seminar and had his first mystical experience. As he was lying on the floor of the lodge where the group met, listening to Gregorian chants, meditating, and looking through the glass skylight at moonlit redwood trees, he felt a deep pain in his chest that left him in an ecstatic state. Heretofore a secular Jew, Stolaroff concluded that the experience was evidence that God had touched him.[4]

It was also at the Sequoia Seminar that Stolaroff, along with Stanford electrical engineer Willis Harman, first met Gerald Heard. An Anglo-Irish writer who had begun his career at Cambridge and Oxford as an academic, Heard had become a committed pacifist in the 1930s and had immigrated to Los Angeles at the same time as Aldous Huxley. In California he'd become a devotee of a Hindu religious order and continued to write books ranging from spiritual essays to science fiction novels featuring UFOs. He also developed a reputation as a mystic. (It was Heard, also a close friend of Rathbun's, who introduced Huxley to Eastern thought.) Heard led a wide-ranging discussion group at one of the Sequoia Seminar retreats, and later Stolaroff, who by then was in charge of instrumentation marketing at Ampex, became a regular visitor at Heard's home in the Pacific Palisades when he was on business trips to Los Angeles.

It was during one of his visits in 1956 that Heard carried on enthusiastically to Stolaroff about a new drug called LSD. The idea shocked the young engineer. He couldn't figure out why such a famous mystic would need a chemical substance to assist him in reaching a heightened sense of illumination. Nevertheless, Heard was fervent about the chemical and told Stolaroff about an unusual man who would occasionally come from Canada and administer the drug to both him and Huxley.

The man was Al Hubbard. Short and stocky, with two passports and a murky history of connections to both law enforcement and intelligence agencies, Hubbard was without question one of the most curious folk characters in America during the 1950s and 1960s. A true believer in the power of assisted enlightenment, he was in effect the West Coast Johnny Appleseed of LSD, avidly scattering psychedelics into a small community of technologists around Stanford, accelerating a swirling set of social, political, and technological forces that would all take shape during the next decade.

By 1961 Stolaroff and Harman would establish a research group in Menlo Park to study the relationship between creativity and LSD. One of their first subjects would be a young Stewart Brand.

O

BRAND'S EARLY FLIRTATION WITH LIBERTARIANISM AND ANTI-communism, coupled with his interest in Eastern ideas and mysticism, had collided with his increasing interest in human ecology. He staked out an intellectual position in which humans had a moral responsibility to care for the species and the Earth, a thread that would echo through Brand's thinking and acting for the rest of his life. Now, sitting with the two dozen Sequoia Seminar participants on the floor of the meeting house amid a redwood grove with a view of the Santa Cruz mountains, he argued that humanity had evolved to the point that, because of human self-awareness, each individual could contribute to and influence the species. It wasn't just a question of the actions of an individual, he said; the question of survival is one that confronts an entire species.

"We are faced continually by choices," he told the group. "Yet we never have complete knowledge of what the alternatives consist or involve. We choose out of partial knowledge and frequently are wrong. This, I maintain, is excusable and certainly educational. The only evil is not to choose. There is our responsibility." He ended by citing *The Little Prince*. In the novel the fox notes, "You become responsible, forever, for what you have tamed."

In the immediate aftermath of the Seminar, on the day he began his senior year, Brand wrote to his parents that he had been transformed:

> I know that the experience, unlikely as it might appear, did occur, for the evidence pervades my actions and thought. At Seminar, essentially, one hauls out one's self and ponders what to do with it. Some people turn the old self in for a new one. I haven't done that, but I made enough discoveries about me, altering in some places, and learned enough possibilities, exciting ideas, and practical methods that perhaps it amounts to the same thing. A money's worth was got, I am sure.

A month later he was still thinking about the relationship between humans, freedom, and their environment. What was unique about humanity was the freedom to choose (which path to follow). "Evolution is the final empiricist," he wrote in his journal. "If a species works it survives; if not, not."

O

AS A JUNIOR, BRAND HAD MOVED OFF CAMPUS TO A COTTAGE IN Menlo Park. At the time, he had proudly written to his mother that he would be the master of his world. "Essentially, I am alone to study and to sleep," he wrote her somewhat defensively. "You can't believe the pleasure it gives to come home (home!) to the serenity of this place after the peopled, frenetic days at campus." Unstated, in the back of his mind at least, was that living by himself would provide greater potential for social conquest. Despite his brief summer interlude in Paris, he remained for all practical purposes a virgin and increasingly unsatisfied with his isolation. All of this had been tremendously amplified by a bitter falling out with Joan Squires, whom he felt had shifted her attention to others. Brand sent her an angry letter, describing her as a "bitch." He quickly realized that while the outburst might have felt satisfying at the time, it was a huge mistake. He tried to justify it as irrelevant, that their relationship had already soured beyond repair—"even good though it was, it was probably lost anyway"—and doomed to circumstances beyond his control. "Love," he concluded, "I am sure, is primarily a matter of timing." Yet recognizing that he had done something stupid and costly, he sat in front of his cottage in Menlo Park and cried. Later it would become apparent that interest in a romantic relationship was more on his side than hers. Joan Squires was looking for a friend, not a lover.

Eventually he would concede that, far from successful, his time living by himself had been "sad and lonely." As a result, during his last year at Stanford he took on the position of student adviser to twenty-one freshmen who were living off campus. With a friend, Eric Field, an engineering graduate student, he moved into a small house on an alleyway on campus near fraternity row. (Jeff Fadiman, who was spending the year in Europe and Africa, was also a close friend of Field's.) Brand's mentoring responsibilities were minimal, and Field, a handsome man with piercing blue eyes and a small-plane pilot's license, jump-started Brand's social life. The two would frequently double-date, and dinners cooked in their cottage for guests were a regular event, though several nascent relationships fizzled out for one reason or another.

As unsuccessful as his social life was that year, his academic fortunes were the opposite. Inspired in part by Fadiman, he continued to think about becoming a journalist and threw himself into a course on writing for magazines. As part of his coursework, he attended a photojournalism conference in Monterey. Wearing a badge and listening to professionals talk about their craft felt as if he were crossing the threshold into a more adult world.

Charles Eames, the famed designer, was the featured speaker at the event, and he came armed with a broad selection of magazines he had purchased at a newsstand on his way to the conference. His dazzling comparisons of the photographs in the magazines awed Brand. The best photography was so evocative you could practically smell the place in the picture, Eames told the audience. Inspired, Brand left the conference certain that he wanted to capture the world in pictures more than words. Nevertheless, he started the other way around. Brand began his search for freelance work by successfully pitching the *New York Times Magazine* an article on the little-known Center for Advanced Study in the Behavioral Sciences, or CASBS, then an independent research organization hidden in the hills just west of the campus.

In an era of dramatic investment and growth for the social sciences, the Center had been established by the Ford Foundation in 1954 with the intent of breaking down the stovepipes that separated the social sciences and creating a more integrated "behavioral science." Brand spent several months making regular visits and conducting telephone interviews with scientists who had previously been involved with the CASBS, such as the University of California at Berkeley anthropologist Sherwood Washburn and Gregory Bateson, the British social scientist once married to Margaret Mead. More than a decade later, Brand's brief intersection with both men would prove to have a defining impact on his thinking.

By the end of the school year, he had a 3,100-word article titled "Fruits of a Scholar's Paradise." The piece gave a half dozen examples of collaborative social science as practiced at the Center. The first of these came from a biology seminar that had been created in 1956 by Washburn, which gave rise to a new theory arguing that it was interaction with tools that drove human evolution. "By the seminar view," he wrote, "the first tool-using creature was not at all like us, and he became what we are through a process of interaction with such cultural advances as tools."

The theory did not endure, but Brand would come in contact with similar concepts in his time spent with a young Stanford professor named Paul Ehrlich, a population biologist. He would become highly influential for Brand during the next decade, partly for his 1968 book *The Population Bomb*, a pessimistic look at the future of humanity, which Brand initially took as gospel.

Brand poured his heart into researching and writing the Center profile, but in the end, the *Times* editors passed. Casting a wider net drew
no better results. *Horizon* sent him a rejection notice, and both *Harper's
Magazine* and the *Atlantic Monthly* declined to offer assignments.

Despite his lack of immediate success, Brand's discovery of Washburn attracted him to the concept that would be first described as "coevolution" by Ehrlich four years later. A little more than a decade later
Brand would also become close to Bateson, and a conversation between
them, titled "Both Sides of the Necessary Paradox," appeared first in
*Harper's Magazine* before being reprinted in Brand's first book, *II Cybernetic Frontiers*, along with his *Rolling Stone* article about computer hacker
culture. Years later, Brand would return to the thinking of several others he encountered during this time, among them cyberneticist Warren
McCulloch, and the MIT engineers Claude Shannon and Robert Fano,
who pioneered theories regarding the measurement and transmission
of information. For the moment, however, he was just another freelancer
without takers.

○

H E HAD FALLEN IN LOVE WITH THE IDEA OF BECOMING A BIOLO-
gist when he read *Cannery Row* and its portrayal of Ed Ricketts. At
Stanford, marine biologist Donald Abbott would become Brand's Ed
Ricketts.

"Classes this quarter look the best ever," he'd written home during the
term he'd first encountered Abbott. "At long last, I'm getting what I was
looking for in biology: ecology, 'the interrelation between living organisms
and their environment. . . .' This is the 'living whole, not the dead parts.'"

In ecology, the competitive exclusion principle, named for Georgy
Gause, the Russian biologist and pioneering ecologist, was the proposition that two species competing for the same resources cannot coexist
as constant populations. However, Abbott told his students, the territory
on the hillside behind campus was shared by two tarantula species, apparently in violation of the Gause principle.

Brand had found an opportunity to do original research.

Fieldwork meant trips to the hills to check on tarantula activity. Brand would lie in fear on a cot perched next to a tarantula burrow, collecting data. On Abbott's advice, he captured samples of the two species and sent them to Bruce Firstman, an evolutionary biologist who specialized in scorpions and spiders. Firstman wrote back, identifying the two species and warning Brand to be careful in handling one of them in particular—an earlier research paper done at Stanford twenty-five years earlier had shown that the venom from the tarantula could kill a mouse more quickly than that from a black widow.

Ehrlich, his senior thesis adviser, was impressed with his research and suggested that he should publish it—a potential path to graduate work in biology. However, as Brand toyed with the possibility, another bit of serendipity arrived.

In addition to becoming a freshman adviser his senior year, Brand was appointed as a student member of the President's Advisory Committee. Shortly before one of the meetings he was chatting with William McCord, a newly arrived sociologist, and asked him if he was teaching anything interesting during the current quarter. McCord responded that yes, he was: a graduate seminar on personality. That sounded interesting to Brand, and two days later he had persuaded the professor to let him join. Soon thereafter he arrived for the class and found that almost all of the personality types for study had been assigned to other students. Only one was left, the anomic personality—"one publicly disapproved of or actively punished by society."

It was perfect! A door had opened giving Brand license to explore the North Beach beatniks Spiegelberg had alerted him to. Despite Brand's nasty letter, he was still friends with Joan Squires, and she helpfully pointed him to Pierre Delattre, a local minister who ran the Bread and Wine Mission on Grant Avenue and who had recently gained national visibility after he appeared in a *New York Times* article on the changing nature of the community.

Four days later Brand knocked on the door of the Bread and Wine Mission. There were no signs indicating he was at the right address, but a young man his age answered the door and told him that Delattre was traveling in Oklahoma.

"What's going on at the Bread and Wine these days?" Brand inquired.

It had largely been converted into the publishing headquarters for two magazines, *Beatitude* and *Underhound*, the young man replied. Brand's eyes lit up. The previous week he had attended a campus lecture by Spiegelberg on "neo-existentialist" poetry, where he had enthused about *Beatitude* as an example of the future of the art form.

Soon thereafter Brand found himself volunteering to help with the production of *Beatitude* #13. Assembling copies of the magazine, he chatted with a photographer who had shot the cover photo for the current edition and found himself talking with a young woman who like himself was passionate about Kazantzakis. Several hours later he left with Al, the young man who had met him at the door. They made their way down Grant Avenue to sell copies of the fresh-off-the-press magazine at the Co-Existence Bagel Shop, which along with the Caffe Trieste and the Coffee Gallery formed the spiritual core of the North Beach community.

It felt a bit as if he had fallen into a well and ended up in a movie. While they were in the bagel shop the editor of *Beatitude* wandered in and asked if anyone knew what the *Today* show was; he had been invited on the program for an interview. At the same time, sitting inside the coffee shop were a reporter and a photographer (both in ties) from *Argosy* magazine, preparing an article on the Beats. Brand watched as a young local couple explained to them how the "Beat" myth got started, what happened on weekends when the tourists showed up, how the police were cracking down, and how they (the locals) thought the whole thing was a laugh. The girl noted that she had recently shown up in a photograph accompanying a *San Francisco Chronicle* article, described as a dope peddler caught in a raid, even though she was a happily married and productive resident of North Beach.

Brand loved it. The kids he was hanging out with weren't Beats except on the weekends when the tourists showed up. They didn't speak like hipsters—hipster lingo belonged to the jazz musicians who were only part of the scene. They were all young and in various artistic fields, "alert, educated people with clear, intelligent, enthusiastic eyes," as he wrote to his parents that night. Beards? Well, he had seen someone who hadn't shaved that morning, but everyone else was smooth cheeked, perhaps making the scene a little more palatable in Rockford.

After dinner in Chinatown, he and Al wandered around the neighborhood with the *Argosy* journalists, and Brand got a quick education on the real world of journalism, photojournalism, and freelancing. They parted company at nine p.m., with Brand getting an invitation from Al to come back the next day for an insider's tour of "The Beach."

As he walked back to his car, he was floating. Passing by a San Francisco Victorian, he heard jazz coming from the garage. He stopped for a while and listened.

He had found his new home.

O

AT THE END OF THE QUARTER, BRAND WROTE A FINAL PAPER FOR his graduate seminar in which he proposed a typology for the members of the "Bohemian Village." "Bohemia isn't a place," he wrote. "It's a state of mind." There were the "rational," who could be further broken down into "seekers" and "rebels." There were also the "irrational," who could be split into "Joiners," "Defiant Ones," and "Kickhounds." It was clear that Brand was preparing to throw his lot in with the rational bohemians.

He continued to explore North Beach during his final spring quarter, which also brought the possibility of a serious relationship with a young woman who was looking for the kind of commitment that Brand was unwilling to give. A troubled, sweet Stanford graduate student who struggled with depression, Marty had already tried suicide several times before she came into Brand's orbit, and he could not figure out how to navigate the chasm between the devotion she was looking for and the enjoyment he got from their connection.

In June his parents arrived for graduation. He wore his ROTC dress uniform to the commencement and that evening celebrated with his parents and Marty. Later he insisted that she share a hotel room and they made love. At the time Brand described it in his journal as "my loveliest graduation present." Later he would come to see it in a different, more realistic light. He came to refer to it as a "charity ball." In the morning, they returned together for a farewell breakfast with his parents.

After graduation, while he waited to begin army basic training at Fort Benning, in Georgia, Brand took a summer job as a greeter for a foreign student exchange program, the Committee on Friendly Relations Among Foreign Students. For $250 a month his job was to meet incoming students at the airport and help them get settled. Much more exciting, he house-sat in North Beach while the owners were away in Mexico. It was summer living in grand style: three bedrooms, completely furnished with a hi-fi, and his only responsibilities were to feed the four cats and water the plants.

Best of all, Peter and Fox King took him under their wing. Not the Johnny-come-lately beatniks on Grant Avenue who were delighting, intriguing, and offending a national media, the Kings were a path to Jean Varda, a central figure in the Beat community, and to Frederick Roscoe, the owner of Discovery Books, a North Beach used bookstore that was the neighborhood counterpoint to Lawrence Ferlinghetti's more famous City Lights Books.

The day after Brand moved in, King invited him along on a boat-launching party organized by Varda across the Bay in Sausalito. To Brand, Varda represented a cultured, intellectual European world that he had only glimpsed at Stanford. Not only a friend of Henry Miller's but also the writer Anaïs Nin, Varda lived on the *Vallejo*, an aging ferry he shared with artist Gordon Onslow Ford, who had been a member of the 1930s Paris surrealist group surrounding André Breton. A year later Onslow Ford would move out and Alan Watts joined Varda on the *Vallejo*. In 1967 the boat would be the location of the "Houseboat Summit" attended by Timothy Leary, Allen Ginsberg, Gary Snyder, and Watts, where they famously discussed the impact of LSD on society.

Varda was a completely social animal who regularly threw memorable parties for a diverse group of Bay Area bohemians and hangers-on. That day's festivities centered on a World War II surplus lifeboat he'd converted into a sailboat rigged in the style of a Chinese junk with a pre-psychedelic, "eye-rattling" paint job. It took almost two dozen people to drag the boat thirty feet onto the mudflats and wait for the tide to float it. There were artists, actors, modern furniture makers, dancers, and writers. Many of the men were bearded and all of the women were pretty and striking, Brand wrote in a letter to his parents. The conversation was fascinating. "Better than the best I found at Stanford," he noted.

Among the first he met were a young couple, Jack and Jean Loeffler. (Loeffler had looked up into the rigging and seen a skinny guy with blond hair curled around the top of the mast taking his picture.) Loeffler was a jazz musician who, when drafted into the army, had become a member of the 433rd Army Band. (At one point, the band was asked to play "The Stars and Stripes Forever" while positioned just seven miles away from an atom bomb test.) After departing from the army, Loeffler made his way to North Beach, where he did a variety of things, including serving twice as a caretaker at Big Sur Hot Springs—once before the legendary journalist Hunter S. Thompson had the job and once after.

That summer, Brand was truly on his own for the first time. It was an uneven learning experience. One evening he invited Stephen Durkee, a young painter, and his wife, Barbara, whom he had met through the Kings, over for dinner. Stephen was a rising star in the New York art scene, creating pop art–style paintings on large canvases. In 1956 he had moved into a Fulton Street studio converted from a small factory that had once been a facility for manufacturing sails for ships. Durkee, who would eventually cofound a commune in New Mexico and ultimately convert to Islam, was at the peak of his artistic popularity, being described as a member of a new generation of painters including Roy Lichtenstein, James Rosenquist, Andy Warhol, Jim Dine, and Robert Indiana. His studio had previously been occupied by Robert Rauschenberg and before that by Cy Twombly.[5]

With a flourish, Brand served them beef tartare. It was only after dinner that Barbara, who was a wealthy Stanford graduate, learned that he'd made the meal from hamburger he had purchased at the local grocery store. She was aghast, but it was too late to do anything about it.

His relationship with Marty sputtered along that summer, framed by the dark places she would regularly collapse into. Eventually, she drifted away, and several months later, after he had left North Beach, Brand learned that she had been found dead in her bed at the San Francisco YWCA from an overdose of potassium cyanide. She left a note leaving her body to medical research, adding "nobody need be notified."[6]

When he read a postcard from a mutual friend passing on the news, Brand slammed his fist into a wall.

In September he took a final solo camping trip before reporting for officer basic training at Fort Benning. His destination was Hidden Lake, a small marshy depression on the north side of Mount Tamalpais in Marin County. As he drove across the Golden Gate Bridge in the late afternoon, the good feelings began to push away thoughts of the struggles he had had during his Stanford years. His clearest impressions were warm Palo Alto days and the smell of grass that stretched in every direction on campus. He discovered a grove of redwood trees, pulled over, and sat down on the matted needles in the sunshine to read *The King Must Die,* Mary Renault's coming-of-age novel portraying the early life of Theseus, the king and founder of Athens and the slayer of the Minotaur. Now it was time to leave his own Athens and face his own trials.

# Chapter 3

# Acid

WITH A BROTHER WHO HAD SERVED IN THE MILITARY AS an artillery officer and a sister who had married a West Point graduate who would become an army general and distinguish himself in Vietnam, Stewart Brand thrived in ROTC at Stanford. He had enjoyed studying military history and the close-order marching drills. He had worn his uniform with pride a decade before the Vietnam War era, when the military would be banished from American campuses by the antiwar movement.

As graduation approached, Brand faced a choice of whether to pursue graduate studies in biology or to try to make a career as a freelance photojournalist. If he had decided on biology, he could have briefly gone into the military and then stayed in the reserve for years and years to come. Instead, he chose to become a young army officer, committing to the military for two years. At Fort Benning, he set out to acquire an army ranger badge and in doing so came face-to-face with his greatest failure, one that would haunt him for the remainder of his life.

In college Brand had briefly flirted with pacifism, attracted by Gandhi's moral purity. He soon decided that while he remained opposed to indiscriminate mass killing, he would be willing to kill another individual with his bare hands. ("I gladly learn the skills," he wrote Eric Field. "One day a hitchhiker may turn on me.") His feelings were derived in part from his early libertarian stance, but they would remain complicated by the pride he took years later from his military training. Despite

a deep hatred for a faceless army bureaucracy, his decision not to reenlist and serve in Vietnam would also trouble him for decades after leaving the military. It created a personal quandary that was ultimately
wrapped up in his failure to become an army ranger.

During World War II the rangers had been constituted as an American force set up along the lines of the British Commandos. An elite
corps of soldiers, they were supposed to be the first to fight and to take
on the most difficult assignments in battle. Brand decided he wanted to
be among their ranks.

Afterward he would tell himself that if the Army Airborne School
had come before ranger training, he would have passed both. And if he
had become an army ranger, his life would have been very different.
The rangers were first deployed in Vietnam as long-range reconnais-
sance patrols in 1965. If Brand had made the cut and stayed in the military, assignment to Southeast Asia would have likely been his path. It
wasn't—because of the vicious cold of a Georgia winter.

The chill began a month before he entered ranger training. The
soldiers spent entire days outside, much of the time moving in weak sun-
light or sitting on bleachers that were hidden from the sun but not a persistent wind. Even with underwear, long underwear, a shirt with a quilted
lining, and a field jacket, the cold penetrated deeply enough to hurt.
Each day was followed by an identical one that was just as unbearable.
As the month went on, Brand was filled with a growing dread that
what he was undergoing would be nothing compared to the ranger
training to come. He would not only be cold, but he would be perpetually wet as well, forced to plunge into icy rivers and climb "frosted"
mountains.

Few things can make a person question their intentions more than
a fear of extreme cold getting even colder, and with each passing day
Brand tried harder to figure out why he was putting himself through the
torture. Obviously the objective was to learn to fight and kill. But was
becoming a killer truly in accord with his principles? That left pride, "but
it has dwindled seriously," he wrote in early December, "and shivers."

Officer training had taken just a month. He was given a brief respite, a trip home for Christmas that underscored the fact that his fervor for the military life was fading. Training was fine for what it was, and he still enjoyed it. But without a war it seemed increasingly pointless. Fantasies that he might become the next Norman Mailer and write great fiction were fleeting without real battle, and this was still peacetime.

"With no war," he wrote in his journal, "it wouldn't do for a book. For that I dismiss it. Not book-worthy, not noteworthy. I am a character in search of an author, situation, plot, and other characters. . . . I scarcely exist."

Briefly at home in Rockford, he felt a new sense of alienation from his parents' life. His father, he decided, was a good soul, but definitely quirky. He was a man with a menagerie of fears—crowds, heights, and being laughed at or belittled. He had never liked his parents' endless round of holiday cocktail parties, and that Christmas he felt completely estranged. Simultaneously he discovered that he had new coping mechanisms for tolerating the banal seasonal small talk that was part and parcel of middlebrow Rockford. "I've found my way and I recognize this other as alien to it," he wrote.

When Brand returned to Georgia for ranger training, both the cold and a new dread were waiting. It was no longer a matter of simply enduring low temperatures and cold winds. There would be freezing-cold water as well. And more runs. Seemingly endless runs. When he broke, after just two weeks of a two-month course, it was his mind, not his body, that failed.

Brand pulled off a surprising initial triumph, somehow persuading the officers running the training program to let him bring his camera on the runs and even to the pool where the trainees were forced to jump into 37-degree water from a high diving board while blindfolded, wearing their gear and ammo belts.

He was fitter now, honed by basic training. He survived the run in formation, wearing a uniform and army boots, by throwing his mind into the struggle, ignoring as much as possible the aches in his legs and the knot in his stomach. During each five-mile run, he could feel his vision narrow, and as the pain approached his absolute limit, he focused on what he would get when it all stopped—chow and mail. Though he always made it to the end, each run again triggered that voice inside asking why he was putting himself through this torture.

A week into training he had to face a round of pitiless water trials. It was not just diving into the water—you had to swim a couple of laps in all your gear. He learned to bite the air while he held his head as far out of the water as possible. He found ways to keep his mind together while doing an agonizingly slow sidestroke. Once out of the water, he would go into convulsions, knowing that in a few minutes he would be back under. He couldn't think of a single reason not to quit.

For a week he made it through each crisis through sheer stubbornness and fear of shame. Then it was announced that later that month training would include swimming across ice-rimmed rivers and sleeping in frozen clothes. Someone mentioned that a squad leader had abruptly quit that morning, presumably because of the knowledge of what was scheduled to come.

Brand was silent, but during a bayonet drill moments later, it dawned on him that all of this was a hardship imposed for no reason. He told himself, "Well, I could just quit. Huh. I guess I'll just quit." This had nothing to do with the learning and lore he had come for—it was simply pain inflicted with no reason, or worse, against reason.

During his map class, he wrote down a list of arguments on both sides, knowing that he had already made his decision. He created two columns. One was labeled "Learning" and the second "Pain." On one side were the survival skills he was acquiring; on the other, a list of things that he was forced to endure—running, cold, wet, lack of sleep, etc. Then he added a series of arguments in each column. On the pain side, he noted, "What is a Ranger in the life of writer-photographer?"

After lunch, he saw his commanding officer and walked up and told him he was leaving. "I don't go along with the principle of hardship for its own sake," he said.

Just like that, he was gone. What Brand worried about most was the phone call to tell his mother he had quit.

O

PARACHUTES ARE AN INTERESTING METAPHOR FOR LOTS OF things in life. They can offer soft landings, they can give the wearer a godlike perspective on the world, they can fail you just when you need them most—or they can bail you out. In Brand's case the parachute bailed him out.

As quickly as he left the rangers, he began parachute training. Just as the dread of the cold had undone him, his love of climbing now became his greatest asset. The parachute courses began with a series of successively higher tower jumps and concluded with five real jumps from a C-123 military transport plane. Sensing a new opportunity to pursue his goals in photography and journalism, once again he persuaded his superiors to let him take his camera along.

Jump school quickly brought his confidence back. When it finally came time to load into the C-123, Brand had a smile on his face, and it wouldn't leave him the entire time—not when he crowded into his seat, not during takeoff, not when the first five cadres jumped out the back of the plane, nor when he followed his group out into the air. He was the last one in his group to go out the door. He felt the side of the plane rush by and briefly closed his eyes, then looked up and glimpsed his parachute opening above him. There was a very strong wind, and the force of air below made him intimately aware of just how quickly he was descending. As his landing point rushed up, he tensed, put his feet together, hit the ground with a bang, and rolled to make it look better.

The five jumps were spread over several weeks, and only on the final one did he perfectly nail his landing. It was a windless day and he touched the ground so gently that he calmly fell backward and watched the bright-green chute settle directly on top of him. He stood up under it, his arms wide open, still grinning.

O

IN NOVEMBER AT FORT BENNING, EVEN BEFORE HE COMPLETED HIS
officer training, Brand received an assignment as a basic training drill
instructor with the Second Training Regiment at Fort Dix, in New Jersey.
The real work of basic training instruction was done by the sergeants—
the lieutenants simply stood around and attempted to look respectable.

Brand would spend the remainder of the two-year commitment attempting to use his skills with a camera and typewriter to garner a more
interesting posting, hopefully to Europe as a military photographer, or
perhaps a public affairs position that would allow him to continue to
write and take pictures.

While haggling with the career management officials about where
he would be posted, he worked on an article describing his experience
titled "First Jump." His efforts to sell it were unsuccessful. Adding to the
hurt, the army censors who read the article were unimpressed with his
prose. A coldly worded edit from the Pentagon information office went
so far as to single out one of Brand's favorite phrases as "ineffective and
should be removed."

He next approached editors at *Argosy* and *True* with a proposal for
an article on the Fort Benning Pathfinder Navigation training program.
"I'm afraid at this time that we are full up on war pieces," an *Argosy* features editor responded. *True* was no more welcoming.

One disappointment followed another: An opportunity to run a
weekly TV program at Fort Benning would involve a return to the South,
where he had felt as if he were living in an American apartheid. He applied to the Army Information School, but the company commander
saw that he had been a biology major and he was summarily rejected.
At the Pentagon he spent two days walking the halls trying to make his
case for a public information officer posting in Germany, only to be told
that an officer with two years' active duty could not get an overseas assignment except to Korea, which he had been warned was a hideous
posting. He offered to commit to three years of active duty to get to Ger-

many, but an army rule prevented that as well. The First Army Headquarters on Governors Island and the Fort Dix public information office
were interested in taking him but were helpless to effect a transfer.

He soon found his days filled with fury at the dull-wittedness of the
army's functionaries, followed by a growing sense of futility. In March
of 1961 he wrote to a friend, "I exist in a limbo: a training company. Neither leading nor commanding, nor held responsible for anything I can
personally control, my job is to sign things, so when fate fucks up, I hang."

It all seemed so distant from his vision of the nobility of service. One
morning he watched two dim and childish sergeants taunting recruits
by thrusting a burning tear gas bomb in the face of anyone slow in putting on their gas mask. One trainee lacerated himself in a barbed wire
fence in which he had become entangled when trying to escape the gas.
Brand later watched as the sergeants ate meat and cake for lunch while
the recruits quarreled over sparse C rations.

That evening his immediate superior pounded on his door and ordered him to get his car off base immediately because he had not properly registered it while running errands. Furthermore, he told Brand
that he had given him the same order the day before. That was news to
Brand, but he was given no chance to defend himself. That night he
went to bed thinking, "To hell with all of it!"

Soon after arriving at Fort Dix, he had won a fifty-dollar US Savings
Bond prize in a *Popular Photography* contest and, searching for a way out of
his funk, he began spending every evening in the base amateur photo lab.
In short order he struck up a genuine romance. Pamela Balme, a regular
at the lab, was a WAC, a member of the Women's Army Corps, who was a
nurse at the base hospital. Their first date was at an Edward Steichen photography exhibit at the Museum of Modern Art in New York City, and it
ended in his bed in the Fort Dix billet. The following weekend they drove
to her parents' home in Poughkeepsie and found their way to a "leafy dell"
on a blanket with a gallon of fresh hard cider. ("The blood moved. And
Spring was," he wrote pretentiously to his friend Jeff Fadiman.)

Attempting to offset his army frustrations, Brand had renewed his
friendship with the Durkees, one of the couples he had met the previous

summer around the Kings' dinner table in North Beach. Soon after arriving in New Jersey, he had helped them set up a painting studio in an abandoned church they had purchased for $6,000 in Garnerville, a tiny hamlet located an hour north of New York City. Two years later the church would serve as a launching pad for a group of artists interested in exploring electronic technologies to create a new kind of multimedia art, and Brand would for several years become an affiliate.

Now, however, he lived for the time he was able to steal away from military duty and began spending his weekends with the Durkees in Greenwich Village.

There were occasional bright spots, such as a three-week assignment to lead troops in a war-gaming exercise at Fort Drum, an army base set along the Canadian border in upstate New York. It was one of the few times he was actually able to serve as a commander. At one point he loaded up with training hand grenades and ran from side to side of the lines behind his troops, tossing the faux bombs as he went. It was the most fun he had had since joining the military.

Back at Fort Dix, however, he continued to run afoul of authority. He was reprimanded after being caught sleeping in a hut on a firing range when he was supposed to be the officer in charge. He also got in hot water for putting his camera in his desk drawer in the orderly room, accused of failing to safeguard his personal gear. His requests for more stimulating assignments only succeeded in further marking him as a troublemaker and someone who tried to avoid responsibilities.

He wrote a bitter letter to his brother-in-law, Donald Sampson, the army commandant, telling him that he was even considering applying to go to Korea and asking him for assistance in getting the army to consider his plea. He was intrigued by Japan, he wrote, and he would get to visit on R & R. Sampson was scheduled to be in Washington, DC, and Brand asked if he could meet him there to get his aid in pressing his case. He signed the letter "Unorthodox, Individual, Troublemaker." Despite the trip, he was even turned down on his request to be transferred to Korea. His brother-in-law was unable to help him. The reason was simple. "Policy." Nothing would come of his appeals for the next half year.

He had been counting on a month-long leave to fly to Europe, buy a VW bus, and travel with Helmi Klee, a pretty young German woman he had met in San Francisco the previous summer. At the last minute his base superiors restricted his leave to just ten days, and then deemed him essential personnel and chose to not let him travel at all, forcing him to cancel his trip. (Klee was gracious enough to purchase the car for him and ship it to the States. She named it Red Skin, which stuck, at least for a while, until Brand repainted it in San Francisco.)

That summer things got worse. He and Pamela Balme were able to get away only infrequently. They especially treasured the weekends they spent in Greenwich Village staying at the Durkees' loft. One weekend they had the place to themselves and were just finishing a corn-on-the-cob dinner when the phone rang. The caller asked if Barbara was there. Brand said she wasn't, and the caller said, "Stephen has a couple of my sketches and paintings he was going to give me advice about. He didn't tell me he was leaving."

Without suspicion, Brand responded that if he wanted to come by to pick up his stuff, that was fine. They unlocked the downstairs door and a few minutes later the caller appeared on the landing of the loft accompanied by two other men, all in street clothes. One identified himself as a major in the army's Criminal Investigation Command and asked if they could look around.

The visitors searched the apartment, going through some of Brand's possessions. It became obvious that they had done their research before showing up. Eventually the friendlier of the three gave Brand some fatherly advice: be discreet with WACs and stay away from condemned Greenwich Village lofts. Brand offered them coffee, but they suddenly seemed as if they were in a hurry to leave. He was left with a creepy feeling that they had left a microphone to eavesdrop on them.

Afterward, Brand's commander persuaded the higher-ranking officers at Fort Dix not to discipline him for "fraternization," and Balme was still accepted into Officer Candidate School. When she left for officer training in the fall, the two continued to exchange friendly letters, but they never met again. The army's obsession with keeping the two separated rankled Brand, however, and deepened his growing animosity toward the military's bureaucracy.

Envious of Balme's challenging assignment, Brand decided that he would give the army's Green Berets a try—new skills, restored pride, risky individual work with foreign partisans, the opportunity to wear a cocky beret, and perhaps a great source of material should he pursue a journalism career. The idea was promptly shut down when he was told that he couldn't apply until he had eighteen months of troop training.

Later during the summer, he saw *Question 7,* a movie that dealt with dissent and political repression in East Germany. Afterward, he reread what he had written about communism and choice as a college sopho-more and decided that it held up well, though he now suddenly saw it in a new light. Right here in America he was experiencing regimentation, pettiness, fear, and institutionalized stupidity—all of the attributes he hated about communism.

"I want to say a curse on both your houses," he wrote in his journal. "I want to simply disengage, retreat, hide, and tell everyone else to do the same because that's the only way left, and maybe if everyone does it, no wars will be fought for lack of soldiers."

A moment later he vacillated. There were good things in America. What he was looking for was a cause worth fighting for. "Pray, what cause is that?" he wondered.

By that fall, Brand was sleepwalking through his day job as an offi-cer and had come to feel like a cooperating prisoner. In October his dis-may had crystallized to the point that he wrote in clear printed letters in his journal: "I detest the Army." Each weekend he would head in a different direction—to Boston to spend time with his brother Peter, then teaching at MIT; to Philadelphia by himself to see a play; to Virginia for a weekend with Jeff Fadiman.

Between Stanford and the army, Fadiman had spent a year in Europe trying his hand at freelance journalism. What that year had amounted to, however, was a running tally of his female conquests in different cit-ies, a number of them described in jaunty letters he sent to Brand. At one point he'd had two girlfriends in Europe simultaneously and one of them was a Danish communist. Serving in the army as an intelligence officer, he was paranoid that his bosses would discover his dalliance. At the same time, now back in America, Fadiman was quite lonely, and so he welcomed the opportunity to double-date for the weekend with Brand in the Virginia countryside.

That fall, after months and months of failing to convince the brass he was worthy of working as a military photographer, Brand finally found a connection who might be able to pull strings in his favor. His savior was Henry Grossman, a young professional photographer who would go on to fame for his photographs of President Kennedy and the Beatles, but who was serving his time as an infantryman at Fort Dix when Brand met him in the base darkroom. Leaving Fort Dix for DC, Grossman said he would see what he could do for Brand. A few weeks later he sent him a note on White House Press Room stationery saying that he had spoken with Vice President Johnson's military aide and several other military public information officers with Brand in mind. It was a single ray of hope in an increasingly gloomy winter.

O

B RAND WAS BACK AT FORT DIX IN JANUARY AFTER A LONG-DELAYED European holiday spent mostly in Switzerland, and in the space of a single phone call his outlook on the world changed. Seemingly out of the blue, he was told to report to the Pentagon on a one-month assignment for a tryout as a photographer in the press office.

Working at the Pentagon was everything that Fort Dix wasn't. As a member of the audiovisual staff in the office of the chief of information, he reported to Hubert Van Kan, a cigar-chomping colonel whose hobby was building models of Civil War cannons and who was also a skilled editor. Brand's assignments were a mixed bag, mostly run-of-the-mill publicity shots. Early on, he was sent to the White House to photograph the president's brief meeting with a visiting group of WACs. When Kennedy arrived, he walked straight toward Brand, who backpedaled furiously, missing his opportunity to shake hands with his commander in chief. (The White House photographer, however, captured a photo of a young Second Lieutenant Brand looking on eagerly while JFK spoke with a group of swooning and giggling military nurses.)

He found an apartment not far north of the White House in one of the city's few integrated neighborhoods. Fadiman was living four blocks away. At work at the Pentagon, Brand set about taking photos and sometimes writing stories as well.

Washington wasn't San Francisco. He found none of the beauty, none of the people, none of the happiness he had experienced there, nor anything resembling a bohemian sensibility. But Camelot was in full swing, and even beyond the Kennedys there were ideas, culture, elegance, and style. To take advantage, he recalibrated and set about "going posh." Brand decided he would frequent the galleries, museums, theaters, and sophisticated homes that he now had access to, and he began to entertain frequently at his apartment. He also began to build up an arsenal of photographic gear and related instructional books.

After a promising start, he mostly sat at his desk, waiting. A photographer's life was much better than that of a basic training instructor, but a bureaucracy remained a bureaucracy, he realized, and often his major occupation was nothing. He seethed. Then suddenly he received an assignment to hop on a helicopter to photograph a session at the Judge Advocate General's Legal Center and School at the University of Virginia. Then to North Carolina to take photographs for a book called *Special Forces*, which was published by the Pentagon, and next a trip to Germany to take part in the production of a propaganda film. Such bursts of excitement were few and far between, however.

Brand had managed to twice extend his original assignment to DC, but in April of 1962 he was called back to Fort Dix. He returned with glowing letters of commendation for his work in the information office but also with a growing realization that he was a bad fit for the military. He was promoted from second to first lieutenant, but while he still occasionally thought about extending his military stay, for the most part his fantasies focused on returning to San Francisco.

Before going back to New Jersey, he had several long talks with Colonel Van Kan about his future in the military. There was a clear sense in the halls of the Pentagon that the war in Vietnam was gathering steam. In January US military helicopters had ferried South Vietnamese troops into battle against the Viet Cong for the first time. Joining the Special Forces would require a return to Fort Benning to repeat the ranger training course—hopefully, this time in warmer weather. Another possibility was to reenlist and serve as a photographer accompanying the troops in Vietnam. The rub was that both options would require a new three-year obligation to which he was unwilling to commit. There was some romance in the idea of going to Southeast Asia as a military photographer, but by the time he was forced to make a decision he had become embittered by his experience as a drill instructor at Fort Dix. He had come to believe that the army as an institution was corrupt, and he rejected it.

Still, he was torn. He had admired what he had seen when he photographed the Green Berets in North Carolina. The idea of Special Forces troops assisting righteous rebels in the Third World appealed to him. However, California was again beckoning, and he walked away from the idea of a military career. In April he headed back to Fort Dix and to the drudgery and mindlessness of the army training regimen for what he thought would be another half year. The "monkey gods" had won again, as he wrote a friend, if only temporarily.

O

FORCED TO RETURN TO HIS LIFE AS A DESK-BOUND PLATOON leader and a basic training instructor, Brand attempted to draw a sharp line between army life and the civilian world he was preparing to reenter. He was supposed to be residing at the bachelor officer quarters, the monastic rooms that contained a cot and a desk for each officer. But he described living on the base as living in an "Olive Drab Chamber Pot." Instead, he rented an apartment in a rural barn three miles from New Hope, Pennsylvania, almost an hour up the Delaware River from Fort Dix. His landlords, the Crookses, were an open and friendly Quaker family who were happy to take in a soldier as a renter. Mr. Crooks was a designer and illustrator, and his wife would frequently drop by with a daffodil or a fresh-baked loaf of bread.

Far from the highway, Brand's place in the country was set among dense hedges overlooking a meadow where Black Angus cattle wandered. A relic of the eighteenth century, the barn was 215 years old with eighteen-inch-thick stone walls. Mr. Crooks had restored it as his studio with an apartment. Soon after arriving, Brand made a quick inventory of the wild-life that surrounded his barn, noting pheasants, rabbits, squirrels, roosters, and hens, as well as birds ranging from cardinals to a brown thrasher.

Time away from the base was precious and he spent some of it with the Durkees, who were completing their move upstate to the converted church in Garnerville. On the weekends when he wasn't away in New York, he took a photography job for a rural playhouse in Bucks County.

On base he tried to put his head down and serve his time without complaint, but his discontent bled through, darkening his last months. He found a Greek poem translated by the San Francisco writer and poet Kenneth Rexroth and used it to sign off on his letters to friends:

> *Will, lost in a sea of trouble,*
>
> *Rise, save yourself from the whirlpool*
>
> *Of the enemies of willing.*
>
> *Courage exposes ambushes.*
>
> *Steadfastness destroys enemies.*
>
> *Keep your victories hidden.*
>
> *Do not sulk over defeat.*
>
> *Accept good. Bend before evil.*
>
> *Learn the rhythm which binds all men.*

—Archilochos

Living in New Hope meant getting up before four a.m. and commuting to the base to take reveille in the dark. Each morning he would stand in front of his troops, his voice calling out, "Present arms! Order arms!" And then his sergeants would respond, "All present and accounted for! All present and accounted for! All present and accounted for!" He would respond, "Sergeants, march your troops to breakfast!" "Sir!" and off they would go. Several times he was late, and on those occasions the sergeants would fake his voice in the darkness.

Brand's reputation as a malcontent eventually reached his brother-in-law, Donald Sampson, who by that time had become the commander of Fort Lewis in Washington. Sampson had fought in Korea and would go on to proudly serve in Vietnam as well. He had made several efforts to help his younger brother-in-law with a favorable posting but drew the line at pulling rank, and now Brand was afraid that Sampson had come to view him with contempt. Nevertheless, he remained disenchanted with the small-minded military bureaucracy that ruled his life.

His growing frustrations with military discipline were compounded by Colonel John Szares, a heavily decorated officer who had fought in both World War II and Korea. He arrived at Fort Dix and took charge of the junior basic training officers. Szares took a dislike to Brand in their first meeting and, after an intense confrontation (partly about Brand's devotion to photography), declared war.

Brand's youthful turn to anticommunism has been described as a source for his later libertarianism and the making of the do-it-yourself attitude that permeated the *Whole Earth Catalog*. However, for Brand communism was always a rather abstract threat. In contrast, Szares was both the real and immediate human face of oppressive bureaucracy.

Decades later Brand would come to have a gauzy view of his time in the military, retrospectively finding the discipline bracing and his training to be among the best learning experiences of his life. While he was a soldier, however, his discontent was palpable. He wrote Stephen Durkee, "The death merchant, the pain merchant, the bully grown bigger, who in his inadequacy and cowardice and sickness turns against man, killing not indiscriminately but with careful deliberation all he knows is best in man. He is the great misuser—using his life to murder life, his intelligence to murder intelligence, his freedom to murder freedom." He concluded, "I learned that in this culture the enemy will often work from behind a committee, which he can manipulate to his ends without taking any of the responsibility—a position of purest hypocrisy."

When he had first returned to Dix in April, he stayed away from photography, indignant that the army seemed so intent on frustrating his passion. Gradually, however, he returned to the base darkroom when he could and also built his own darkroom in his apartment in New Hope.

Feeling the need for professional education, Brand attempted to use a short vacation leave to sign up for a photography workshop with Ansel Adams in Yosemite Valley but was told that the class was full for the year. Instead, he was accepted into a weeklong summer workshop in Denver run by Minor White, another well-known photographer who, along with Adams, was helping create a photography department at the San Francisco Art Institute. Brand asked the base higher-ups for permission to attend and was immediately crushed, his application promptly "returned without action due to shortage of CO grade officers during this period."

Driving home to New Hope after his application was rejected, he was full of bottled rage. He felt trapped. But perhaps there was one last thing to try. The army's bureaucracy had a soft underbelly—you just had to find it. And now Brand did.

There were, in some cases, allowances made for education. With nothing left to lose, he applied for early release from the service to begin classes at the San Francisco Art Institute and San Francisco State College in the fall. Magically, this time he succeeded. He was freed from the army to start classes and to register for the reserve at the Presidio in San Francisco.

This might have ended much differently. If he had finished ranger training or if the army had let him work as a photojournalist, he might have stayed on. And in that case, in all likelihood, he would have found his way to Vietnam at the very beginning of America's war in Asia.

Instead, at the end of August Brand set out to return to the Promised Land for good. He packed his VW bus and headed back to California.

O

B RAND'S ADMISSION TO SAN FRANCISCO STATE COLLEGE AND the Art Institute had been aided by Gerd Stern, the Beat poet, who wrote a series of introductory letters. Stern had left San Francisco for rural New York and was living near the Garnerville church where the Durkees were setting up shop. He was busy creating elaborate mechanical/electronic sculptures and Brand had connected with him.

Stern owned a converted navy laundry barge that was moored at Waldo Point on the Sausalito waterfront. The barge, named the *Holland*, was thirty feet by one hundred feet, in ill repair, and Stern was actively trying to sell it; in the meantime, he offered to rent it to Brand. (The *Holland* later became the stuff of legend when Allen Ginsberg claimed that he'd lent Stern a forty-thousand-word, single-spaced letter written by Neal Cassady that was supposedly the inspiration for Jack Kerouac's *On the Road*. In a 1968 interview in the *Paris Review*, Ginsberg accused Stern of losing the letter by tossing it off the deck. For years Stern denied the charge, and eventually the letter was discovered in an attic and sold at auction in 2017.)

Though it was infested with fleas, the *Holland* was an ideal home for Brand. It had a romantic fireplace, a fine kitchen, a huge bed, and a view out into Richardson Bay. The deal was that he would rent his friend's houseboat for several months while he looked for an apartment in San Francisco. Soon after arriving, he battled the fleas to a standoff by piling eucalyptus branches in all of the corners of the barge.

Brand leaped back into California living. He never became a great sailor, but he refused to cease trying. He bought a Sunfish, a small single-handed sailing dinghy that he knew well from his childhood at Higgins Lake, and launched it directly from the deck of the barge.

But San Francisco Bay was not Higgins Lake, and he occasionally found himself in trouble. On a beautiful fall morning, he decided to go out for a quick sail before breakfast. Exhilarated by the sight of the Golden Gate Bridge, he rode a fifteen-foot swell under it and kept going. When, hours later, he turned to head home, he was greeted by an epic ebb tide that swept him farther and farther away from the bridge. The sun was setting and he began to worry. He turned around and with a good wind he figured that he would soon sail to shore, but while he sailed fast over the water, the tide kept him from moving forward.

He hailed a nearby sailboat and asked for a tow. They threw him a line and he took down his sail and ran the line through a loop of metal on the bow of his boat. To his horror, the nose of his boat dived into a wave and kept going under. He released the line and came back to the surface and watched the other boat sailing off and waving "good luck!"

Now his sail was down, and the current was still carrying him toward Hawaii. He saw one last boat motoring in, and he hailed it. This time he got aboard the other boat before he tied a line to his Sunfish. When he finally got back to the barge, he collapsed. He was exhausted and sunburned, but he had survived.

**B**RAND'S CLASSES AT BOTH SAN FRANCISCO STATE AND THE ART Institute were electrifying. He took a black-and-white photography course on the view camera and another on color photography, a design course, a class on medieval art, and a poetry course taught by Kenneth Rexroth, the translator of the poem Brand had used to sign off on his letters.

Before he left the East Coast, he had presented himself as a potential freelance journalist both to the *New York Times* and to an editor at *Horizon*. Now he began pitching Bay Area publications with ideas for photographic spreads and accompanying articles. He found no takers.

At the same time, he began exploring Northern California, taking long hikes and wandering the coast in his VW, still named Red Skin. Lynda, a friend from DC who had ended up being a regular romantic visitor at his New Hope farmhouse retreat, showed up on her way back from a trip to Japan. He had looked forward to the idea of a serious relationship, but post-Asia they weren't able to easily reconnect, and she left after just five days. Shortly afterward, the cat he had acquired ran off as well.

All lives are shaped by serendipity, but in California Brand seemed to have an uncanny sixth sense for being in the right place at the right time. Henry Miller's *Big Sur and the Oranges of Hieronymus Bosch* had been published the same year Brand discovered Big Sur with the foreign students during his sophomore year at Stanford. Now, five years later, he arrived just two weeks after the publication of Jack Kerouac's autobiographical novel *Big Sur*, in which he describes several visits to Lawrence Ferlinghetti's cabin set above the Pacific.

Certainly Brand didn't know what to expect when he wheeled his VW bus south from Sausalito to Big Sur just two weeks after the publication of Kerouac's novel. Nonetheless, his timing was remarkable. Whether it was "consciousness expansion" or "intelligence amplification," something was afoot in Northern California at the beginning of the 1960s that would be instrumental in both the creation of the sixties counterculture and, in the 1970s, the formation of Silicon Valley. The spectrum extended from the spiritual, mystical, and chemical—"instant mystic"—paths to mind expansion, to the pragmatic access-to-tools philosophy that Brand pioneered in the *Whole Earth Catalog* and that would be best expressed by Steve Jobs in the 1980s when he described the personal computer as a "bicycle for the mind." Brand's Big Sur weekend would point him in a radical new direction, a path that ultimately contributed not only to the emergence of the counterculture in Northern California but also to the birth of a new environmental movement.

The human potential movement can be dated to the fall of 1962 in Big Sur, at what would soon be named the Esalen Institute. (The name derived from the Esselen, who populated the Big Sur region before Spanish colonization.)

Esalen's founders, Michael Murphy and Richard Price, had distributed the first brochure for their new institute immediately before Brand arrived on his visit. Its cover featured an infinitesimal calculus equation from Bertrand Russell and was titled "The Human Potentiality," a notion that came from a 1960 lecture in which Aldous Huxley had argued that it might be possible to unleash the latent 90 percent of the untapped human intellect.[1]

That was very much the theme of the first Big Sur Hot Springs seminar that Brand stumbled into in September of 1962. At that point, the weekend was confined to an academic discussion led by Willis Harman, who had the previous year helped launch the International Foundation for Advanced Study, to explore LSD and human creativity, in Menlo Park. It was one of several spin-offs from the Sequoia Seminars, in which a number of the founders had participated. The foundation had begun offering an LSD experience as part of the experiment they were running to determine whether the psychedelic drug fostered creativity. For $500 (the equivalent of more than $4,500 in 2021), the foundation would take you through a carefully guided LSD trip.

Harman's assistant in organizing the Esalen seminar in September was Jim Fadiman—Jeff's younger brother, who had briefly met Brand when visiting DC. Fadiman had been an undergraduate at Harvard, where he studied social relations, but he focused most of his energy on being an actor. After graduating in 1960, he spent a year in Paris, and while he was there, Timothy Leary and Richard Alpert (who would later become Ram Dass), along with Huxley, passed through on their way to deliver an academic paper on psychedelics in Copenhagen. In Paris, Alpert, who had been Fadiman's professor at Harvard, told him, "The greatest thing in the world has happened to me and I want to share it with you," and he proceeded to pull a small bottle out of his pocket and introduce his former student to LSD.

Forced back to America a year later by the threat of the draft, Fadiman moved to California and began graduate work at Stanford. Ultimately he would become the bridge between East Coast and West Coast groups who were independently experimenting with psychedelic drugs. He would also go on to carry the torch for LSD for many years, becoming the public face of the microdosing scene that swept through Silicon Valley beginning in 2011.

Fadiman arrived at Stanford as a distinctly unhappy graduate student in 1961. In school only to avoid the military, he was feeling that it was a waste of his life; he would rather have been in Europe. Having recently been introduced to psychedelic drugs, he now saw the world as a much different place. Full of self-pity, he began leafing through the Stanford class catalog looking for something that might be interesting to study. He found a small section of cross-disciplinary classes, including one being taught by Harman, called "The Human Potential." The class was to be a discussion of what was the highest and the best that human beings could aspire to. Fadiman was intrigued.

That morning he walked across campus to the electrical engineering department to visit Harman. The man he introduced himself to looked like a straight and conservative engineering professor. When Fadiman asked if he could enroll in the interdisciplinary course, Harman replied that it was already full for the quarter and perhaps he should think about it for the next quarter.

"I've taken psilocybin three times," Fadiman said quietly.

The professor walked across the room, shut his office door, and said, "We'd better talk."

In the end, Fadiman became Harman's teaching assistant. He also soon became the youngest researcher at the newly founded International Foundation for Advanced Study.

Brand had been intrigued when he initially encountered the younger Fadiman on the East Coast. Now he was wild with enthusiasm. "I went into the seminar clothed in all of the scientific skepticism I could muster," he wrote to Stern after the weekend, "listened to the nine or so people who had had LSD enlightenment, and came away with only one doubt: where could I scare together the $500 necessary for the month-long process in Menlo Park."

He'd likened it to skydiving with his body, but he added that these people were skydiving through the universe with their minds. "When they spoke about what they saw and felt," Brand rhapsodized, "it was poetry they spoke."

O

STERN'S BARGE SOON PROVED TO BE LESS OF A PARADISE THAN IT had first appeared. For one thing, a faulty sewage system was not easily repairable; and for another, the fleas proved daunting. When Fox King's mother moved back to Europe, and her apartment, located just down the hill from Peter and Fox's, opened up in the heart of North Beach, Brand jumped at his chance to move to the city. The Kings had been an important avenue from Stanford to a much wider world, including among many others Jean Varda, who continued to hold his Sunday afternoon parties and cruises on the Bay. With a sweeping view of the city from their window facing north, the Kings held regular dinner parties that might include four people or twenty.

Brand's apartment at 570B Vallejo Street was just around the corner from the Kings and a half block up the hill from Grant Avenue, which was the true heart of North Beach in the same way Haight Street would become the center of the Haight-Ashbury district just a few years later. The apartment was within walking distance of the San Francisco Art Institute. Jacques Overhoff, a sculptor Brand had met through the Kings and Jean Varda, lived nearby, and a year after Brand moved in, Gary Snyder came to live close by in North Beach.

To be at the corner of Grant and Vallejo just steps from Caffe Trieste and just up the hill from Lawrence Ferlinghetti's City Lights Books was to be at the center of the cool part of San Francisco—even if the prime of the Beat era was past. Pierre Delattre's Bread and Wine Mission was now gone, as was the Co-Existence Bagel Shop (which had never actually sold bagels), but there were beatniks, bums, businessmen, Chinese, Italians, WASPs, grocery stores with vegetables out front, old buildings, a hill, good weather, and a park. For Brand, best of all it was an affordable neighborhood that was deeply multicultural.

Returning to North Beach in the fall of 1962 proved to be remarkable timing. The Beats were fading, and it would be four years before the emergence of Haight-Ashbury and the hippies and five years before the Summer of Love, which broadcast the arrival of the counterculture to the world. Brand had placed himself at the center of one of the most creative places in the country just at the moment when a great rupture from mainstream culture was about to occur.

The Vallejo Street address was actually two apartments, and he rented them both for forty dollars a month, affording him a guesthouse on the third floor. The building itself was dilapidated, and it was soon condemned, making his stay there continually precarious. Downstairs there was a storefront fish shop named Friscia Fresha Fisha, run by an Italian who, when washing down the street in front of his shop, would gently harass any attractive woman walking past with a squirt from his hose. Brand could walk to Adolph Gasser's, the city's premier photographic equipment store, and he immediately turned one of the apartment's rooms into a darkroom.

As soon as he moved in, he carved a low window in the wall at floor level that looked out into the bushes from the bedroom/sitting room. There was a sofa left by Fox King's mother that he disliked, and so he built a high platform bed above it. He painted the main room black for effect. He added a light in the garden that permitted him to sit and watch the cats play and occasionally catch a glimpse of the ankles of the stewardesses who lived in an apartment behind his.

In addition to regularly joining the Kings at their table, he hosted his own dinners. One night he invited his high school friends—now married—Joanna and John Edson to supper. The Edsons had moved to San Francisco so John could complete his residency at San Francisco General Hospital, and Brand had suggested that they look for housing in the Mission District.

"Isn't that all Mexican?" Joanna had replied. Despite having left Rockford and studied at college in the East, her midwestern provincialism was still intact. (She and John ended up renting a house in the Sunset District, then a largely white working-class neighborhood.) In response, on their second night in the city, he invited them for a Mexican

dinner at his apartment that he prepared with his girlfriend of the moment, a young Mexican named Rose, who had become a regular visitor
first to the barge and then for several months afterward in San Francisco.

Joanna was a Republican, she had just had her first child, and when
she walked into Brand's apartment, she was dumbstruck. Right next to
the door, half as high as a fireplace and touching the floor was a window
that looked out on a garden lit with spotlights. And the room! She knew
he was a photographer and had an inkling of the aesthetic of capturing
lights and darks, but she had never been in a completely black room
before.

Michael Murphy became a regular visitor at Brand's guest apartment in North Beach. After Stanford, where he had undergone a mystical conversion in Spiegelberg's class, Murphy had taken a vow of chastity,
traveled to India, and spent the next eleven years in meditation and
study as a committed virgin. Now he was reentering the world, trying to
fuse East and West.

That fall, shortly before an Esalen seminar on "Human Potentiali-
ties," Jack Loeffler, Brand's friend from the Varda cruises who was then
working as caretaker and morning chef at Slates Hot Springs, introduced Murphy to peyote. During his peyote experience, Murphy met
Erica Weston, photographer Edward Weston's granddaughter. The two
soon began a love affair they carried on as guests in Brand's second
apartment. Murphy would later recall that Brand would on occasion politely ask them to keep the racket down. The two men would become
good friends, and Murphy would be a frequent guest when he came to
San Francisco, where over Irish coffees they would fantasize about Murphy's plans to build Esalen.

R UN BY PSYCHOLOGISTS AND ENGINEERS WHO WERE COLLECT-
ing data to find a link between the drug and creativity, the LSD experiment in Menlo Park included both an extensive preparation and a

debriefing. The first step was to write an extended personal biography. Brand had gone through similar exercises at both Exeter and Stanford, and now a month before his daylong initiation he threw himself into an explanation of where he had come from and where he was going. He described his family and his time at school. His army experience had left him with still-raw emotions. He called it his first strong personal experience with depravity. "I learned that bureaucracy is terrified of the truth, any truth, and I learned to use it as a weapon," he wrote. "I learned of the rage and shame that despotism brings, and I learned the joy of resistance." He explained that he was interested in taking LSD as a learning experience in the hope that it would enhance his judgment and appreciation of beauty, especially as the latter related to his photography. He added that his philosophy was rooted in biology, principally evolution and ecology, and could best be described as "pragmatic." He didn't pray or think of religion often, although he noted that he had gone through the Sequoia Seminar several years earlier and thought the anecdotes about Jesus were "good stuff." He had recently dived into Zen Buddhism, and he had been reading about its influence on art; he wrote that his increasing interest in Zen assumptions, approaches, and values was one of the reasons he wanted to try psychedelics.

When, in March of 1961, Stolaroff and Harman set up the International Foundation for Advanced Study on a quiet street in downtown Menlo Park, just two miles from the Stanford campus, they weren't the only ones exploring the therapeutic uses of LSD on the San Francisco Peninsula. Experiments were going on at the Veterans Administration Hospital in Menlo Park, and the Palo Alto Mental Research Institute had already begun introducing local psychiatrists and psychologists and even poets such as Allen Ginsberg to psychedelic drugs.

But the foundation was something new. Engineers rather than medical professionals led the project, and the clinic was intent on charging a fee for each experience. They worked with several psychologists, including Fadiman, as well as the mysterious Al Hubbard, who was a member of the board of directors. Fadiman, finishing his PhD in psychology at Stanford, was teaching at San Francisco State, and his research at the foundation focused on the changes in beliefs, attitude, and behavior that resulted from taking LSD.

Other participants included Charles Savage, a physician who had conducted medical experiments for the US Navy in the early 1950s, exploring the use of psychedelics as a truth serum, and Robert Mogar, a psychologist at San Francisco State College who helped design and administer psychological tests. Toward the end of the studies Robert McKim, a professor of industrial design at Stanford University, joined the project to help explore the relation between technical design and psychedelic drug use. Don Allen, a former Ampex engineer, and another man worked as "counselors." A significant portion of the experimental subjects were engineers who worked at different companies on the Midpeninsula—a group of people who were believed to be very "uptight."

Before long the group published a glowing research report based on a survey of its first 153 subjects. The results were in the realm of the kind of advertisements found on late-night TV. Fully 83 percent of those who had taken LSD found that they had lasting benefits from the experience, including an increase in their ability to love and to handle hostility as well as enhanced communication skills, interpersonal relations, self-esteem, and understanding of themselves and others. They also reported a 66 percent decrease in anxiety, and 83 percent believed that they now had a new way of looking at the world. The researchers also noted a high correlation between "greater awareness of a higher power, or ultimate reality," and a general claim that the benefits were permanent. What wasn't entirely clear—and was the ultimate point of the experiments— was whether psychedelics could be used in a directed way to improve rational cognitive abilities.

Among the first 153 subjects (ultimately more than 350 people would participate) was Stewart Brand. (Not long after Brand's original LSD experience, the US government began to add restrictions limiting experiments, and so the researchers specifically chose scientists, engineers, architects, and other researchers as their test subjects.) Simultaneously promoted by another group of researchers in Milbrook, New

York, including Leary and Alpert, LSD and other psychedelics would eventually have a dramatic impact on a societal scale. They also had an equally dramatic personal impact on Stewart Brand.

Brand's clinical exposure to LSD was a very different process from what would become commonplace several years later when acid became a recreational drug for college students. He showed up at the foundation at eight a.m. on December 10, 1962, for an all-day session. He had been told to bring photos of his family and other objects of interest to him. During the day the two guides who led him through his LSD experience would ask him to look at certain things and then record his reactions.

At the start of the day he was given a dose of carbogen, a mixture of carbon dioxide and oxygen that was used as a benchmark at the clinic to determine how the subject might react to psychedelic drugs.

The effect was immediate: as Brand later described it, he went to a "very interesting" other universe for what he thought must have been "seven eternities." When he came back, everyone who was watching him was still sitting there, but their cigarettes were just a little shorter. Brand thought carbogen (which he presumed forced his brain to take in too much oxygen and "flame out") was just great, and he later concluded that, in comparison, LSD was a disappointment.

Brand took a goblet containing LSD at 8:41 a.m. He then lay in a quiet, darkened room listening to classical music through headphones. He was given a second goblet at 10:00 a.m. and a final dose by injection at 2:00 p.m. He didn't know it at the time, but the injection was an effort by the researchers to generate a reaction in a patient who seemed particularly blocked or unable to open up.

Subjects' reports of their psychedelic experiences tended to be, well, psychedelic—and sometimes arbitrary and inane, particularly to an outsider. In his report Brand broke the session down into different periods, which he described as "purple attics," "purple helixes," "vacuum cleaners," and "cement."

In the first period there were cartoonlike pictures that played through his mind in sync with the music playing in his headphones. "I recall the notion of gaily pursuing cobwebs through a succession of angular attics, of feeling the music was too spectacular and superficial, and of intimations that Being was large and take-able for granted but out of my then range of vision," he wrote. "Bodily sensations were pleasant chills and a neck-ache. I recall chuckling with feelings of things that had no humor."

After the second goblet of LSD, the experience changed. He asked for simpler music. He looked at a rose that had been placed in the room to create a peaceful ambiance and found it enjoyable but not profound. He became talkative. He began to race through various "scopes of being" and imagined his body at various scales at different locations on Earth.

In the afternoon he was asked to sit up, a change that made him very uncomfortable. He began to feel he could separate people from their masklike faces. Staring in the mirror, he saw his visage as battered and tough. He was asked to look at murals and yin-yang symbols, but he found nothing interesting or very deep. He walked to the bathroom and found the experience dizzying and humiliating—it appeared to him that he was holding a child's penis.

After the injection of LSD, everything transformed into what he described as "vacuum cleaners" and "cement": a roiling series of images that passed through his head before he began to feel as if he could barely move. Asked how he felt, he replied, "Very 'thing.'" He was shown a picture of Christ and suspected that he was being manipulated.

In the afternoon Jim Fadiman joined the session and asked Stewart to look deeply into his eyes. Brand did and then vomited; when he looked at his vomit it was purple.

When the session ended, he was taken to Fadiman's house, which he greeted with pleasure and the relief of escape. After Brand sat down, Fadiman displayed a series of images: an indistinct woman on a record album cover, a statue of the Buddha, a transparent picture that in his head turned into a mask made of two stones and a carrot, and yet for inexplicable reasons reminded Brand of himself. Fadiman showed him several more pictures, including a dark, hellish scene with a satanic child silhouetted against the backdrop. As Brand peered at it, it dissolved into a peaceful valley.

Dinner turned out to be a bizarre experience in which he was intensely aware of his chewing and swallowing. Brand found that he was traveling down into the plate among the potatoes. He watched as a potato crumb lit by the candle on the table became a heroic version of himself.

Later that night, after he thought the effects of the drug had worn off, he walked outside and looked up at a full moon. He stood frozen as it receded, transforming itself into three separate dancing images.

To the Menlo Park foundation researchers, Brand had proved a tough nut to crack. Their analysis was that he was stuck in here-and-now concepts and resistant to fully "letting go." They saw him as the model of the uptight, intellectual guy who depended on logical analysis for emotional defense. Even the LSD injection "booster" had been unable to shake his defenses. Although he had been pushed into more inner-generated symbolism, he still kept "one foot on the ground."

It would only be later, during several follow-up sessions, that the Menlo Park psychologists noticed he softened, becoming less defensive, less concerned about how he was perceived, and more accessible and easier to be with.

Not surprisingly, the next morning Brand was in an odd mood. By the time he returned to the clinic for debriefing, that mood had turned decidedly blue. He stayed deeply depressed for several days until he accompanied Fadiman to a Japanese dinner in North Beach prepared by Fox King for a small group of friends. Over the meal, he said to Fadiman that he wished he had tried to look into the therapist's eyes again after he had vomited.

"Try it right now," Fadiman said.

As Brand stared at Fadiman over the single candle set on the table, he found that tears were forming in his eyes. Fadiman told him to let them come. Finally he told Brand to close his eyes and to "stay with it."

Brand continued to focus on his feelings, and then he realized that

Fadiman was crying too. Their eyes locked for a few more moments, and when Brand rejoined the party, he felt rejuvenated.

At the end of the evening, with the other guests watching, Brand took off his clothes and dived into the spooky blue light of a backyard swimming pool.

# Chapter 4

# American Indian

COMFORTABLY SETTLED IN NORTH BEACH, STILL DIGESTING his LSD experience while taking photography and design classes, Stewart Brand quickly got two big breaks from Dick Raymond. The Palo Alto architects who had designed Stanford's new student union building were unhappy with the publicity photos that had been shot. Raymond recommended Brand, and as a result, he received his first professional photography assignment in California. The fee was just $140, but it put Brand over the moon. "Dick Raymond for Knight-hood," he wrote in a thank-you note.

His Stanford assignment also paid off in a more subtle but ultimately more dramatic way.

At the same time that Mike Murphy was concocting Esalen, two Stanford computer scientists, Douglas Engelbart and John McCarthy, were embarking on pioneering technologies that would completely transform the world during the next half century. While McCarthy, who had coined the term *artificial intelligence*, was trying to build a thinking machine at the Stanford Artificial Intelligence Laboratory (SAIL), Engelbart was attempting to create a system that extended the human intellect—intelligence augmentation—on the other side of campus. It was AI versus IA. One man worked to replace computers with humans, the other to extend humans' intellectual reach. This philosophical divide would create the defining paradox in the world of computing that persists to this day.

One day, while rambling among his old haunts on campus, Brand wandered into the computer center in Pine Hall. There he saw two young men playing an early version of *Spacewar!*—the first video game. *Spacewar!* had been written the previous year at MIT by Steve Russell, a young computer hacker who worked with McCarthy and had followed him to Stanford to help establish the AI research laboratory. At that point, it required a mainframe computer and a rare computer graphics terminal. The program displayed two small joystick-controlled spaceships capable of firing projectiles at each other, all while whirling around a sun that generated gravitational pull toward the middle of the monitor.

Perhaps Brand's recent LSD experience had left him more highly attuned, but whatever the case, it was obvious to him that the two young men were having what he would later describe as an "out-of-body experience." There was something to that: *Spacewar!* was quite likely the first glimmer of what would later be known as cyberspace—that virtual landscape that would begin to emerge at the end of the sixties with the creation of the ARPANET, the Pentagon-funded network of computers. Brand filed the "out-of-body" idea away for a possible story. Almost a decade later he would return to explore what he'd noticed.

STRUGGLING TO BEGIN HIS CAREER WITH GREAT AMBITION BUT few resources and no job, Brand battled intermittent bouts of depression that would become increasingly challenging during the next decade. In the fall of 1962 he accompanied Fadiman back to Big Sur Hot Springs to attend a weekend seminar given by Gerald Heard, the British-born writer who had turned Stolaroff and many others on to LSD and who had been a significant influence on Murphy and Price when they were beginning to organize Esalen.

Brand, however, would have none of Heard's "human potential" proclamations, nor what seemed like Esalen's deep embrace of them. He left before the weekend was over, afterward writing Murphy, "I fled Heard the other weekend because I picked up the old-lady musty smell of a cult. . . . To be surrounded by true believers is murderous. If your retinue (and revenue) take your every fingernail clipping seriously, it's all too easy to join them in their error." He also issued a stark alarm about the coming wave of LSD, warning Murphy that its looming impact on the culture would range from subverting politics to the "biggest baddest witch-hunt since McCarthy."

Brand's second break came in the spring of 1963 when he attended a Sierra Club wilderness conference with Raymond, who took his protégé around and introduced him to all of the players in the national wilderness movement, including Ansel Adams. It was the second time Brand had met the famous photographer, having once dropped in unannounced at Adams's workshop in Carmel to ask his advice on conservation photography. Adams had then told him, "As we grow older, mirrors become windows." More than a decade later Brand would prominently print that insight in an early edition of the *CoEvolution Quarterly*.

In the early sixties Raymond had established a consultancy in recreational land use, and with Zach Stewart, an up-and-coming San Francisco architect, he obtained a small contract with the Warm Springs Indian tribes in Oregon to develop their reservation for tourism. Stewart had grown up in the state of Washington and gone to Reed and Yale before studying architecture at Stanford. He was close to the San Francisco Zen scene, and some years later he would design a home for Gary Snyder at Kitkitdizze in the Sierra foothills. He was a charismatic, razor-smart, ultraconfident, fast-talking creative design type whose ideas went far beyond conventional architecture. He was also prickly and controlling, characteristics that would eventually undermine his partnership with Brand.

Established by treaty in 1855, the Warm Springs Indian Reservation occupied more than a thousand square miles along the Deschutes River in north-central Oregon. Raymond introduced Brand to several of the tribal elders at the conference and suggested that they hire him to photograph an upcoming wild horse roundup on the reservation to help create a brochure marketing Warm Springs to tourists. Raymond, perpetually thinking big, encouraged Brand to consider the idea of producing a photographic book documenting the lives of the three tribes (Warm Springs, Wasco, and Paiute) that occupied the reservation.

One of the representatives at the conference was Delbert Frank, a member of the tribal council that represented the three tribes on the reservation. When Frank extended an invitation to make the pilgrimage to the Oregon reservation, Brand quickly accepted.

American Indians were then struggling against federal policies that were the culmination of centuries of being perceived as obstacles to national progress. In the 1950s there was a growing push to extinguish all remnants of Native life in the United States.[1] With *Let Us Now Praise Famous Men* as a template, Brand envisioned attacking the stereotype of the Indian "problem," replacing it with a vision of the Indian "contribution," which stressed that Indians understood how to live in harmony with their land and that their values resonated with his childhood conservation pledge—to protect the land, water, and wildlife.

At the beginning of May he drove to Oregon in a recently acquired pickup truck, camping along the way. He arrived at the reservation after two days on the road, carrying with him a signed copy of Ansel Adams's masterpiece *This Is the American Earth* as a gift. It was still several weeks before the roundup, and he occupied himself with meeting the tribespeople and taking pictures of the eastern Oregon landscape. He had arrived on the reservation with a set of preconceptions about "Indians," but he soon realized that the people he met were not like anything he was familiar with from his midwestern upbringing.

After several days, Frank returned from a business trip and the two men quickly bonded. Frank was bitter about the college-educated pro-development Indians who opposed him in the tribal council, pointing out that while he had been in San Francisco at a conference, they had struck a deal with a Portland-based architectural firm for the design of a posh hot springs resort catering to white tourists.

Frank gave him his charter—to produce a brochure that would, by highlighting the Warm Springs Indians and their land, advance his notions of "wise" development of the tribal land.

Brand fell increasingly under Delbert Frank's spell as they toured the reservation. They drove to a meeting about Indian fishing rights on the Columbia River at the Dalles Dam, which had been completed a half decade earlier, in 1957. The village of Celilo, once a legendary fishing community, was now submerged in the twenty-four-mile-long reservoir created behind the dam. Lost with it were ancient petroglyphs; as they drove, Frank shared memories of a large mysterious face carved in the rock.

When the Warm Springs roundup began in mid-May, one of the first things Brand realized was that while the mythic icon of the American cowboy was as a loner and an individualist, these cowboys were part of a collective endeavor. They were also a window into a world of contemporary Indians that Brand hadn't realized existed. For several days he tracked them as they went about roping and branding the wild horses. With Frank's blessing, the Indian cowboys gradually opened up to him. They were suspicious about the plans to develop the reservation for recreation, worried that this might just be the first step in taking their land away. "A white man wants to conquer everything," one of them told him. "If white men had this land, there wouldn't be any wild horses. There wouldn't be much of anything else."

Brand was transformed when he returned from Warm Springs; everything he looked at now had a different context. He came away with an understanding that America needed the Indians—and he had the images to prove it. The brochure he talked about with Delbert Frank would never be produced, but for Brand that was irrelevant: he had found a calling.

WHEN HE ARRIVED BACK IN NORTH BEACH, A YOUNG POET WHO was staying in his apartment thrust a copy of Ken Kesey's landmark first novel, *One Flew Over the Cuckoo's Nest*, into his hands. Brand saw that many of the things he had found on the Warm Springs Reservation were echoed in Kesey's story of an Oregon psychiatric hospital as narrated by Chief Bromden, a half Indian who pretends to be deaf and mute. It also was an echo of his frustrating struggle against the irrational army bureaucracy at Fort Dix. "As Kesey writes it, the battle of

McMurphy versus Big Nurse is identical to Indians versus Dalles Dam or me versus the Army," he noted in his journal.

Published only a year before Brand discovered it, *One Flew Over the Cuckoo's Nest* had made Kesey famous. He had come from Oregon to Stanford for graduate studies and entered the school's prestigious creative writing program. Like Brand, he had recently taken LSD. Kesey, who was three years older than Brand, had been introduced to the drug when, to supplement his income, he had participated in government studies at the Menlo Park Veterans Administration Hospital.

On a whim, Brand mailed some of the photographs he had taken in Oregon to Kesey, who was in the process of moving over the hill to the small mountain community of La Honda because his cottage on Perry Lane in Menlo Park was being torn down. Brand added a short note: "Pictures were taken this May on the first stage of a brochure job for the Warm Springs Indians. They're spelling out what they plan to keep, not lose any more of. See you sometime." Several days later Kesey called and invited Brand to visit.

Dorothy Ostwind, a Stanford graduate student, was hanging out with Kesey and several of his friends one afternoon when a red VW camper drove across the bridge leading to the property. She watched a tall, thin blond man get out of the bus, struck immediately with how serious he seemed, that he was clearly on a mission. The next thing she noticed was that his van was full of books—in fact, it was like some kind of 1960s bookmobile.

Kesey's band of Merry Pranksters, which would later be immortalized in Tom Wolfe's *The Electric Kool-Aid Acid Test*, was just starting to form. An eclectic mix of outsiders, poets, journalists, and camp followers would come out to see Kesey in the redwoods to hang out, get stoned, and play. In Ostwind's eyes Brand was different. He came across as "quiet, self-contained, and intense." She was immediately attracted to him and they would soon begin dating.

Brand's mission that day was to explore a joint photojournalism project with Kesey for *Esquire* magazine. Days before, news had exploded about an "LSD Paradise" that Timothy Leary and Richard Alpert were creating in Zihuatanejo, Mexico, intended as a research center under the auspices of the International Federation for Internal Freedom (a group Leary and Alpert also had created). The plan was to bring together a small number of physicians, psychologists, artists, and writers to study the effects of psychedelic drugs. The Mexican press got wind of the community, and the news filtered into American newspapers.

*Esquire* had written to Kesey asking for an article about the experiment. He was already planning to make the trek to Zihuatanejo in July, but because he was in the throes of finishing his next novel, *Sometimes a Great Notion*, he hadn't responded to the magazine editor's note. He told the young photographer he was open to the idea of a "word-picture" collaboration.

Brand then used his connection with Jim Fadiman to introduce himself to Leary and get his own invitation to join the Mexican experiment. Leary wrote back telling him he was quite welcome but that he should expect to sleep in a hammock—the hotel was jammed.

Brand then wrote his own query letter to *Esquire*, concluding, "About LSD nobody is objective. Ditto experimental communities. From inside the experience, I think will come the only pictures—and words—worth much. (How much?)"

*Esquire* loved the idea, but the authorities were not cooperative; within weeks the Mexican government had evicted Leary and Alpert from the country. Soon thereafter Brand learned that his apartment on Vallejo Street had been condemned and the landlord was putting the building up for sale. He avoided immediate eviction, but it would make his stay there increasingly precarious. For solace he turned to the I Ching, which fortified him: "The town may be changed but the well cannot be changed." It counseled: "Strength in the face of danger does not plunge ahead but bides its time. . . . Thus, the superior man eats and drinks, is joyous and of good cheer."

As it turned out, the condemnation order was rescinded and the potential sale temporarily delayed. Brand would ultimately find three more years of grace and nominal rent in a space that he could make over as he wished and that served as a crash pad for a steady stream of friends.

O

**B**RAND AND MINOR WHITE BOTH SHARED A FASCINATION WITH the I Ching. White had previously studied with Alfred Stieglitz and become close to Edward Weston as a student and a friend when he moved to San Francisco, which put Brand within range of one of the greatest American photographers of the period. As he went deeper into the craft, his interest began to drift from photojournalism toward artistic photography.

During the summer of 1963 he began a long and circuitous process of finding ways to avoid his army reserve duty, which was supposed to be fulfilled with regular exercises at the Presidio military base. In August he wrote the commanding general a letter asking to be moved to inactive reserve status because of the nature of his civilian job—freelance photographer.

It took the army a month to respond that he was not eligible for transfer to an inactive reserve unit. Thus began a multiyear cat and mouse game in which the military always seemed to be a step behind and unwilling to enforce any of its threats. The reality, he would later learn, was that the military lived in dread of the possibility that veterans might ignore a call-up.

Brand continued to take advantage of the Northern California outdoors, sailing on the Bay, kayaking in the surf near Bolinas, and picking up sport parachuting, jumping from a small airfield in Calistoga, a village nestled in the upper Napa Valley. He began working on a script for a film about skydiving.

At the end of the summer his interest in design and his class at the San Francisco Art Institute appeared to pay off. For a class project he came up with a new approach to packaging a six-pack of cans. He thought he might have a breakthrough idea and turned to his brother Mike for advice on how to commercialize it. Mike's polite response was that he might not want to leave his day job. At that point, of course, Brand didn't have a job, but he would soon get one.

His instructor referred him to Gordon Ashby, who was also teaching at the institute as well as running a design practice from an office near Fisherman's Wharf. Ashby was a young designer who was a protégé of Charles and Ray Eames. A model maker and a specialist in three-dimensional design, Ashby in the early 1960s had worked in their Los Angeles design studio, where he had led the creation of a pathbreaking museum exhibit for IBM, originally installed at the California Museum of Science and Industry in Los Angeles. *Mathematica: A World of Numbers . . . and Beyond* revolutionized museum design with its interactive exhibits, setting a standard that was later emulated by museums like the Exploratorium in San Francisco, the Tech Interactive Museum in San Jose, and the Museum of Science in Boston.

IBM had contracted with Ashby to create a sequel titled *Astronomia*, which was scheduled to open at the Hayden Planetarium in New York in June of 1964. He was looking for someone to go to libraries and observatories to do the basic research for an exhibit that would capture the history of modern astronomy.

During the summer Brand interviewed at Ashby's studio. The designer's first impression was of an intense young man who had the "potential to be somebody" and someone who "wanted to know about everything." Brand's photographic expertise also seemed a good match, and so Ashby hired him to do the background research, write the exhibit copy, and collect artifacts as well as contribute with design ideas, eventually paying him a part-time wage of $5 an hour, or $600 a month.

*Astronomia* allowed Brand to set his hours and travel the country while also doing his Indian research. He applied for a library card at the University of California at Berkeley and began reading everything he could find about the history of astronomy. He was particularly affected by Arthur Koestler's *The Sleepwalkers*, which traced the history of modern cosmology back to Mesopotamia, with an emphasis on serendipitous discovery. Brand decided Koestler's framework was a good metaphor for his own life—not a carefully plotted arc of ambition, or even a narrative, but rather doing one different thing after another, each of which seemed like a good idea at the time and which, hopefully, would evolve into something profound.

He also read Koestler's *The Act of Creation*, which concluded that invention and discovery shared the pattern of blending unrelated concepts into a new one, a process Koestler termed *bisociation*. This, too, made sense to Brand: true progress was made not by pursuing straight lines but by unanticipated incorporation. Koestler had a powerful influence on Brand, long before Gregory Bateson's thinking on cybernetic theory would further his understanding of the power of paradox.

Having IBM and the Hayden Planetarium as sponsors opened doors. Brand was able to fly around the country negotiating for loans of artifacts that would take them away from their home institutions for five years. He visited the Adler Planetarium in Chicago, the Smithsonian in DC, Caltech, Mount Palomar, and others. He designed and built an elaborate portable photographic copy stand intended to fold into a suitcase to allow him to safely work with fragile materials.

Brand's function was not merely to visit archives and libraries. He was charged with commissioning a replica of Galileo's first telescope. Another exhibit he designed offered a cosmic interpretation of the importance of perspective. When he wasn't traveling, he would spend the weekends sport parachuting until a jump in which nearly everything went wrong—his chute didn't open as planned, and he barely missed a power line as he approached the ground—spooking him enough to put the sport aside. (He was saved by his reserve parachute.) A daredevil who was fond of heights, he would have an even closer call three months later, permanently ending his life as a thrill seeker.

THREE BLOCKS AWAY FROM FISHERMAN'S WHARF AND JUST SIX blocks north of Vallejo Street, Water Street was a tiny block-long alleyway of small homes, offices, and the warehouse for Friscia Seafoods, the firm that ran the storefront shop below Brand's apartment. Off the beaten path, Water Street, where Zach Stewart had his design studio, was still a magnet for North Beach bohemian creativity. When Brand wasn't haunting the library in Berkeley, he became a regular at Gordon Ashby's studio, just a block away on Vandewater, another alleyway.

Brand was also actively pursuing photography and enjoying the North Beach Beat scene. When Gerd Stern returned to the city to prepare a new kind of artistic happening at the San Francisco Museum of Modern Art—which was then located at the Civic Center—Brand was pressed into service.

Gerd Stern, whose endorsement had helped Brand get to California to pursue his photography studies, had been a flamboyant character in San Francisco during the 1950s. His second wife was Ann London, whose father managed a luxury hotel in San Francisco, and as young socialites they regularly appeared in Herb Caen's gossip columns in the *Chronicle*.

San Francisco was a breath of fresh air for Stern, who had completely struck out in New York, where the art community ignored him. He had been reading an early manuscript of Marshall McLuhan's *Understanding Media*, which he had borrowed from the poet M. C. Richards, who had gotten it from the composer John Cage. The notion that it was the medium rather than the content that defined society fascinated him, and he was intent on moving away from his poetry of words to create collages of dynamic objects, images, light, and sound.

Stern had persuaded the San Francisco museum's director to let him use the museum auditorium to put together a performance to help raise funds for *Contact Is the Only Love*, an octagon-shaped, seven-foot-high, neon-lit piece of art. When it was complete, it had a speaker on each side, and an endless audio loop would play what Stern called the Verbal American Landscape, or VAL. A motor-driven cam switch mounted on a truck tire would force the sound to ping-pong back and forth from speaker to speaker.

Stern knew Ramon Sender, a young composer who the previous

year had launched the San Francisco Tape Music Center with Pauline Oliveros and Morton Subotnick to explore electronic music. Sender introduced him to Michael Callahan, a local college student and electronics wizard who was working as a technician at the center, and together they began assembling a crazy quilt of televisions, projectors, and audio equipment to produce two evenings of a twin bill: *Who R U?* and *What's Happening?* Stern enlisted Brand and Ivan Majdrakoff, an art instructor at the San Francisco Art Institute, to wander the streets of San Francisco taking photographs of "words"—every street sign and billboard they could find.

Stern dubbed the San Francisco event "an experiment in simultaneous communications," and he quoted McLuhan in the show handbill: "Patterns of human association based on slower media have become overnight not only irrelevant and obsolete but a threat to continued existence and sanity." It would be for all intents and purposes the birth of modern multimedia.

The event itself was a remarkable San Francisco happening. Attendees included Herb Caen, Allen Ginsberg, and the mayor of San Francisco. Stern used four projectors, one of which was operated by Brand. In addition to the four screens, Stern had placed a panel of sociologists onstage with four telephones. They would listen to and interact with four separate groups that were holding conversations in "isolation booths" in different parts of the exhibit space, those conversations amplified, mixed, and transmitted to the stage in syncopation with images while the entire event was simultaneously broadcast via closed circuit and displayed on a television set.

Critically, the event bombed, to put it mildly. The next day the *Chronicle*'s music critic declared: "Some colossal flops have taken place in the public halls of San Francisco in my time, but Gerd Stern's show called 'Who R U?' and 'What's Happening?' is the only one I can recall that laid so vast and horrendous an egg as to score a kind of success."[2]

After the Stern disaster, Brand took off for a monthlong stay on the East Coast over Christmas and New Year's. There for *Astronomia* work, he also wanted to visit New York friends, such as the Durkees and Henry Grossman. Having organized his Warm Springs Reservation photography into a portfolio, and eager for funding for his book project, he planned to show it to United Nations officials. After Christmas he headed to DC, where he hoped to make contact with Stewart Udall, the secretary of the interior.

Out for a walk one chilly January afternoon on the towpath near Georgetown, he decided to do what he frequently did—climb up and take a look around. In this case his vantage point was the Chain Bridge, which spanned the river between McLean, Virginia, and DC. At first he slid effortlessly, his palms resting lightly on the steel on the ledge created by the bridge's supporting girder surface at his back. Every five yards or so there was a brace to grab on to and swing around. He simply inched along with his heels rocking over the riveted surface of the ledge. Then, after he had swung around the second brace, he stopped to rest and consider his situation. He looked down at the brown water covered with floating ice, and his legs began to vibrate from foot to hip. He forced himself to make a start and edged away from his handhold, then lost courage and his legs started trembling even harder. Suddenly he felt like he couldn't see. He barely made the next handhold and stopped again, still shaking.

He saw people walking on the towpath but realized there was no way they could save him. In despair, he tried singing, tried talking to himself, tried incantations—"Light," "love," "look," and "here and now, boys."

His shaking diminished, he edged out again, this time with his eyes fixed on the icy trees on the horizon, his mind focusing on "doing it," to keep moving.

And he made it to the next brace. There was a rhythm to this, he realized, a sort of chant, and, playing it in his head, he swung around the next brace without slipping. Keeping his momentum and then repeating until he didn't need to edge any farther, he finally lowered himself down and then jumped onto the wet path.

His skydiving mishap several months earlier had been a premoni-
tion, he decided. After his encounter with the bridge, his love of climbing permanently vanished, replaced by panic attacks in the face of even
modest exposure to heights, such as on ski lifts. Yet not all was lost.
Brand decided that even though he had been paralyzed in fear, he had
learned something: keep moving—he who hesitates suffers. Momentum
builds and will help. This was his bridge, narrow and high, and if he
fell, he died.

O

BACK IN SAN FRANCISCO AT THE BEGINNING OF 1964, BRAND TOOK
a research trip to Southern California with Gordon Ashby. While
they were there, visiting Caltech and the Mount Wilson Observatory,
they paid a visit to the offices of Ashby's mentor Charles Eames.

Eames had been an inspiration since Brand's student days, when he
had heard him speak about magazines and design at Stanford. His studio was consumed by a deadline for the design of the IBM pavilion for
the 1964 World's Fair. As Brand wandered around the office the designers grabbed him. Because the theme was the IBM marketing slogan
"Think," they were eager to collect photographs expressing the idea.
The result was a cameo in the exhibit, a prominent picture of Brand,
deep in thought and nibbling on a pencil.

When he wasn't researching for Ashby, he threw himself into the
American Indian world. His photographic portfolio from the Warm
Springs Reservation became his calling card. He made phone calls and
sent off a stream of letters to possible funders ranging from publishers
to foundations. He also burrowed into the San Francisco Public Library,
exploring Indian culture in tandem with pursuing the history of astronomy. Not long after he was back, he stumbled across an odd volume titled *America Needs Indians!*. Published in 1937, it consisted of a hand-lettered
cover with a roughly drawn abstract graphic of something that might
adorn an Indian woven blanket. Written under the nom de plume of either Iktomi Lila Sica or Iktomi Hicala—both author names appeared in

different places in the book—it assessed (sometimes sarcastically) the condition of modern American Indians in a manner that spoke directly to Brand's efforts. (Eventually it became apparent that Iktomi was a man named Ivan Drift, likely one of a small group of white Americans who were so taken with the Indian experience that they went native.)[3]

In any case, the name *America Needs Indians!*, so evocative of what Brand had brought away from his visit to Warm Springs, stuck. Determined to capture the contrast between an older civilization and his modern one and at the same time find a way to create a new fusion of the two worlds, he enlisted the Loefflers as his research assistants and began planning a summer to capture the Indian experience in pictures and words on the reservations in the Southwest. The project would take many forms, ranging from a multibook set of photo essays to an immersive museum exhibition, before crystallizing as one of the first true multimedia slide shows.

At the end of January, Brand, Dorothy Ostwind, the Loefflers, Kesey and his wife, Faye, as well as Mike Murphy and Erica Weston, assembled at Big Sur Hot Springs, where Jim Fadiman was helping lead a seminar, Existentialism & Mysticism. During the weekend Brand spent several hours relaxing and chatting with Kesey in the pools. Although Brand would orbit around the edges of Kesey's world for almost three more years, becoming close friends with several of the Merry Pranksters and taking part in a number of their adventures, ultimately he would hold Kesey at arm's length and never become a full-fledged or "on the bus" Prankster. It was a standoffishness that would be reciprocated. The two men would share a role as bridges between the Beat era of the 1950s and the hippie culture in the 1960s, but Brand was alarmed by Kesey's insistence that you had to buy into the notion of the revolution in consciousness that he saw happening. If you agreed, you had to be there with your entire mind and your body. Brand decided that if someone told you that you were either on the bus or off the bus and that meant total commitment, you should get off the bus.

Eventually he did.

The two men would remain friends until Kesey died in 2001, but

Kesey shared some of Brand's reticence. After Brand had left the Pranksters and had spent time working in Governor Jerry Brown's office in
Sacramento as an informal adviser, Kesey was quoted as saying, "Stewart recognizes power. And cleaves to it."[4]

It was probably unfair; after all, Brand would remain on the periph-
ery of the Brown administration, just as he had with the Pranksters.
However, the remark stung, and decades later he would still fret that
there might be truth in the assertion.

Part of the plan for the Big Sur weekend had been to take his portfolio to show Ansel Adams at his Carmel studio, but Adams had been
called away, so Brand spent the day with the Loefflers. A practicing Zen
Buddhist, Loeffler was a deep mystic and a confirmed nomad who had
found his way into the world of peyote several years earlier. Peyote can
be consumed in a variety of ways, as tea, mush rolled into balls, dried
buttons; in this case, the Loefflers chose the edible variety. (Throwing
up was not uncommon, but the Loefflers had no such distress.) The trio
then wandered around Cannery Row looking for a studio space for the
Durkees, who were then considering returning to California.

At the end of the day, Dorothy Ostwind and Brand went back to
North Beach, where they discussed the idea of her moving in upstairs
on Vallejo Street. Although they had had a passionate weekend, Brand
was deeply hesitant about a stronger commitment and was actively
scheming to set her up with his friend Jim Fadiman. It worked. His two
friends were soon madly in love, and the following year Brand would
serve as their wedding photographer at Esalen.

The next week Loeffler took Brand's Warm Springs photographic
portfolio to Adams and asked for his endorsement. Afterward, he called
Brand to say that the legendary photographer had been impressed with
his work and told Loeffler he would approach Stewart Udall to ask for
$5,000 for their project as well as suggest that the Sierra Club consider
publishing Brand's book.

After he got off the phone, Brand was ecstatic but also frightened
by the weight of responsibility in measuring up to the praise. In any

case, there was no turning back. The next week he received a letter from Adams apologizing for missing him over the weekend and adding a rave review.

"Your pictures are extraordinarily good!" Adams wrote. "I am very happy about them! You have accomplished something!!! And there is every indication that this is only the beginning."[5]

O

GORDON ASHBY'S STUDIO ON WATER STREET WAS A LIVELY SCENE, and the street would occasionally be closed off for block parties. He intentionally cultivated an open-door policy at his studio and kept boxes of colored chalk around so neighborhood children could use the alleyway as a free-form canvas. He kept the glass door open to let the public walk in and out and even permitted them to use his equipment—including cameras and projectors and a wide range of design tools.

Although Brand was a novice, he had quickly impressed Ashby as having an unusual kind of creativity. As they worked together, Ashby gave Brand more responsibility and let his young protégé create his own exhibit ideas.

As the *Astronomia* project progressed, Ashby began acquiring almanacs. Some of them were intended to be used as artifacts in the exhibit itself and others were simply reference tools. He gathered a range of books that were used by laypeople to learn about everything from the weather to the stars. One afternoon, as Brand and he stood contemplating his growing array of historical documents, Brand mused, "What we need now is a new kind of almanac." It would be the first inkling of the *Whole Earth Catalog*. It was also an example of how many of Brand's notions would bubble for years before they emerged.

As Brand schemed with the Loefflers to begin *America Needs Indians!* research, his plans for a summer expedition began to take shape. They would "go native," living in tepees they built. They would start by visiting the Washoe in Nevada. Brand was particularly attuned to the peyote scene and was planning to make contact, with Loeffler's aid, with the Native American Church, a religious movement that used peyote as a sacrament. From there they would head south, visiting the Hopi, Navajo, Zuni, Jicarilla Apache, and Papago Indian reservations. There was also a tantalizing rumor to be chased down: Loeffler had heard of an Indian tribe, somewhere in northern Nevada or perhaps southern Utah, that had had no contact with whites.

Brand kept searching for funding, with no tangible results. He tried to hold a seminar at Esalen on the project, bringing together big names, including Adams, Udall, and even Buckminster Fuller; he tried to exploit a Ford Foundation grant Raymond had gotten to explore the possibilities of suburban utopias; he wanted to display a roomful of his relevant photographs to kick-start his *America Needs Indians!* research project. But nothing worked.

In the spring he and the Loefflers created Brand & Loeffler Design as a cover to permit them to buy equipment and supplies wholesale. Brand's VW bus was a reliable expedition vehicle, and they added a 1951 International Harvester truck to their fleet and began to convert it into a camper. Conceived by Zach Stewart and fashioned over several weeks in the alleyway outside of Ashby's studio, it had a bit of a covered wagon feel to it, all the more so because the resulting camper top pivoted on an axis, making it possible to unroll it and sleep under the stars.

One small irritant was that the US Army still failed to realize that Brand had no wish to serve in a reserve status. They rejected his letters asking for deferrals and instructed him that his reserve assignment was scheduled to begin on May 7. In response, a guest who was living upstairs wrote back that she had received the letter in his absence and that he would be out of town on a work assignment for the next five months. The wolf was once more held at bay.

It would be four years before Carlos Castaneda published *The Teachings of Don Juan: A Yaqui Way of Knowledge*, but already in the spring of 1964 Brand was intensely interested in the world of peyote and the expanded consciousness that could result from it. He had begun experi-

menting with marijuana and mescaline while he was in the army, and
Jim Fadiman had plunged him into the world of LSD. Now Loeffler, who
had also introduced Michael Murphy to peyote, was offering him a gateway to a deeper understanding of the American Indian experience.

Unlike a generation of college students who were to follow in the late
1960s, he never really became a "stoner." He rarely purchased psychedelics, and he walked away from the scene completely just as the baby
boomer generation was entering it. For a time, however, he was consumed
with the notion that psychedelics might offer some deeper understanding of himself and the world around him. Several years later he wrote a
friend, "I vote Yes on drugs. Yes with huge capital letters that spill over
into everything else. Yes with my eyes open." But by the time psychedelics had become a recreational activity on campuses across the nation, he
had explored them thoroughly and moved on to different tools.

The previous summer he had gone to Berkeley and introduced himself to Michael Harner, who was then the acting director at the Lowie
Museum of Anthropology on the UC campus. In 1960 and 1961, Harner
had begun conducting personal experiments with ayahuasca, a powerful Amazonian traditional spiritual medicine. Indeed, he had delved
deeply enough into the herbs that he decided he would become a shaman himself, leading others in chemically augmented rituals.

Brand plied Harner with questions, including: 'Was there a relationship between psychedelics and cultural evolution?" In the discussion that
followed, a bond quickly developed, and he and Harner became close
friends.

That spring the anthropologist had a PhD candidate named David
Perry, who was a Pomo Indian. His grandfather had lent the Lowie Museum his shaman's kit, which included two large quartz crystals. It was
Harner's responsibility to return the kit to his student to give back to
the Pomo tribe. Harner was, however, a stickler for Native traditions and
he decided it was necessary to perform an appropriate ceremony to return the "power" to the crystals.

He enlisted the support of the Loefflers and Brand in conducting the ritual, to take place at sunset at a spot where a spring flowed into the ocean at Drake's Beach, a lonely stretch near Point Reyes in Marin County, almost two hours north of San Francisco. At the last minute the group piled into two cars and raced over to the coast. The ceremony required a variety of herbs, including bay leaves and angelica root. Because they were in a hurry, they reached out of the car windows as they drove, grabbing bushes, tearing their hands in the process.

When they arrived at the coast just before sunset, the beach had an eerie sense about it. There was a strange light, a strong wind blowing debris off the cliffs, and the surf was phosphorescing. Their tardiness meant everything being done at a manic, double-time pace, giving the event a *Monsieur Hulot's Holiday* feel. Just as the sun set, Perry rushed out into the surf and began striking one of the crystals on a rock. Back on shore, Harner whispered to Brand, "If it breaks, he dies."

Brand and Loeffler were assigned the task of building a fire. As they began to stoke it, the wind picked up, racing through the driftwood, creating a potential disaster as it ignited an expanding circle of nearby debris, spreading the fire all over the beach. Soaked up to his waist, Perry returned to shore and said ominously, "Bear Doctor is here, and we've got to get out of here." Jean Loeffler had already begun to feel increasingly uncomfortable, and with this warning, she completely freaked out.

At that point in his life, Brand was still prone to suggestions about the existence of magic and decided that if this guy told them to get the hell out of there, then they should get the hell out of there. They struggled to put the fire out and then fled back down the beach to where the cars were parked. They drove north to a country inn, feeling as if they had just escaped certain death.

As they sat trying to recover, the two anthropologists asked Jack Loeffler to come outside. In the parking lot, they told him that David had seen an "owl shaman" on the cliff above the fire and he had shot a dart at David, but that it had missed and struck Jean, putting her at risk of death as fog entered her veins.

"Well, for chrissake don't tell Jean!" Loeffler replied.

O

TWO WEEKS LATER BRAND ATTENDED A SYMPOSIUM AT SAN JOSE
State College that featured most of the current star LSD researchers. Alpert, Bateson, Frank Barron, Sterling Bunnell, Robert Mogar,
Charles Savage, and William Van Dusen all spoke, at times with abun-
dant fervor, on the state of the science and the present legal situation for
the drug. Alpert described the drug as like handing someone the key to
a jet airplane but argued that it should be given away for free—to applause from the audience.

The view from Rockford, Illinois, however, was less rosy.

Brand's enthusiasm for avant-garde experimentation and bohemian
life in San Francisco did not sit well with his parents, who continued to
subsidize him with occasional checks to help meet expenses. Shortly
after the LSD conference, which he had described with great excitement
in a phone call home, he received a letter from his father. Bob Brand
began by noting that his son seemed to be a candidate for a "Letter." His
mother and he had been disturbed by their conversation, "and I think
it is no less my responsibility to give you some words of experience on
the subject of what I call the Half People and the consequences of asso-
ciating with them," he began. These were the people who congregated
in places like Greenwich Village, Beacon Hill, and North Beach and visited watering places in Provincetown and Big Sur. They have talent in
diverse fields, such as art, music, and theater, his father admitted, and
as a result, their groups tended to have more than the average percentage of "homos, perverts, lesbians, and the like." He cited Van Gogh as
an example—someone whose work would command a fortune but sadly
only after his death. Largely penniless, they tended to become expert
moochers—charming and flattering "but leeches all the same," he continued. On the other hand, his son already had a substantial amount of
property and would have considerably more in the future, making him
a capitalist, with responsibility for "the protection, good use, and growth

of this wealth." Anything less would be unfair and ungrateful to those who had passed that wealth on to him and those who might receive it in the future.

"I would shuck the Big Sur crowd and get with people more worthy of your talents because I suspect your present friends are playing you for a Live One," he wrote, just getting warmed up.

The real message was about narcotics, hallucinogens, "and all that jazz." Brand's father pointed out that a local youngster had just blown his hand off trying to make a rocket. "Sure, it is exciting, but so is setting your building on fire, or throwing stones at summer-house windows," he added, indicating that he had still not forgotten a twenty-year-old infraction. What was he to say when his friends asked, "What is Stew doing now?" He wanted to be proud of his son, not making up explanations that didn't quite explain things. He closed on a sad and defensive note: "Now you and your friends can read this letter aloud, have great fun over it, which will degrade me thoroughly. 73, Dad."

The problem, Brand decided, was that he and his father lived in different parts of the country. The next day he sent his father a five-page single-spaced letter, standing his ground and defending the honor and integrity of his friends. It was a full-throated rebuttal, an argument for his "creative" lifestyle that included an itemized list of his friends and their achievements. California was simply the future and you had to be there to understand that, he wrote. A recent issue of *Look* magazine on California had pointed out that what was happening today in the Golden State would appear a year later in the rest of the country.

He refused to give an inch on drugs either. He cited the research from the symposium he had recently attended, detailing how knowledgeable authorities were no longer discussing total control of psychedelics on the grounds that they should not—and in fact, could not—be controlled. Moreover, unlike other drugs, psychedelics were not addictive.

"If," Brand wrote, "the nature of my friends, and my work, and my writing do not give evidence that my judgment is similar to yours, and based on the same values, and worthy of trust where mutual knowledge is lacking, then I'll be sorry to see the communication diminish."

He was telling his parents that he hoped they understood where he was headed, but that whatever they felt, he was on his way.

O

IN THE SECOND HALF OF MAY 1964, ASHBY AND BRAND WENT TO New York to install the *Astronomia* exhibit at the Hayden Planetarium. It was an opportunity to spend time with the Durkees and Gerd Stern at the Garnerville church as well as visit DC, where he had struck up a friendship with Sharon Francis in Stewart Udall's office. An ardent environmentalist and an outdoorswoman, Francis loved the idea of Brand's Indian project and arranged for an audience with the secretary. Eventually, she became a dedicated supporter of Brand's *America Needs Indians!* crusade. Thanks to her backing, Brand spent the evenings in Udall's offices after everyone had left, sending out fundraising letters for the project. (He took advantage of a giant eight-foot-long bathtub that had been installed by a previous interior secretary and took to cooking his evening meal in the office as well.) Eventually he was discovered and thrown out, but not before he was able to get Udall's enthusiastic backing and a small financial commitment as well as interest in his project from Udall's wife.

During the evenings he also began work on the draft of an essay sketching out his case that "America needs Indians." It would remain unpublished, but it would be the first inkling that Brand was playing with the ideas that he would publicly articulate two years later. Indeed, while eventually he would claim that the inspiration for the *Whole Earth* photo first came to him during his 1966 rooftop LSD trip, in reality, the fundamental concepts were already fully developed during the time he spent in DC that spring.

Indian culture, he wrote, was linked to the idea of wilderness, which was both an ecological condition and a state of mind. He felt that Kennedy's assassination—and the haunting image of the riderless horse leading his funeral procession seen by millions of Americans on television—marked a transition into a new era. The shape of that epoch was still unexplored: "We will see what has been entered with the publication of the first hazy image from the moon of our Earth—glowing disk, mandala of unity, inescapably ourself," Brand wrote.

Where did this come from? Most likely it was a mixture of what he had learned from his time in Oregon on the Warm Springs Reservation, reading Marshall McLuhan—and quite possibly the draft text of an inauguration speech that would be given by President Johnson in January of 1965. Brand had read an early copy in Udall's office, and he had underlined this passage:

> Think of our world as it looks from that rocket that is heading toward Mars. It is like a child's globe, hanging in space, the continent stuck to its side like colored maps. We are all fellow passengers on a dot of earth.[6]

O

THE MILLBROOK MANSION WAS LOCATED ACROSS THE HUDSON River and about an hour north of Garnerville, where the Durkees were setting up shop in their converted church. Brand would visit Millbrook on at least three occasions. After his first visit, he described it as "a lovely glimpse of how it might be," an intentional psychedelic community composed of twelve adults and ten children "without romanticism." Ultimately, however, he became disillusioned. In July the Pranksters had visited the imposing Victorian manor but departed quickly with an overriding sense that the Millbrook crowd was boring and elitist. Brand was ultimately put off by the way Leary treated women. While at Millbrook he heard about how Leary had callously abandoned a series of women and concluded that he was a cad.

Garnerville offered something more positive. USCO—Company of Us—was an artists' collective started by Durkee, Stern, and Michael Callahan that focused on the brand-new multimedia technology and art. In the early days, the trio referred to themselves as "Weirdo," "Beardo," and "Techno." During his East Coast research trips, Brand spent several weekends in Garnerville. At one point, after playing word games with Durkee and Stern, he wrote in his journal:

**Contribution of Stephen, Gerd, and myself.**

*Take the NO out of NOW*

*Then*

*Take the OW out of NOW*

*Then*

*Take the THEN out of NOW*

*Then*

Later Stern would use this "electronic mantra" in a variety of art pieces, first for USCO and, after various other mechanical and painted forms, even refrigerator magnets. It would also resonate with Ram Dass's book *Be Here Now*, which Durkee and Brand would assist with several years later.

Secretary Udall had recommended Brand to Alvin M. Josephy Jr., a leading historian of Indian history, who was an editor for American Heritage Publishing. When they met in New York, Josephy, who would become a supporter and a friend, offered Brand a long tutorial in American Indian history and culture, punctuated by a note of optimism. Numb and conquered for more than a century, the Indians were currently going through rapid change, Josephy told him. They were developing a nationalistic spirit, exemplified by the National Congress of American Indians, which represented eighty-five tribes and was beginning to be a potent and interesting force. Before they parted, Josephy alerted Brand to the next National Congress of American Indians convention, which was scheduled to be held in Sheridan, Wyoming, at the end of July.

It would prove a fateful detour later that summer.

The installation of the *Astronomia* exhibit was marred by an incident that would embarrass Brand for decades. The exhibition showed many historical artifacts that Ashby and Brand, using IBM's influence, had been able to borrow from museums and laboratories all over the world. Ashby had made certain that everything was insured.

One of the gems of their collecting efforts had been a collection of original glass photographic plates that had been created in the early nineteenth century by Joseph Ritter von Fraunhofer, a Bavarian physicist who had invented an astronomical instrument known as a spectroscope that splits a light source into its constituent wavelengths. Fraunhofer discovered the dark absorption lines in the spectrum of the sun that were known as Fraunhofer lines.

As Brand worked on the exhibit one afternoon, he carefully set the plates on the floor and turned around to work on the glass enclosure where they would be displayed. He then took a step back, and as his heel bore down, the stack shattered; a shock went up his spine as he realized what he had done. They were irreplaceable and he was mortified. But there was nothing to be done. They were gone.

In the end his misstep was not fatal to the project. The *Astronomia* exhibit, which opened on June 30, 1964, earned favorable commentary from luminaries such as Isaac Asimov and Carl Sagan. IBM was also pleased, and so toward the end of his time in New York City, Brand used its sponsorship of the project as a pretext to visit one of its marketing executives to see if IBM would be interested in supporting his Indian photography. He took Gerd Stern along and predictably got nowhere. "IBM bombed. I could see him saying no no, Indians scruffy, IBM clean: the subject stopped him," Brand wrote Ashby afterward.

In early July the Loefflers met Brand at Pyramid Lake in northwest Nevada and the trio spent several days exploring the hot springs, including a creek that ran with warm water; you could sit under a thundering waterfall in a pool that was a perfect temperature. For Brand it was his first visit to the desert, and he quickly fell under its spell.

A week later, following the Loefflers, who had gone ahead, he made his way through the Black Rock Desert. (It would be more than a quarter century before it was discovered by the people who created Burning Man.) The Black Rock Desert playa is one of the flattest places on Earth, covering approximately two hundred square miles with no inhabitants but "Basques and buckaroos." There was an extraordinary quality to the playa, an "extreme planetary surface weirdness," as he thought of it. It was in wandering this desert that Brand discovered a new kind of solitary freedom. In a sense, it was like Big Sur, which itself was an extreme version of the West Coast that he had set out to immerse himself in. The Pacific Ocean and the high desert equally captured his spirit. Here was Brand's version of "lighting out." The smell of rain on sage in the deserts of the Southwest was the smell of freedom.

O

LOEFFLER HAD INTRODUCED HIM TO PEYOTE NEAR PYRAMID Lake, and then later in the summer Brand joined peyote meetings with the Navajos in New Mexico and Arizona. Within a short amount of time, he would delve deeply enough into the world of peyote rituals to attain the status of a roadman, the leader of the peyote meeting. (Several years later his first published article would be "The Native American Church Meeting," which appeared in a quasi-academic journal called *Psychedelic Review* in November of 1967.)

To Brand, peyote was more manageable than LSD. It felt as if it increased his ability to perceive the world around him without hallucinations. It also altered his view of technology. The way Indians viewed time, he decided, was different from the way whites experienced it. After the near-calamitous beachside ceremony to return power to the crystals of the Pomo shaman's kit, Brand had written in his journal, "If you would know an Indian feeling, try this: Move slowly."

Now his peyote experiences ratified his conclusion that for whites, time was a rapid sequence, whereas Indian time was land time. It was geological and astronomical. It was much slower.

Brand found that he loved car camping in the desert; like Edward

Abbey, for Brand it was a way of being solitary and—for a bit, at least—away from all human contact. Now, spending a week in the desert largely alone, Brand thought about the great technological strides that America had made in the past century and decided that he had no idea what the country was "progressing" toward. He concluded that there was no meaning in technological advancement as an end in itself. He had been reading McLuhan, and here, far from the noise of the city, he realized that Indian time is livable only to the extent that it was possible to avoid or transcend the barrages of electronic culture. Things were becoming clearer.

O

THE YOUNG EXPLORERS ROAMED IN THE DESERT FOR WEEKS. They made it as far north as a small Mormon outpost, Kanab, Utah, before turning south and heading to Santa Fe, where they stayed with friends of the Loefflers'.

In Santa Fe, Brand had a series of epic dreams involving a quest that involved several men and women in a forested mountain wilderness searching for something lost. What it might have been he couldn't quite put his finger on, though he had been reading Tolkien in the desert, so the theme was perhaps unsurprising. Both Jack Loeffler and the actor Max von Sydow appeared in the dream. Von Sydow, who as a young actor had a striking resemblance to Brand, was with an attractive woman. In one dream the actor gestured at something as it emerged for an instant large and white in the distance and Brand joined in the quest.

Toward the end of July, Brand left the Loefflers in Arizona and drove north to Wyoming just prior to the National Congress of American Indians convention. He camped on the edge of town and set about photographing those gathering for the event.

The convention received a smattering of national attention, being described in newspapers as a "powwow."[7] More than a thousand attendees from tribes around the country had been expected, but actual attendance was relatively sparse. On the first day only twenty-two delegates representing eleven tribes, from the Dakotas, Montana, Washington State, Oregon, Utah, Oklahoma, North Carolina, and Washington, DC, registered.

It was in the midst of the presidential election campaign between Lyndon Johnson and Barry Goldwater, and Stewart Udall used the convention as an opportunity to flay Goldwater for his position on Indian welfare. Brand was able to briefly reintroduce himself to Udall and set about making contacts with the various Indians who were attending.

He also met a young Indian woman there, Lois Jennings, a half-Ottawa mathematician employed by the Office of Naval Research who was working at the registration desk when he arrived. Her mother, who was active in Indian politics, had persuaded her to attend during her summer vacation. Short and bespectacled, she had raven hair and what Tom Wolfe would later describe in *The Electric Kool-Aid Acid Test* as a radiant smile.

Jennings was in many ways the perfect "Indian maiden," as Brand would later describe her. She was also a genuine "hidden figure," the description for the women who did mathematical calculations by hand who were essential for the early space program. The daughter of a federal government economist, with a brother who would become a nuclear physicist at MIT, Jennings had majored in math in college, even taking an introductory course in programming high-speed computers.

She was someone who did not fit neatly in any single category. She would say, "Among whites, I'm an Indian, among Indians I'm an Ottawa, among Ottawas I'm Thunder clan, among Thunder clan, I'm Wasaquam [her mother's birth name], among Wasaquam, I'm Lois."

When she came to the Sheridan convention, Jennings was a relatively unhappy and underpaid specialist in debugging software, who in addition to doing calculations for a classified US Navy missile program was tasked with making the office coffee when she came in to work in the morning.

Her first impression of Brand was a young man who had still not completely found himself, but she was attracted to him and made several visits to his VW encampment on the outskirts of town for evening trysts. After Brand left Wyoming to head back to Arizona, they would stay in touch, occasionally by letter and more frequently by phone. A mutual infatuation grew slowly; it would be almost nine months before Jennings would visit and almost two years before she would move to California to live with Brand.

After Sheridan, Brand drove back to the Southwest and rendezvoused with the Loefflers. Udall, with the prompting and assistance of Sharon Francis, had sent letters recommending that the Sierra Club consider publishing Brand's Indian book, arguing that the Indians had a great deal to teach whites about conservation.

O

BRAND HAD MADE A PARTICULARLY GOOD CONNECTION TO THE Native American Church in meeting Hola Tso, a church vice president, who invited him to participate in an all-night ritual. "There is not really so much irony . . . in my contemplating Marshall McLuhan's super-up-to-date *Understanding Media* in a windblown ponderosa forest several hours before participating in an Indian peyote meeting," Brand wrote. "Indians munch peyote, sing, and listen more attentively to the voice that print culture's din has made nearly inaudible."

The Native American Church ceremony was more devout than anything he had ever encountered. There were strict rules about when you could stand and when you could leave and in which direction you walked. Brand found himself praying with great emotion.

After the peyote was consumed, singing commenced and continued until dawn. The leader leaned on an ornamental staff and with his free hand shook a gourd rattle in time with a drummer. To Brand, their singing reminded him of Gregorian chants.

Afterward, still high on peyote, Brand drove the Indian roadman home. At his house he accepted Brand's request for membership in the church. It would eventually come complete with a membership card.

O

THE *AMERICA NEEDS INDIANS!* PROJECT WAS BEGINNING TO TAKE
clearer shape in Brand's mind; it was becoming less of a traditional
book and more of an exercise in multimedia. Brand had seen a preview
at the Eames's studio in Los Angeles of their multimedia exhibition for
the 1964 World's Fair, which encompassed fourteen large and eight small
screens and a seating arena that was hydraulically lifted into a domelike projection screen. He was also aware of Zach Stewart's vision of
a "sensorium"—an elaborate immersive visual experience inside an en-
closed dome. In the desert Brand had imagined a domelike inflatable
"airhouse," much like an oversize Navajo hogan, with a ten-foot-high
tunnel entrance. On one side of the tunnel would be a photo of a cigar
store Indian and on the other side would be an "Uncle Sam Needs You"
poster. Inside the tunnel a sound collage of pop music, advertising, news
reports, and traffic noises would assault the senses as an unsettling contrast between traditional Indian and urban white cultures.

Once inside the dome, with a high ceiling fifty or sixty feet overhead,
the visitor would be guided through a maze of eight-foot-high curved
panels presenting a stream of murals accompanied by localized looping
soundscapes. In a peyote exhibit, there might be sounds from a ceremony:
drumming, rattling, and singing. Stewart Udall's or Margaret Mead's
voice might accompany another photograph.

His idea was to complete the traveling exhibition first and then pub-
lish a book of the photographs. He imagined that he might persuade the
Department of the Interior to cover the costs of travel in the United
States and maybe even overseas as well.

He talked about the idea with Zach Stewart by phone from the desert. Stewart had partnered with Dick Raymond on the Warm Springs
Reservation project because he had a deep interest in Indians and Indian rights. His great-grandfather had been the Indian commissioner
for the Oregon Territory. But he also tended to blow a fuse, and when
he learned that Brand had submitted a bill for his photos to the Warm

Springs Indians, he had tossed him out of his office. He was deeply skeptical of the growing San Francisco drug scene, and his tendency to micromanage, coupled with his antidrug stance, would eventually drive a bitter wedge between the two men. Early on, however, they more or less papered over their differences, and Stewart agreed to partner with Brand on the new project.

By September, *America Needs Indians!* was in full swing. Brand left the Loefflers and drove back to San Francisco, where he was welcomed at the home of Dan Osborne, Zach Stewart's business partner. But the next day he was feeling out of sorts, once again in an urban world, far away from Indian lands and the cactus of the Southwest.

"I found myself upside down in my own skin," he wrote to the Loefflers in early October. He said he felt uncool. A meeting with Stewart was full of hostility, even though they basically agreed. Eventually, they began to outline the *America Needs Indians!* presentation on a long scroll of architectural paper. Their project would ultimately become one of the first multimedia exhibitions defining an emerging American counterculture, but their partnership wouldn't last.

## Chapter 5

# Multimedia

IN JANUARY 1965, STEWART BRAND HEADED EAST ON YET ANOTHER fundraising tour for *America Needs Indians!* (now an ambitious traveling sensorium). As before, he made little progress. An approach to Stewart Udall's wife, Lee, for $5,000 to underwrite a multimedia gallery aimed at undoing Indian stereotypes was met with a polite handwritten note: "Stewart, I'm afraid you are slightly out of our class. Our yearly operating budget is $8,000!" (The secretary of the interior did offer a consolation prize: a $1,000 contract to photograph an Indian Jobs Corps camp near Winslow, Arizona.) The Bureau of Indian Affairs had initially shown interest, but it turned out it wasn't as interested in countering Indian stereotypes as it was in marketing the bureau's work. His approaches to New York publishers, made with the blessing of Josephy turned out to be a distraction and a dead end.

Being in Washington also meant that he spent more time with Lois Jennings. What had begun as a casual affair had continued as a series of extended phone calls, and Brand was waking up to the fact that there was much more to this young woman than he'd initially realized. In March, before heading west, he wrote both in his journal and in a letter to a friend: "Lois was disappointed today about the Atlas-Centaur exploding on its pad. She feels that each one of those shots brings us closer." He added, "She is right. She is an unexpected amount of girl. I salute Lois Jennings, 23, half-breed Ottawa, for being the least hung-up girl I have loved."

Brand returned to the Southwest in the middle of March and began eagerly interviewing Indians about their herbal pharmacies. He also plunged into archaeology that predated the modern Indians, planning a trip to the remote Crack-in-the-Rock Pueblo north of Flagstaff, which had seen human habitation for a thousand years before being abandoned in the thirteenth century.

In Santa Fe, he again met with the peyote church Indians he had studied with in the previous year. In Taos, he watched as a local medicine man diagnosed a pinched nerve in a tribe member's back. "Purified his hand in cedar smoke, and reached in and un-pinched the nerve," he wrote Durkee.

On the advice of Minor White, he visited Walter Chappell, a photographer who had previously been part of the Big Sur circle of photographers that included Ansel Adams and Imogen Cunningham. In Taos, Chappell had begun to photograph Indian ceremonial life, and his work would become associated with the Taos Pueblo. Brand spent an afternoon lost in his photographs. That night an Indian dancer and drummer arrived, and he sat entranced while Chappell improvised on the piano as they sang and danced until dawn. Several days later Brand gave Chappell a ride to Oraibi, a Hopi village on a daylong drive west of Taos. That night, under a full moon, he left the photographer on the mesa and drove on for miles in the moonlight with his headlights off, held spellbound by the desert and the moon.

When he arrived in Winslow, however, the spell was broken. He discovered he was unexpected, and it took the better part of the day to learn that his prized consultancy had been abruptly canceled by a remote and faceless Washington bureaucracy. For several days he despaired, forlornly wondering, "What do I do with my life now?"

Trying to make sense of his situation, he decided he would spend a week in the desert anyway. Jack and Jean Loeffler had established themselves with the Navajos in northern Arizona in the fall of 1964. They had found an older abandoned three-forked stick hogan, a traditional Navajo dwelling, and asked the tribe if they could occupy it. Someone had probably died there, which was why it was abandoned, so the answer from the Indians was "We wouldn't do that but take your chances." The Loefflers were elsewhere, so Brand headed for the abandoned dwelling, tucked away in one of the most remote areas of the reservation, where in the early 1960s few spoke English. Once there he took the mattress out of his VW van, laid out his camera gear, lit the interior with a kerosene lantern, and set up for a weeklong retreat.

For the first several days he simply did chores and luxuriated in the silence. He heard nothing but the wind against the hogan's walls, the occasional snowflake hissing on the stove, and outside, the wind moving through stubby trees. Finally, prompted by an unusual mixed message from the I Ching one night, he took the "medicine"—LSD.

Suddenly not only did the hogan feel like some kind of protective being, but Navajo Mountain, to the north, transformed itself into an immense deity with outstretched arms—a kind of Jolly Green Giant, very big, strong, kind, and laconic. For a while he became the mountain; later he just appreciated it from his distance, watching as it aged eons in a short time. He strolled outside, feeling like a pilgrim of some sort. Later he returned to the hogan and made coffee and walked naked around the stove, waving the cheese he was holding as if conducting an imaginary orchestra.

He had what he had long sought in psychedelics—a transformative experience. In Menlo Park he had felt claustrophobic restrictions; now, release. The experience was sharpened, he wrote in his journal, because his situation felt so dangerous. He was far away from help and on his own.

In the morning he looked up at the smoke hole and saw several inches of new snow, which felt redemptive. The loss of Udall's support now seemed as if it were only a speed bump. Things would be all right in the world. He quickly packed his van and, skittering on the slushy mountain roads, hightailed it back to California.

O

IN THE SPRING THERE WAS AN EARLY SCREENING OF *AMERICA NEEDS Indians!* at the home of Zach Stewart's architecture partner, Dan Osborne. Brand's stated goal was to give his audience the experience of peyote without actually taking the sacrament. The presentation itself, however, had more breadth, making a visual argument that was highly critical of his own white culture. He had gathered his photographs from his travels to Indian reservations during the past year and juxtaposed them with images he had drawn from middle America. Brand had been introduced to the photography of Edward Curtis by Josephy, and some of those images were integrated into the presentation. Inspired by his USCO friends, *American Needs Indians!* played on multiple slide projectors and two screens married to a soundtrack that included Indian chanting and drumming as well as music. Ultimately, the performance would expand to include a movie projector and slide carousels holding seven hundred slides.

In *Hippies, Indians, and the Fight for Red Power*, historian Sherry L. Smith points to *America Needs Indians!* as the seed of a shift in the American perception of the value of preserving American Indian culture. She describes how Brand's visual essay touched off a political campaign that changed government policy—"As more and more people accepted Brand's premise, political support for policy transformation followed."[1] *America Needs Indians!* would serve as a bridge between the cultures. Behind his simple message was the unstated assumption that the Indians' situation was fragile, their existence was threatened, and the nation should both learn from Indian values and do something to protect them.[2]

Brand could barely contain himself. He had added the Beatles' "A Hard Day's Night" to the soundtrack. "We wound up dancing with the projectors—images popping out of the darkness all over the room," he wrote Stern. Afterward, Mike Harner joked that the Irish coffee must have been spiked with LSD.

Fresh off his opening success, Brand took the multimedia presentation public. A round of last-minute telephone calls offering invitations to a performance on a chilly evening in the courtyard of the San Francisco Art Institute drew more than two hundred people. The follow-up show, sponsored by the UC Berkeley anthropology department, generated another full house, with people turned away.

*America Needs Indians!* would be shown almost a dozen times during the remainder of 1965, but as Brand and Stewart tried to work together, their partnership frayed. Brand continued to find Stewart to be a controlling micromanager with an out-of-scale ego. And Stewart in turn remained outraged by the fact that Brand dabbled in psychedelics. The architect frequently chastised his younger partner for what he belittled as offbeat ideas.

"A peculiar and unpalatable attitude keeps creeping into *America Needs Indians!*," Stewart wrote Brand while traveling in the Southwest. "It is a cross between pontification and unreality—a kind of lecturer's 'I know something you don't' approach." He concluded by suggesting that Brand memorize Antoine de Saint-Exupéry's *The Little Prince*—a children's book that conveys the lesson that life is only worth living when it is lived for others, not for oneself.

Despite continuing friction, the high point of their collaboration came in September 1964 when the show (now including Indian dancers and Chappell accompanying them on piano) had a four-night run at the Committee, a popular North Beach nightspot, for a three-dollar admission charge. Several months later, the show received a rave review in the *Chronicle*: "The picture show was terrific. The slides and films were used at the start primarily to bring out all manner of startling parallels, contrasts, and switches between the culture of the American Indian and the white majority."[3]

Since leaving the army, Brand had driven across the country a number of times in his trusty VW bus, and that summer he and Jennings lived as nomads, often camping in an Airstream trailer, making their way as far east as Oklahoma and stopping at various encampments and celebrations. They headed first for northwestern New Mexico to meet up with the Durkees. Their rendezvous point was the mountaintop where Jack Loeffler was working as a fire lookout. The night they arrived almost everyone ate peyote, except for Loeffler, because he was trying to remain a conscientious employee, and Jennings, who had been put off by the peyote rituals that had taken place in her family basement when she was a teenager.

The Durkees had driven in from New York, where they were living because Stephen felt he had to be there to make it commercially as an artist. However, Barbara hated it there and wanted to return to the West. New Mexico was a revelation, and they would return the next summer with big plans to begin building a communal new world.

O

WHEN BRAND CAME BACK TO SAN FRANCISCO AT THE END OF August, he was still under Kesey's spell. He had missed the legendary August 7 La Honda party orchestrated by journalist Hunter S. Thompson, where the Hells Angels motorcycle gang shared LSD with the Pranksters. However, he was present at other events with Angels and Pranksters that fall and at one point wrote in his journal: "I feel more freedom and joy nowadays with Kesey and the Hells Angels than anywhere. Perhaps in part because it's not given, provided there. You have to find it."

After the North Beach shows, Brand got wind of a Me-Wuk Indian festival being held for the first time in the foothills of the Sierras. He drove to the town of Tuolumne on the outskirts of Yosemite Valley to attend what would be the first of the Me-Wuk Acorn Festivals, since then held annually for the next fifty years and counting.

When he attended the festival, he was planning to capture more material for his multimedia show. He found exhibitions, food, music, and dancing. However, what caught his eye most was a series of games, including some very physical rugby-like games involving a ball being played by teenage Indians of both sexes.

He also learned about the hand game, played widely by American Indians since before recorded history. It pits two tribes over a pot of money, and one side attempts to guess the location of a marked "bone." He was even more impressed watching the teenage boys and girls play together. In white America teenage sports were almost universally separated by gender. At the festival he saw boys and girls playing, grappling, touching each other, and having a great time. The sight sparked an insight that would germinate several years later.

The following month, Kesey and the Pranksters rolled out of La Honda, heading for a big rally in Berkeley organized by an antiwar coalition known as the Vietnam Day Committee and a planned march to the Oakland Army Terminal. Kesey, who was not on the list of announced speakers, was invited onstage by *Realist* editor Paul Krassner and deflated the audience by arguing that the demonstrators were wasting their time. Tom Wolfe captured his address in *The Electric Kool-Aid Acid Test*: "They've been having wars for ten thousand years and you're not gonna stop it this way. . . . Ten thousand years, and this is the game they play to do it . . . holding rallies and having marches . . . and that's the same game you're playing . . . their game. . . ."[4]

In the crowd, Brand found himself in agreement. Although he had left the army disillusioned by its faceless bureaucratic irrationality, his ongoing anticommunist outlook continued to keep him separate from the early antiwar movement, which had a significant socialist and communist participation in the mid-1960s. As a former army officer, he felt he had knowledge the protesters didn't have. Moreover, his brother-in-law, Donald Sampson, had gone to fight in Southeast Asia. Brand wasn't for the war (he considered himself to be on the "psychedelic side" in the political dispute over Vietnam), but he had a basic sympathy for the enlisted men, and he bridled when he heard that protesters were calling them baby killers.

Shortly after the Vietnam Day protest in October, he wrote to the Loefflers approvingly that Kesey had "sicced" the Hells Angels on the Vietnam Day protesters. During the October march, the Angels had attempted to disrupt the protest, leading to the arrest of some of the gang members.

Brand viewed the protests through the prism of the ecological theory he had studied at Stanford, pointing to the vigor and even chaos found at the edge of a tide pool in his letter: "The labels blurred into reality and Vietnam is here at last. The borderline of living paradox is the new frontier and the domain of the new mountain men."

This made sense. Brand's understanding of his progress as being the result of chance and collision, of unplanned navigation, was rooted in libertarian preferences, not a concern with day-to-day, or even year-to-year, social concerns.

As the Vietnam War raged on and protests tore the country apart, Brand did his best to stay above the fray. When he set out to become a publisher in the fall of 1968, just months after Chicago police beat and tear-gassed protesters, he decreed the new publication would have nothing to say about the Vietnam War, and he stuck to what he believed was a no-politics editorial policy for the first three years he published the *Catalog*.

O

B RAND INVESTED TWO YEARS INTO CREATING *AMERICA NEEDS Indians!*, but in the end it remained a largely unfinished effort. Other photography projects fell by the wayside. His partnership with Stewart would come to an ugly and abrupt end that winter when Brand came to Stewart's architectural studio to claim the slides to show *America Needs Indians!* at what would become the first full-scale public Acid Test held by the Pranksters at the Big Beat nightclub in Palo Alto. Stewart, enraged at the idea of Brand going off on his own to participate in a drug-based sensorium, exploded, and the two men came close to blows. Brand took the carousels and left. While the men would collaborate once more a month later, they would hold deep personal grudges for decades. (Stewart sent angry letters for almost a decade, even to Brand's mother, demanding repayment of a small amount of expense money the architect claimed he had advanced the project. Brand would later say Stewart was one of only two people he had worked with in his entire life for whom he retained an active hatred.)

In Palo Alto the Big Beat nightclub, which planned to target the "go-go dancing scene,"[5] was scheduled to open in mid-December. Just before it did, Kesey rented it for one night as a kind of shakedown cruise for something radically different—a new kind of experience designed to erase the line between the audience and band.

The first Acid Test had been held just after Thanksgiving at the Soquel, California, home of Ken Babbs, a Marine Corps veteran and a Merry Prankster friend of Kesey's. Members of a band named the Warlocks, soon to become the Grateful Dead, performed. A second test followed soon after in San Jose. Brand brought his *America Needs Indians!* extravaganza to the third Acid Test, and it left a lasting impression. Phil Lesh, the bass player for the Dead, would remember it decades later: "To many of us—white kids who had grown up watching Westerns in the fifties—these revelations struck like lightning bolts."[6]

In the fall of 1965, Mike Hagen, one of Kesey's young Pranksters, arrived at Brand's place in North Beach and excitedly told him the Pranksters were planning to put together a supersize Acid Test that would eventually be named the Trips Festival. Brand thought it was a great idea. He also knew that the Pranksters were disorganized and that, left to their own devices, they would never make it happen, so he got on the phone and began making calls.

There was already a nascent music scene in San Francisco at the Matrix nightclub, and a group known as the Family Dog had put on some early rock concerts at the Longshoremen's Hall—a cavernous, six-sided building with a high ceiling—earlier that year. Brand was able to reserve the venue for a January weekend for a $300 down payment on the condition that the hall would be completely cleaned up after each evening's show.

The immediate challenge was how in the world he was going to advertise his happening. His second call was to Jerry Mander, a partner in Freeman, Mander & Gossage, a high-profile San Francisco ad agency with an office in a converted firehouse near the San Francisco waterfront. Mander and Brand had crossed paths frequently in North Beach, and Mander had watched several performances of *America Needs Indians!*. He had been impressed when Brand ended the performances by picking up a slide projector and waving it around while chanting wildly, mimicking the sound of a Native American chant he had learned from attending peyote gatherings during his travels in the Southwest.

In Mander's mind Brand was a pioneer in exploring the new electronic communications and entertainment world that was being touted by Mander's advertising partner Howard Gossage—a close friend of Marshall McLuhan's.

"We have something we want to talk to you about," Brand told Mander, and the advertising exec invited him over. Brand, Kesey, Ramon Sender, and a young man named Ben Jacopetti, who ran an alternative theater group in Berkeley, showed up at the agency a couple of days later, explaining that they had no idea how to market their event.

"I know a guy who knows how to sell tickets," Mander said. He turned and called Bill Graham.

Graham, a German-born refugee from the Nazis who had grown up in New York City, had come to San Francisco in the early 1960s to be closer to his sister. He had worked briefly as a statistician for Southern Pacific Railroad before going to work as a promoter for the radical theater group the San Francisco Mime Troupe. Mander knew him as an aggressive guy who would drive around town in his Karmann Ghia convertible putting up event posters in bookstores and record shops. Graham showed up, loved the idea, and agreed to come on board.

That was how, several weeks later, while on Christmas vacation from a teaching job she had taken on a reservation in Montana, Lois Jennings would barrel through the financial district as a passenger in Graham's Karmann Ghia as part of an unsanctioned parade led by the Pranksters during the afternoon before New Year's Eve. The caravan tossed out handbills as they rolled along. (In San Francisco, getting a parade permit was virtually impossible, but if you kept moving and passing out handbills, it was possible to cause quite a stir while staying one step ahead of the authorities.)

Brand was in full Prankster mode. He maneuvered his VW bus—a sticker on its side read "Love Generator"—through the confetti and toilet paper and old calendars that, per San Francisco tradition, were raining down on the last working day of the year, giving a running commentary from a loudspeaker on top of his bus to the financial district crowd, whom a reporter described as "secretaries and vice presidents and clerks."[7] When the caravan turned the corner onto Montgomery,

Brand's loudspeaker blared: "You're urban folkniks. Help us litter up this street. There's a message for you on each of these snowflakes. Read them. Hey, there's a pigeon landing on a window up there! What's wild is we're all in this parade. It's all a big trip."[8]

The group arrived several blocks away in Union Square, where they took out three giant helium balloons with a long cloth banner spelling "N-O-W." One of the Pranksters, a young woman whose Prankster name was Mountain Girl, spray-painted "Trips" on the balloons, and they were launched into the sky.

Two years later, in 1968, Brand and Jennings would achieve international celebrity when they appeared in the opening pages of *The Electric Kool-Aid Acid Test*, Tom Wolfe's adrenaline-fueled re-creation of Kesey's and the Pranksters' exploits. Wolfe described a mythic figure, the "half-Ottawa Indian" with her hair thrown back, a radiant smile with a jumble of poorly formed teeth. (After being recruited into helping promote the event, Jennings would miss the actual event because she had to return to teach school after the holidays.)

The Pranksters were known for their costumes, and Brand and Jennings had taken to wearing Native American regalia as they went about advertising the upcoming festival. "I remember that we went around to all the hotels with promotional literature for the festival, and the people in the lobby kept saying, 'Oh, the travel desk is over there,'" Brand would recall later.[9] In a way, they weren't altogether wrong.

O

THERE WAS A SERIES OF DRESS REHEARSALS FOR THE TRIPS FEStival in December and early January. Bill Graham had organized a benefit for the San Francisco Mime Troupe at the Fillmore Auditorium in December, the night before the Pranksters held the second public Acid Test at Muir Beach. Then the Pranksters held an even larger Acid Test at the Fillmore on January 8. Two thousand people showed up to hear the newly minted Grateful Dead play. At the time, the Fillmore was simply a rundown music venue in a black working-class neighborhood that had been built a half decade after the San Francisco earthquake as a dance hall.

That night, amid all the cacophony, Stewart Brand had an unusual encounter with Neal Cassady. Iconic as the quintessential Beat during the 1950s, Cassady was, like Brand, also a bridge between the Beats and the Hippies. The film documentary *The Other One: The Long Strange Trip of Bob Weir*[10] documents how Cassady was essential in shaping the Grateful Dead's worldview and by extension the values of the emerging counterculture. His life captured the free spirit ideal that Brand would later express succinctly in the *Whole Earth Epilog*: "Stay hungry. Stay foolish." Already a larger-than-life character in the fifties, portrayed as Dean Moriarty in Jack Kerouac's 1957 novel, *On the Road*, he was the towering figure of the Beat generation. He was twelve years older than Brand, and two years later he would die in Mexico, alone, from a drug overdose after being found comatose on a lonely stretch of railroad tracks on a cold and rainy night.

In the sixties, with Kesey and the Pranksters, Cassady was a speed freak, a wild man, never stopping, speaking in scat verse, carrying a small sledgehammer that he would hypnotically flip in the air over and over. He was the default driver of the Pranksters' bus, and was known for shifting the gears manically as he drove.

The night Brand and Cassady crossed paths at the Fillmore, among all the chaos, they found a quiet alcove where it was calm. When Brand approached him, Cassady was standing motionless, staring down at the band and the dancers. Brand discovered a Cassady who was not dialing his way wildly through multiple personalities, who was sensible, who was interactive. He was surprised as he realized that there was a "reflective Neal" hidden inside all the time, something other than the manic hipster the Pranksters celebrated. In between the music, the two men chatted quietly. What Brand found was a friendly, thoughtful man who was aware of what was happening.

"It looks like the publicity for your Trips Festival is going pretty well," Cassady said.[11] It was indeed.

O

A FEW WEEKS LATER, PETER COYOTE WAS RIDING HIGH. THE YEAR 1966 was only three weeks old, but the twenty-five-year-old Grinnell College English-lit major had already arrived in San Francisco, had joined the Mime Troupe, and was dating a stripper from the raunchy section of the city's North Beach neighborhood. Like many youthful hipsters in the Bay Area, he had seen the flyers advertising the Trips Festival, a multisensory event that was supposed to offer a simulated LSD experience. There would be music and dancing. And so, on a chilly Friday evening, he showed up with his date at the Longshoremen's Hall to discover a long line of young people waiting in the parking lot. Once inside, he found his way to a darkened corner and under a table he reveled as his stripper friend "taught me 600 new things that skin could do."

Afterward, high on LSD, he wandered through the crowd, bombarded by a cacophony of sound and light. Attire ranged from scruffy to occasionally topless. He spied Brand, thin with already receding blond hair, shirtless, with a large Indian pendant around his neck, a circular diffraction grating affixed to his forehead, wearing a black top hat capped with a prominent feather, sitting high above the floor on top of the flag-festooned "Tower of Power" that served as the Pranksters' command center for controlling the projectors, speakers, black lights, and strobes that were intended to simulate the psychedelic experience. "That motherfucker knows how to throw a party," Coyote thought.

It was, in Brand's later retelling, the first time the Bay Area's ten thousand hippies realized that there were ten thousand hippies. The three-day party opened a cultural fault line in America that defined the 1960s, which in many ways still resonates in the political, cultural, and environmental battles taking place in America today. The Trips Festival became a catalyst, igniting the counterculture that would soon take root in San Francisco. Indeed, there is a straight line from the Trips

Festival to the creation of the Haight-Ashbury scene and the Summer of Love. The weekend also led directly to the establishment of the Fillmore Auditorium as the catalyst for the San Francisco music scene, etching a path that led to Woodstock. It was a spark that helped ignite a global youth movement that reached a peak just two years later in 1968.

O

AS SUCCESSFUL AS BRAND WAS AT PROMOTING THE TRIPS FESTIval, it was Kesey who inadvertently put it over the top when, late one January night several days before the festival, he and a young woman, Carolyn Adams, known to the Pranksters as Mountain Girl, were arrested while tossing pebbles off the roof of Brand's apartment building. It was front-page news; Kesey was already in trouble with the police over an earlier drug bust in La Honda. Freeing him to await trial, the judge publicly admonished Kesey not to attend the upcoming festival, vastly increasing awareness of the event. (Camouflaged in a silver space suit with helmet, he would show up anyway.)

Afterward composer Ramon Sender observed that he and two other people were the only ones there who were straight on the Friday night the Trips Festival opened. One of the others was Bill Graham, who clearly understood that he had stumbled onto a gold mine. He raced around with a clipboard, pausing to express his sympathy for Jerry Garcia, the Dead's lead guitarist, when he discovered that the neck of his guitar was broken.

The first evening Brand began taking money from people at the door, but when he was confronted by the huge line waiting to get in, he quickly realized he was out of his element. He asked Michael Phillips, a young banker working in the Bank of America's research department, to take the gate. (It seemed reasonable to ask a banker, who obviously must know something about money.) Phillips undertook the task for about half an hour and then, studying the scene in the hall, said to himself, "I've had enough. This is too much fun," and persuaded Graham to take his place.

At the time, LSD was still legal, and the drug wars that would tear America apart in the late 1960s were just getting underway. For many of the young people who attended that weekend, the Trips Festival would be not only the first time they'd been stoned or tripping but also a life-changing event, their consciousness chemically altered or not. One example was Peter Calthorpe, a precocious Palo Alto sixteen-year-old who built geodesic domes in high school with Lloyd Kahn, later the editor of the Shelter section of the *Whole Earth Catalog*, who went on to live in a commune and produce light shows for several years. He studied architecture and became an urban planner internationally known for the concept of New Urbanism, walkable communities, while he lived as Brand's neighbor on a houseboat in Sausalito a decade later.

Until then, at rock shows put on in places like the Matrix, the audience had remained in their seats as they watched the performers. Now Kesey and the Pranksters were tearing down the walls. For good or ill the audience was going to be part of the show. The Pranksters had draped parachutes and other paraphernalia around the floor to damp the echoes in the cavernous hall. Ron Boise, a local sculptor who had come to prominence in San Francisco two years earlier when some of his art pieces depicting the Kama Sutra had been seized by the police, brought along an odd instrument called a thunder machine. It was made out of large formed metal pieces taken from old cars and strings that could be played simultaneously by ten or twelve people to make a wild variety of odd noises. Exhibits and art stations were scattered around the floor as well as concessions. One table sold Trips Festival sweatshirts while another was selling books about insects. A champion UC Berkeley gymnast had been tasked with diving off a balcony onto a trampoline; under the strobe light his act added to the surreal ambiance.

The event would prove bittersweet for Brand. On the first night, he set up a tall Indian tepee covered with canvas off to one side of the hall. Inside, Brand and Michael Harner played the hand game that Brand had discovered the previous fall. The problem was that the audience couldn't see what was happening in the tent. The projectors were playing images taken from Brand's multimedia slide show off the tent's canvas, but it didn't make a lot of sense to the festival attendees. Both men were wearing microphones, and after a while Brand began to chant: "Hey! Now we're flying away! Any day now we're flying away! Way-oh-may-ha! Su-na-na-na-yi-ha! Su-na-na-na-yi-ha!" Unfortunately, the game and the participants were hidden from the audience, who had no idea that Brand was trying out a new form of presentation for his multimedia show. It was a nice idea, but it made very poor theater. It would be the last time Brand would show *America Needs Indians!*.

Separately on Friday, Kesey was using one of the overhead projectors as a giant proto-Twitter account, writing compact witticisms such as "WHAT FANTASTIC MARVELS HATH OUR KILLOWROUGHT" on one of the giant screens.

The whole event was exasperating to Zach Stewart. He was angry at the Pranksters and had no interest in experimenting with LSD. He kept projecting architectural slides over Kesey's handwritten messages. And he was continuing to quarrel with Brand, who in turn was furious at the architect for acting like a control freak.

Although the Trips Festival would be mythologized as the event that marked the dawn of the sixties counterculture, Brand and several others considered that Friday night a bust. The audience, it turned out, was not there to take in an avant-garde experience—they wanted to dance. In the middle of the evening, someone walked up to the microphone and said: "This is a bore, even on acid." Confused and increasingly uninterested, a good portion of the audience left halfway through the evening.

It wasn't until Saturday night, when the Dead mended their instruments and found their rhythm, that the Trips Festival finally took off. People had indeed come to dance, not stand outside a tepee watching what seemed to be a random slideshow. Afterward, Brand would say, "It was the beginning of the Grateful Dead and the end of everybody else."[12]

On the Monday after the Trips Festival, the original organizers met at a small café across the street from the Fillmore Auditorium to divide up the $14,500 that had been raised from the long weekend. Graham showed up at the meeting having just signed the lease on the music hall. His portion of the take from the weekend was just $900, but he had seen clearly that there was a fortune in the new music being played that weekend by Big Brother & the Holding Company, the Grateful Dead, and the Loading Zone. Jacopetti's theater group took $900. Brand came away with $1,400. In the end, most of the money from the weekend went to the Pranksters, although Mountain Girl would later say that much of it disappeared with Kesey, who, under the threat of jail, had split for Mexico.

The next month, when Graham launched the Fillmore with a three-day "dance" concert, the poster noted, "With Sights and Sounds of the Trips Festival!"

The sixties were beginning in earnest.

O

**B**RAND WASN'T IMMUNE TO THE URGE TO DANCE AND, WITH THAT in mind, went down to Los Angeles for one of several Acid Tests being held there. Unfortunately, the event seemed to be more about making films, and at midnight he decided to head back to the Bay Area. As he was leaving, Neal Cassady accosted him and asked if he could hitch a ride north. They took off with Brand at the wheel.

In the early hours of the morning, as the VW bus began to labor up the Ventura Freeway into the hills, Cassady suggested, "Why don't I drive?"

Brand pulled over and they swapped places. Soon the bus was flying faster than Brand had ever driven it and Cassady casually began playing a frightening mind game in which he would swerve toward each passing telephone pole, seeing how close he could get the car to the edge of the road at speed before yanking the vehicle to a safe distance.

Brand was paralyzed.

"Would you roll me a J?" Cassady asked nonchalantly as the VW veered toward an oncoming telephone pole. As the bus lurched back and forth, Brand did his best to comply.

It was the most frightening ride he had ever experienced, and he breathed a sigh of relief when they finally pulled into Salinas and Cassady got out. Alone, Brand headed over the mountains to Big Sur, euphoric. He had survived close calls before—his chute not opening, the near disaster on the bridge over the Potomac—but this was different, and the relief he felt was different too. Far from feeling the fool, he was left with a sense of liberation. It felt like he could do anything: he could walk to China on foot or even marry Lois Jennings. He hadn't just escaped an accident—he had *escaped*. He was that free. It was a serendipitous approach to life led without a grand plan, keeping an open mind, and doing whatever seems most promising at the moment.

In February he wrote his father, again trying to calm his fears. He assured him that despite reports that the new counterculture was taking a dangerous socialist turn, he had no beef with capitalism. He made the case that the Trips Festival had been a business success. His financial investment had been only $250, he noted. He had received $1,400 of the $14,500 gross, and he had paid off everyone in just four days.

While he had once been interested in Ayn Rand, he added, he increasingly viewed her laissez-faire capitalist worldview as old thinking. Now the ideas of Buckminster Fuller—pro-technology, with a deep faith that the coming of computerization and automation would result in an infinite abundance that would arrive shortly—were increasingly appealing to him. Fuller was an iconoclastic inventor and futurist who had popularized the geodesic dome and whole-systems thinking, and Brand fell under his spell while attending a series of lectures he gave in San Jose in early 1966. He would become a devoted adherent to Fuller's whole-systems thinking approach; it would become a framing principle for the *Whole Earth Catalog*.

In his letter to his father he pointed to the power of cooperation: "Synergy—the unique behavior of whole systems, unpredicted by their respective sub-systems' events," he explained. He chose not to emphasize one implication—that systemic behavior might also generate unexpected consequences, and there was no reason to assume that they would all be positive.

Brand also explained to his father that he was equally influenced by

Marshall McLuhan and his message that the world was in the midst of a communications revolution. He argued that his father's passionate ham radio hobby was the harbinger of this much deeper revolution: "Thanks to your field of electronics, implosion, and detribalization is where we're at," he wrote.

It was an emotional argument that went on at a distance through letters and phone calls. It was heated enough that the senior Brand angrily condemned his son for bringing shame to the family name. Brand wrote back, telling his father that he was welcome to change his name if he wanted to, but that the younger Brand had no intention of changing his name: "I'm proud of it, of my parents, and of myself. Love, Stewart."

He soon learned that his father had begun a battle with cancer that would ultimately claim his life. He had no sense of how to talk to his father about his illness, and so he sent him a letter asking what advice he would give on the subject: "Should I bite my lip in uncommunicative dismay? Or should we consider dying as a significant topic?" Brand had had his brushes with mortality, but this seemed different. This was the first time that someone he was close to faced death, and it put Brand in a more serious frame of mind. It would be almost two years before his father died.

Brand was close to three men, Jack Loeffler, Steve Durkee, and Ken Kesey, who each seemed like a potential model for how he might live his life. He thought of them as Lao-tzu, Buddha, and Nietzsche, respectively. Loeffler had chosen a rural life, Durkee a communal one, and Kesey was basically urban. Loeffler's approach to life would be best for personal advance, Durkee's for creativity. Oddly, despite Kesey's rejection of activism, Brand viewed him as leading a life that would have the biggest impact on society. Brand realized he wanted to go down all three paths simultaneously, which meant that he would have to find a fourth—his own—path.

Brand had grown closer to Lois Jennings, but she was still teaching on the reservation. That winter she became pregnant and then miscarried, which, in their pain, brought them even closer.

Before the miscarriage he wrote Jerry Mander a letter asking about

hiring him in his advertising company. His family, he noted, was about
to grow by one and possibly two Ottawa Indians, and he was considering "modes of regular work." Although Mander liked Brand, he didn't
have a position available, but he offered him a temporary job repairing
some furniture.

Without gainful employment (his photographic work was enjoyable,
but it wouldn't become a profession), his success as a minor impresario of
the Trips Festival was more or less his only career option, and he would
spend the rest of 1966 and 1967 organizing events, without a clear direction in mind. What he had, just as he had had as a child, was an unending stream of notions. Every once in a while, one of them would be worth
acting upon.

His father's illness was one of the things on his mind when at the
end of February 1966 he decided to spend the afternoon tripping on
LSD on his rooftop in North Beach. He came down off his trip with the
idea of posing the question he had spent the afternoon wondering about—
why the space program had not publicized a photograph of the whole
Earth—to a broader audience. The next day he went to the library and
copied a list of addresses from the *Congressional Directory* and sent a hand-
made button with his slightly conspiratorial question to every senator
and representative. He found another directory and sent buttons to
NASA officials, United Nations officials, Russian scientists, and even
members of the Soviet Politburo. He sent buttons to Buckminster Fuller
and Marshall McLuhan for good measure, and to his surprise, Fuller
responded that his campaign was in vain because it would never be possible see more than a little less than half of the Earth at any one time.
(Sometime later, after the two men met at Esalen, Brand was able to clar-
ify his concept, and Fuller would become a supporter.)

After Brand came down from the roof, he fashioned a plywood sandwich board painted Day-Glo blue with Day-Glo red letters that read "Why
haven't we seen a photograph of the whole Earth yet?" He added a tray
to hold his buttons, which he planned to sell for twenty-five cents each.

With encouragement from Jennings, who arrived on the scene in
time to help overcome his stage fright, he outfitted himself in a white

jumpsuit, adding a top hat with a crystal heart and a flower, and headed over to Berkeley to stand at Sather Gate amid the passing student crowds.

Despite appearing just a year after the Free Speech Movement, he wasn't a student and the dean's office tossed him off campus. Before he was forced to leave, however, he encountered astrophysicist George Field, who took five buttons and told him he was going to visit NASA the following week. He also encountered a *Chronicle* reporter, who wrote a modest article about his quest.

Next, he repeated his vigil in front of the Student Union building at Stanford, where he ran into a group of NASA scientists and traded gossip about the space program with them. Finally, Jennings and Brand headed off to the East Coast, where he made similar appearances equipped with his feathered top hat, sandwich board, and buttons at Columbia University, Harvard, and MIT. At Columbia he ran into a guy who insisted that the reason they wouldn't let a photograph of the Earth be shown was that the Earth was shaped like a doughnut. Later at MIT Brand's brother Pete, who was teaching in the engineering school, joined the crowd he had gathered. ("Who the hell is that?" asked an MIT dean as he walked by one afternoon. "That's my brother," Pete responded.)

In hindsight, Brand's whole Earth insight has been said to represent a seismic shift in our thinking about the relationship between nature and technology. "The image of the whole Earth forced all who saw it to think holistically," wrote Andrew G. Kirk, the author of *Counterculture Green: The Whole Earth Catalog and American Environmentalism*.[13] In the moment, it was never clear whether Brand's campaign triggered any response within the space agency. However, years later he would encounter a NASA security officer who told him that he had been tasked by headquarters to investigate an odd fellow in California asking about why there was no photograph of the whole Earth. He had done a thorough investigation and concluded that Brand was harmless. At the end of his report he inquired, "By the way why haven't we seen a photograph of the whole Earth yet?"[14]

After his campaign ended, Brand and Jennings extended their stay on the East Coast, moving in for a while with the USCO artists in Garnerville. Gerd Stern was living in nearby Woodstock and spending most of his time at the USCO church, and he introduced Brand to John Brockman, another young army veteran, who was still wearing his uniform because, unlike Brand, he was regularly showing up for his service in the army reserve.

While studying for an MBA at Columbia University, Brockman had been on his way to a conventional business career when he discovered the *Realist* at a New York City newsstand. Paul Krassner's underground newsletter convinced him that he wasn't crazy, or at least that others shared his view of the establishment. In many ways, Brockman was an East Coast mirror of Brand. When the two of them met, Brockman had just organized an avant-garde film series he named the Expanded Cinema Festival, which echoed many of the same themes being explored in the Trips Festival on the West Coast.

In another sense, the two men were complete opposites. Brockman was a Boston Jew while Brand was distinctly WASPish. Brand had explored psychedelics, as had many of the USCO artists, while Brockman studiously resisted any mind-altering substances. The mantra of the USCO artists was "We R All One," but when Brockman became a fellow traveler, the group modified the slogan to read "We R All One, except for Brockman."

After Brockman got USCO a gig putting together a video disco in an old aircraft hangar at Roosevelt Field on Long Island, the artists shifted their focus toward producing an exhibition at the Riverside Museum in New York that was described as a "be-in," a term Stern had coined earlier that year.

*We R All One*, which completely took over the museum, opened in June and became a brief media sensation. *Contact Is the Only Love*, Stern's infamous creation for the San Francisco Art Museum in 1963, was shipped from the West Coast, and other USCO artists contributed pieces, including Durkee's recently electrified paintings that included flashing lights among their panels. Brand's contribution was an ambient audio mix called *Phase Raga*.

The exhibition garnered enough attention that Steve and Barbara Durkee appeared on the *Today* show with Barbara Walters, outfitted with diffraction gratings on their foreheads. *Life* magazine put the exhibit on the cover of its September issue, for a story on psychedelic art. "We try to vaporize the mind by bombing the senses," *Life* quoted a "psychedelic artist."[15]

O

BACK ON THE WEST COAST IN JUNE, BRAND AND JENNINGS HELPED Stern and Michael Callahan produce several more showings of *We R All One*, including a performance at a high-profile San Francisco conference on psychedelics. Brand continued to look for venues to exhibit his work, without success, and to put on multimedia happenings, and he began proposing a series of events designed to blend the USCO multimedia and Kesey Acid Test experiences. Eventually, he approached Glide Memorial Church, which under the Reverend Cecil Williams was becoming a progressive institution, with an ambitious proposal to bring together a range of folk and rock musicians like Joan Baez and the Grateful Dead with USCO artists and other counterculture celebrities like Allen Ginsberg for a Saturday evening extravaganza he dubbed "Metamorphosis." An electronic evening at the church would attract a new generation, he wrote to the church elders. "To pay a visit to the Cathedral on our own spiritual terms would, gentlemen, blow our minds," he wrote.

The blown mind learns, he added. Despite his enthusiasm, his message failed to resonate with the church and the plan died a quick death.

The act of learning was about to become one of the central themes in Brand's life. At Stanford, he had experienced an epiphany: "I'm not really actually here to learn French, or whatever, I'm learning how to be able to learn anything, and then I can go forth and have a life. . . . I don't need the class." Learning was an end unto itself.

He was coming to see learning as an essential element in keeping both individuals and organizations vibrant. You needed to learn skills and acquire tools. His romance with tools—the *Catalog* would be subtitled *Access to Tools*—came in part from his 1966 encounter with Fuller, who was legendary for claiming: "If you want to teach people a new way of thinking, don't bother trying to teach them. Instead, give them a tool, the use of which will lead to new ways of thinking."

Brand came to believe that tools were inherently democratizing and would serve as agents of social change. Further, his passion for tools would translate into a general stance as a technophile with a deep faith that human progress was dependent on technological advances. It was a perspective that would be reinforced the following year when he fell into the orbit of Doug Engelbart, the iconoclastic computer scientist who saw the computer as the most universal of tools.

In July, Brand and Jennings headed to Big Sur for an Esalen seminar. Afterward, they hiked back into the wilderness and spent three weeks exploring what it was like to live off the land. Despite his affection for San Francisco, Brand remained captivated by the romance of primitive living. Jennings was significantly more skeptical, but she went along for the adventure.

To get fifteen miles back into the wilderness, they crossed a 3,600-foot-high ridge, looking directly out at the clouds hanging over the vast Pacific Ocean. In the first years of their relationship they would spend weeks on the road, much of it camping in an Airstream trailer. When they hiked together on the mesa in Arizona, Jennings, befitting her Indian heritage, was generally fearless in the wilderness and would ignore the lightning. At the same time, she wasn't naive, and she would always put Brand first on the trail. Rattlesnakes!

They took cornmeal, a copy of Euell Gibbons's classic guide to wild foods, *Stalking the Wild Asparagus,* and a rifle. For a while they made do with just the cornmeal and some lily roots they discovered growing by a deep pool hidden beneath a seventy-foot-high waterfall. Things improved considerably after Brand poached a deer. It was dusk and he saw the deer in thick brush on the other side of a stream. He tracked ahead to a clear shot where the deer was heading and killed it with a single bullett to the head. He field dressed it, ate the liver raw, and slung the carcass over his shoulder to head back to camp, where they cooked and lived on it for several days.

Their flirtation with rural life continued as they headed to the Southwest, visiting with the Hopi Indians and then driving to a commune in New Mexico, where the Durkees had returned with radical dreams to found a commune.

It was called Solux-Lama, later just Lama. The commune was founded that summer by the Durkees and their friend Jonathan Altman, with the help of Timothy Leary's former Harvard colleague Richard Alpert. (Alpert taught at Stanford and lived for a while in the Los Altos Hills south of the university, and Brand would occasionally pitch his tepee and live in his backyard.) Brand would be with them when they first discovered the property in New Mexico. He erected his tepee on the site where the commune was being built and lived off the land for several weeks. It was the first structure on what would become one of the defining communes of the back-to-the-land movement. Alpert, the LSD guru who became Ram Dass that year, would write his bestselling *Be Here Now* in a collaboration with Durkee, who designed the original graphics and typography for the book. Brand contributed photographs to the first edition.

Ironically, though in a few years the *Whole Earth Catalog* would play a key role in the commune movement, Brand never embraced the idea of rural life. He and Jennings stayed at Lama for only a couple of weeks and then moved on. Durkee was committed to building a new world, and Brand was fascinated by the project but not committed in a personal way. Jennings had even less interest in the communal hierarchy imposed by Steve Durkee, which she thought was overly regimented and verging on being a cult. They quickly decided to go back to California and settle on the Midpeninsula.

This didn't, however, prevent Brand from continuing to dream about creating an ideal community. He fantasized about the features his utopian collectives might have. (In one case, he fixated on the idea of separating the community in halves, divided by a volleyball court—a level of detail that was both meticulous and decidedly utopian.)

In August at Libre, an artists' commune in the Huerfano Valley in southern Colorado, Brand came up with the idea of creating a mountain community with access only by foot. It would be governed by an executive council and regular town meetings, with a voting age of sixteen. Revenue would be generated by taxing cultural events. Population would be controlled by "disadvantaging" families with more than two children. Innovation in community activities such as games would be expected, but how things might be set up to encourage entrepreneurship was a question left hanging in his journal.

Perhaps better than going back to the land completely, he decided, would be a backcountry retreat to get away from the urban world occasionally. He proposed the idea of having a "back forty" to Lou Gottlieb, who had been the bassist for a folk-music group called the Limeliters. Gottlieb happened to have some land he had purchased in Sonoma County. The following year, 1967, Sender, who had co-organized the Trips Festival with Brand and whose first wife was a descendant of one of the members of the original Oneida community, a nineteenth-century religious commune, suggested that the land be used to create a new communitarian experiment, which led to the creation of the Morningstar commune.[16] It would be precisely the opposite of what Brand had suggested.

All of this left Brand feeling out of sorts as the summer wound down. He had a sense that he might be in the wrong country. The commune world was no utopia, to say the least, yet he was finding that he grew easily bored "in the woods." His perception of the growing commune movement was further complicated by the fact that they were often created around charismatic leaders, leading to the cult behavior that he was anxious to avoid. What could he do that was "wholly of use"? he wondered.

BEFORE LSD SPREAD INTO THE GENERAL POPULATION AS A REC-reational drug, it had been explored by a small community of academic and military researchers who pursued its uses, from mind control and mind expansion to spirituality. It became illegal in California on

October 6, 1966. Ken Kesey had fled to Mexico and attempted to fake his suicide, but on the weekend of September 30, his disembodied voice wafted from a campuswide audio network at San Francisco State that the organizers of a "happening" known as *Whatever It Is* had created for their event. Although it was unofficial, what would be the penultimate Acid Test took place one weekend night, sponsored by the student-run Experimental College.

At the start of the school year, Brand had contacted a group of San Francisco State student leaders who hired him to organize an event modeled on the Trips Festival. The poster advertising the event noted that the weekend would be spent researching the question "How can we get higher?" There would be a full menu of psychedelic rock and roll (Brand had booked the Grateful Dead, Mimi Fariña, and a variety of other Bay Area bands) as well as light shows, a dance workshop, elaborate "sound sculptures," and a McLuhanesque television crew to feed the event back to the audience as it happened. Designed by Brand, the poster conveyed an image of the Earth as seen from the moon as envisioned by his "Why haven't we seen a photograph of the whole Earth yet?" campaign earlier that year. In the poster, the surface of the moon was populated by naked figures from Hieronymus Bosch's *Garden of Earthly Delights.* Anticipating trouble from school officials, he had printed eight thousand of the posters before he showed them to anyone. Many more than the planned ten thousand tickets were sold, almost matching the entire sixteen-thousand student enrollment at SF State. The weekend proved to be a proto-Woodstock. The Jefferson Airplane wasn't part of the official program, but after the Dead had been playing for a while one evening, members of Airplane got up onstage with them, and the two bands jammed for a while.

Brand had put the Pranksters on the bill, and they all wanted to be there, including an incognito Ken Kesey. On their way from Mexico, the Pranksters' bus broke down repeatedly—the temperature in the desert went well above 100 degrees—and their pet parrot died on the trip, but in the end, they pulled into San Francisco at sunset on the Friday night *Whatever It Is* was set to open.

Brand would later say that his antipathy for the New Left began when he recoiled in horror in response to the Leninist politics the San Francisco State students displayed while he worked with them. The reality was more nuanced.

That *Whatever It Is* happened at all was thanks to James Nixon's election as student body president in the fall of 1966. Nixon had been living in the Haight, and he thought of himself as part of the hippie movement; he had been among a handful of students (including his wife, Cynthia) who in 1965 created the university's so-called Experimental College. Self-styled revolutionaries who were heavily influenced by the Cuban Revolution, they had decided to begin their own insurrection by collectively seizing control of the school and organizing their own system of instruction intended to undermine the established educational hierarchy. Like many of the alternative education efforts that blossomed in the Bay Area in the 1960s, it would come and go quickly (in this case lasting only until the end of the university's student strike in 1968). Despite Brand's antipathy to the politics behind the students' insurrection, he would take many of their ideas and their community organizing approach with him both to the *Whole Earth Catalog* and afterward in his various learning-oriented projects.

As student body president, Nixon found himself in control of a huge budget: a half million dollars from the annual revenue of the Associated Students organization and a million dollars a year from the cafeteria and bookstore. When Brand approached the students with the idea for a cultural extravaganza, Nixon jumped at his proposal, viewing it as a way to merge the civic and cultural politics that were afoot in the Bay Area with the educational revolution they were trying to bring about.

The students set Brand and Jennings up in an office on campus to plan the event. Brand hired Jerry Mander's firm to generate publicity and was able to get local TV and press attention ahead of the event. Despite riots in several San Francisco neighborhoods (including the Fillmore District) the previous week following the police shooting of a Black teenager, and the university administration's fears that San Francisco's Black citizens were going to take over the campus, the event came off without a hitch. Although private security had been hired, Nixon kept them out of view and, in a nod to the growing influence of the feminist movement, controlled the event with an all-female security team.[17]

Although the event itself was a success, it would foretell the deep divisions that would emerge between the New Left and the counterculture, as well as a more personal fracture between the Nixons and the Brands a year later.

For several weeks afterward, Kesey, who had returned from Mexico, where he had been a fugitive from the California marijuana charges to secretly attend *Whatever It Is*, played a cat and mouse game with the police before he was arrested driving south from San Francisco while Brand and several other Pranksters followed in a second car. Two agents who were driving home spotted him accidently and then chased him after he got out of his car and ran. Miraculously, despite Kesey having faked his death and spent six months as a fugitive in Mexico, the FBI chose not to pursue charges against him, and his lawyers were able to persuade a local judge not to throw the book at him. As LSD was now illegal, the author convinced the judge that he could use his stature to persuade a young generation to move on past the drug culture that he had helped create.

An Acid Test "graduation" ceremony, meant to convey the idea that the psychedelic era was ending and the Pranksters and their friends were moving on, was scheduled for the end of the month. Originally it was supposed to include the Grateful Dead and be held at the Winterland Ballroom, but Kesey had lost the trust of Bill Graham as well as the *Chronicle*'s influential music critic, Ralph Gleason, whose endorsement would have carried weight among those who might be interested in attending. Also, the psychedelic community itself was fractured, and many members felt that by renouncing psychedelics, Kesey was betraying their movement. Brand stepped in to help the Pranksters organize the event, now to be held in a dingy warehouse south of Market Street near the police headquarters. While the press showed up in force, there was a relatively small turnout at the event, at which Cassady handed out Acid Test graduation diplomas.

At the end of the evening, in the cold morning air, Brand and Jennings went up to the roof of the warehouse, where he asked her, "Isn't it time that we get married?" Two weeks later they held a ceremony on the beach in Santa Cruz officiated by Paul Lee, a University of California at Santa Cruz philosophy professor who was a friend of Kesey's. It was a foggy morning in November and the ceremony was a small one, attended only by the Loefflers and by Brand's Prankster pal artist Roy Sebern.

Brand realized that it was time for his own graduation from Kesey and the Pranksters as well as his attachment to mysticism. He had concluded that Kesey was running a cult; he bridled at the control the novelist asserted over the Pranksters. Years later, after seeing the same process around Baker Roshi, abbot at the San Francisco Zen Center, he would come to describe the use of such charisma as "theft."

Brand had also gone through the doors of perception to the cosmic place as described by Huxley enough times to realize that it wasn't news anymore. Also, in meeting the Nixons, he was becoming intrigued with ideas in alternative education while simultaneously catching a glimpse of the technology on the horizon of what would become Silicon Valley. It occurred to him that education and technology could be combined in intriguing new ways.

It was time for his next notion.

## Chapter 6

# Access to Tools

**W**HILE THEY BRAINSTORMED, THE TWO COUPLES DROVE all night, winding their way on and off highways and into neighborhoods and back alleys around the Bay Area. Stewart and Lois Brand and James and Cynthia Nixon spent the night rapping while Brand drove, exploring ways to blend the grassroots education work the Nixons had been doing at the Experimental College with the new computer technologies the Brands had learned were emerging in the technology world south of San Francisco.

Nixon and Brand represented polar opposites of the sixties counterculture. Brand, like Ken Kesey, identified with the "psychedelic side" of the antiwar movement while Nixon found inspiration in the Cuban and Chinese revolutions. Nixon wanted to take power while Brand wanted to circumvent it. In the end their differing philosophies would be the undoing of their partnership, and their clashes would leave Brand permanently embittered toward the New Left. The failure of their project would also lead directly to the *Whole Earth Catalog*.

The Brands roamed the country freely during the summer of 1967, intermittently returning to the Bay Area, where they parked their trailer in Menlo Park, just off the Stanford campus, at the home of Diana Shugart, a former Stanford student then working as a social worker, who became a close friend of the Brands'. When they arrived, Brand wrote in his journal that he had returned "to let my technology happen here."

Buckminster Fuller had opened his eyes, and he was spending his time reading widely with the intent of giving himself a "world technology education."

The consumer electronics industry was not yet anything like the force that it would become, but there were hints that something was afoot on the Midpeninsula. Brand would be one of the first people to catch wind of the remarkable technology transformation that was just over the horizon.

The future of computing and what would become the internet were being designed at two labs located on either side of the Stanford campus: the Stanford Artificial Intelligence Laboratory (SAIL) and the Augmentation Research Center at the Stanford Research Institute. Many of the young researchers in those labs were close to the counterculture, and through Dick Raymond, Brand became friends with a number of them.

One day he wandered into the office of Dave Evans, a young Australian computer researcher who was a key member of Doug Engelbart's Augmentation Research Center, and discovered a Janis Joplin poster on his wall. Brand decided that he felt quite at home in the emerging computing world.

He announced to Raymond that he was "sort of" looking for a job in business technology and hoped that Raymond would have either a position or a pointer to something interesting. Raymond had recently left his consulting position at the Stanford Research Institute. Caught up in the alternative education movement, he had created the Portola Institute the previous year as a haven for freelance entrepreneurs in the field of education. In an early profile of the *Whole Earth Catalog* that appeared in *Esquire* in 1970, Ed McLanahan and Gurney Norman, two young writers who had been fellows in the Wallace Stegner creative writing program at Stanford, described Raymond as conveying "an almost tangible aura of Goodness, a beatific quality that is part sincerity, part innocence ... a truly virtuous man."[1] His decision to step away from the high-pressure business consulting world was emblematic of the impact of the counterculture movement during the 1960s. The best and the brightest were leaving the establishment and striking out on their own to create alternative institutions.

The Portola Institute became an umbrella for a wide range of projects and would become Silicon Valley's first "incubator." Raymond pioneered the idea for nonprofits in the sixties, and as a result of his philosophy, the institute would have a significant impact. Both the *Whole Earth Catalog* and the Homebrew Computer Club, which gave rise to several dozen companies that forged the personal computer industry—including Apple—emerged from the fertile ground that Raymond created. He would also help pioneer the entrepreneurial culture that is still at the heart of Silicon Valley—where failure due to having taken risks is a badge of honor rather than shame.

Another Portola Institute founder was Bob Albrecht, a refugee from the mainframe computing world who, like Brand, had moved to San Francisco, attracted by the city's bohemian culture, with a passion for teaching children about computing. When Raymond invited Albrecht to join him in creating the Portola Institute, he moved to relatively bucolic and suburban Menlo Park.

The institute was housed in an unassuming commercial building in downtown Menlo Park, just off El Camino Real, the original route of the Spanish padres that runs from San Jose to San Francisco. Initially Raymond put in some money and Hewlett-Packard donated some more, but it didn't add up to much. The eclectic board of directors included Richard Baker, who would soon become the abbot of the San Francisco Zen Center, and Fannie Shaftel, professor at the School of Education at Stanford University. People would walk in off the streets with ideas, and the principal control mechanism was that the institute kept careful books and knew exactly what it was funding.

Several years after it opened, the Portola Institute served as the umbrella for Albrecht's publishing company, Dymax, a for-profit spinoff that took its name from Buckminster Fuller's term Dymaxion—the blend of *dynamic* and *maximize*. It was a word coined by Marc LeBrun, a teenager who had grown up in Portola Valley and who had been initiated into the world of computer hacking at the Stanford Artificial Intelligence

Laboratory. Several years later the publishing firm spawned a newsletter called the *People's Computer Company*. The cover of the first issue had a hand-drawn sketch by LeBrun. Across the top was written: "Computers are mostly / used against people instead of for people / used to control people instead of to <u>free</u> them / time to change all that— /we need a . . . / People's Computer Company."[2]

It was an eclectic scene that had its own magnetism. LeBrun had an office at the Portola Institute, and Shel Kaphan, one of his close friends from high school, spent more and more time hanging out with him. Kaphan became completely fascinated by the Whole Earth Truck Store and the *Catalog,* both located in the building next door. He had just graduated from high school, and his father, who was in the aerospace business, had gotten him a messenger job that he intensely disliked. Still, he was getting ready to go to college and he felt that he needed to work, so he walked next door and asked J. D. Smith, who was working as the store manager, if he could have a job.

"OK," Smith responded.

That would become his exposure to the mail-order business, as well as to a bookstore. It also gave him a front-row seat as a parade of interesting people would file in to visit Brand that summer, including Richard Brautigan, Ken Kesey, and Hugh Romney Jr. It was a great summer job. Kaphan would go on to become an early computer software hacker and would be hired by Jeff Bezos as the first Amazon employee.

And so it was. A growing group of hobbyists who had emerged from the institute discovered that computers could be used for more than just crunching numbers and were captivating even in their most primitive state—even machines that had to be laboriously programmed by toggling switches to enter individual instructions to start. The secret was beginning to trickle out into the world, no longer obvious only to the priesthood of engineers and programmers who had access to corporate computers. When Dymax moved to a tiny shopping center in Menlo Park, a People's Computer Center was created in the adjacent office, and soon terminals were connecting to a time-sharing computer service. People could walk in, program, and play games. The system, without even the blocky graphics of the first personal computers, was a powerful fantasy amplifier. Little more than text printed on paper by the teletypewriter terminals, the games were still remarkably compelling.

LeBrun was one of the hundreds of kids whom Bob Albrecht turned on to the power of computing. Albrecht had a clear vision of taking computers—at the time viewed by most people as cold, controlling machines—and liberating them so they could be used by anyone. It was a pure sixties vision: he became the Pied Piper of personal computing, intent on bringing the power of computing to the people.

That world was just over the horizon when Brand wandered into the institute in the summer of 1967. His friends were still heading back to the land, but Brand was one step ahead, looking toward a different future. Not long beforehand Raymond had met Michael Phillips, the San Francisco banker who had helped sell tickets at the Trips Festival. Phillips had organized an Educational Innovations Faire that had taken place on the San Francisco State campus as an early effort to explore the intersection of education and technology.

Raymond asked Brand if he wanted to extend Phillips's original educational fair and help organize a larger version. At first, he demurred, telling Raymond that he was through with public events for a while. The Vietnam War was just beginning to generate a national protest movement. Earlier in the year, the War Resisters League had asked him to organize a series of events at San Francisco State, UC Berkeley, and Stanford that he had dubbed "World War IV." His idea, in tune with Kesey and the Pranksters, was to promote "friendly violence." He was quoted in the San Francisco State student paper the *Daily Gater*, arguing that "a person's natural aggressive feelings are diverted harmlessly in mock fighting. You get the fun and satisfaction of fighting without inflicting pain—to any serious degree—on others."[3]

Afterward, Brand felt that his events, which were based on the idea of competitive games to work out aggressions—rather than war—had been failures. At Stanford he had a run-in with David Harris, then the student body president and a committed pacifist, who had rejected Brand's idea of "nonviolent" violence.

Spurred on by his need for work, however, he gradually warmed to the education and technology exhibition and several weeks later took Raymond up on his offer. Next, he brought in the Nixons.

As the project picked up speed, Brand invited a range of his Prankster friends to visit the institute, asking them to share their fantasies about the future of education. Together they decided they would hold a multiday event showcasing the best educational ideas and technologies at the San Mateo County Fairgrounds, tentatively planned for October of 1968. Albrecht, their most experienced computer industry contact, told them they could expect to attract roughly a hundred equipment makers, each willing to spend $50 and $300 for booth space.

It would also spotlight the ideas of two men in particular. One was Buckminster Fuller, whom Brand and Nixon approached at Esalen about the idea and who agreed to become a featured speaker. The other was Douglas Engelbart, the idealistic computer scientist who in 1962 had embarked on a plan to give humans powerful new intellectual tools by developing a system that would "augment" the human mind. Engelbart had assembled a small group of young computer hackers just across the railroad tracks from the institute's offices at the Stanford Research Institute.

Between the two men a complete vision of the role that tools could play in aiding an individual to remake his or her world came into view for Brand. A year later he would make their ideas the foundation for the *Whole Earth Catalog.*

IT WAS THE MIDPENINSULA FOLK DANCING SCENE THAT INITIALLY connected Brand to Engelbart. Every weekend there was a spirited gathering on an outside deck at the Stanford Student Union. Both Albrecht and Engelbart were feverish folk dancers (Albrecht hosted combo Greek dancing/computer classes at his home in San Francisco), and they soon became friends. This drew the computer researcher into the Portola Institute community as an informal adviser, and it opened a window into his computing research for both the Brands and the Nixons.

Engelbart had a grand vision of a "bootstrapping community"—small teams of knowledge workers supplemented by powerful computers—which he believed would accelerate human progress. Called the oNLine System, or NLS, while it ran on an early mainframe computer that connected workers at terminals using a suite of custom programs for collaboration and productivity, it was the forerunner of the Apple Macintosh, Microsoft Office, and the World Wide Web, several decades ahead of its time. Indeed, the original design papers for the ARPANET, the computer network that was the forerunner of the internet, specified that the first test of the network was to permit remote access to Engelbart's NLS software, partly because that software was used for the ARPANET Network Information Center that hosted all of the standards documents that defined the ARPANET.

Engelbart was a quiet evangelist for the idea that the power of computing would transform the world. While Brand was trying to recruit the Augmentation Research Center to participate in the fair, Engelbart was in turn recruiting Brand and Nixon to join his bootstrapping computing community. Finally, when the fair collapsed, largely from its inability to obtain foundation support for a proposed $138,000 budget and in part because of increasingly strident quarrels among the Nixons, Brand, and Raymond, Engelbart would briefly fold Nixon and Brand into his team in different ways.

Inspired by Engelbart, Brand, immediately after the failure of the education fair, began thinking about an idea he called the Electronic Interconnect Educated Intellect Operation, or E-I-E-I-O. Initially a fleeting idea, it would reemerge a decade and a half later as the Whole Earth 'Lectronic Link, or WELL, a computer conferencing system that Brand launched in 1985. Meanwhile, after his falling-out with Brand, Nixon worked at writing manuals for the NLS system from his home in San Francisco.

Ultimately the political chasm between Brand and Nixon could not be bridged. The Nixons thought of themselves as revolutionaries, and they were immersed in campus Marxism as well as the growing antiwar movement. Neither Brand nor Raymond agreed with them on almost anything political and, increasingly, on almost anything else. The project was further complicated by more intimate fragility. The Nixons would separate soon after the project ended. Jennings and Stewart were not doing well either. One night in bed Jennings complained that Brand's monologues were growing harsh and boring. She asked him to please "gentle up."

Their partnership on the brink of collapse, the two couples managed to produce and circulate a funding proposal for the education fair. They had a maximalist vision: "architecture, booths, exhibits, bands, lectures, student guides, wandering barkers, food, rides, art environments, electronic hardware, signs, souvenir stands, political harangues, movies, children, innovational demonstrations, dancers, newsmen, seminars, flags, computers, and people with cameras." Brand hoped to bring in everyone from IBM, RCA, GE, and the US Army to the Mime Troupe, a custom car show, and a steam calliope run by the Merry Pranksters. In 2008, four decades later, a Maker Faire, which felt very much like the idea Brand had initially envisioned, was held at the San Mateo County Fairgrounds.

After Raymond pulled the plug on the project, the Nixons left, while Brand, with Raymond's blessing, began to think about what he might do next. The experience had convinced him that rather than be the operations coordinator for the fair, he wanted to focus on new technologies. One night Brand and Jennings invited Engelbart to dinner, and over their meal he described a coming vast increase in computing power that would make the computer unique among all tools. Earlier in his career, Engelbart had given a talk on the subject at an electronics conference in Philadelphia, describing a coming exponential increase in computing power—an idea that would become known as Moore's law, based on predictions made by Intel cofounder Gordon Moore in a 1965 article.[4]

It was an idea that the futurist Herman Kahn was also spreading widely. A week earlier, at an Esalen talk that Brand attended, Kahn argued that computers increased in power tenfold every two years. That night at dinner Engelbart pointed out that a ten-fold increase would be a qualitative change, and that such changes made predictions about future technologies and their impact difficult. It offered Brand a hint of the world that Engelbart was inventing and a head start in understanding the transformation that Silicon Valley would bring to the world. Engelbart was busy developing what would become a universal intellectual tool, and his vision was very much on Brand's mind when he began to think more deeply about the importance of tools and learning.

At the same time, he was hesitant about completely accepting Engelbart's vision. The prematurely silver-haired engineer was experimenting with both electronics and "social" technologies. He was trying to draw on all of the ideas that were in the air on the Midpeninsula during the 1960s—everything from est, the "personal growth" movement, to LSD and even Maoism—to help build his community.

"Does the Bootstrap community have a chance anyway?" Brand wrote in his journal. "Is encounter groups etc. going to solve their problems? I'm afraid I doubt it. I Ching?"

He threw the I Ching and it returned "tread carefully." Then its message changed into "difficulty at beginning."

O

FROM HIS WINDOW SEAT, BRAND GAZED IDLY AT THE DARK EXpanse sliding beneath the plane. He had returned to Rockford to see his father before the end of his long struggle with cancer and then attended the funeral.

He took turns with his mother and his sister, sitting as they watched his father fade away. He didn't feel grief. Their relationship was a puzzle. He had never felt the same bond that he had had with his mother. Now Brand kept hearing his father's voice from just a few days earlier as he thrashed and tore at his various restraints and tubes: "I'm free. Hey, I'm free." Perhaps, in a sense, Brand was too.

Flying home after the funeral, somewhere over Nebraska, between chapters of the book *Spaceship Earth* by Barbara Ward, an early advocate of sustainable economics, he fell into a reverie about his friends who were building communes all over the rural countryside. He wondered what he might be able to do to assist them. He remembered growing up looking through the L.L.Bean catalog, and he mused about the value it provided—blankets, boots, sweaters, binoculars. It was a celebration of a somewhat mythic lifestyle that married tramping through New England wetlands during hunting season with sitting around the dinner table as cranberry sauce was served. But the catalog's fundamental value was in what it sold, some of which could not be easily acquired from other sources.

He was quietly musing about his family inheritance. It was something he had largely ignored throughout his twenties, but now that his father was gone, it struck him that he had a responsibility to use it wisely. Now he thought about it in the context of what he might do that would be helpful to his friends who were busy going back to the land and creating communes. He had already been involved on the edges of three such efforts, including the Garnerville Church, Lama, and Drop City, another early rural commune.

Brand realized that many of the challenges that his friends were facing came down to a question of access. What they needed, he decided, was a means to acquire a particular tool or service or knowledge about how to perform some task—not via piecemeal tips passed on from person to person but in a coherent, systematic, easily transferable manner. The *Whole Earth Catalog* would be subtitled *Access to Tools*, a bold statement implying the intent to enable skills—to reinvent, initially, the individual and, ultimately, civilization.

How to create such a service? He and Jennings had spent a portion of the past two years roving the West in Brand's VW bus, and he began to fantasize about a truck that could travel the countryside with information and samples of items that were worth having. It would be almost like a bookmobile, but instead of lending, it would be selling, and it would be spreading not just physical items but also information on how to use them. He added the notion of a catalog that would be continually updated, largely by its users.

Recalling Fuller's admonition that you have about ten minutes to capture an idea "before it recedes back into dreamland,"[5] he began scribbling furiously on the inside covers of Ward's book: "What I'm visualizing is an Access Mobile (accessory?) with all manner of access materials + advice for sale cheap. Including performances of stuff, books, dandy survival and camping equipment, catalogs, design plans, periodical subscriptions, copy equipment (+ other gathering equipment—some element of barter here). Prime item of course would be the catalog."

As an afterthought—in a riff on a then-popular backpacking catalog with a cult following, he added: "Notion: every catalog item pictured is held by a naked lady."

He would never finish reading *Spaceship Earth*.

When he arrived back in Menlo Park, once again something had changed in the way he viewed the world. Early the next morning he drove to the Portola Institute and pitched the idea to Dick Raymond. Raymond listened while Brand sketched his vision for the access catalog idea and then asked a couple of questions for which Brand had no ready answers: "Who do you consider as the audience for this 'catalog'? What kind of expenses do you think you'll have in the first year? What will be in the catalog? How often would you publish it? How many copies?"[6]

After some back and forth, Raymond said OK to the idea that Brand would move into the scheme gradually, bootstrapping the project from Portola's office and using its phone, stationery, and financial support.

Several weeks later Brand was ensconced in his new office when Raymond leaned his head into the room and asked, "By the way, what do you think you'll call it?" At this point, Brand hadn't even gotten that far. His 1966 Whole Earth campaign still echoing in his mind, he responded, off the cuff: "I dunno, *Whole Earth Catalog*, or something."[7]

In history, some years acquire what cultural historian Jackson Lears describes as a "numinous" quality.[8] He points to 1789, 1861, 1914, and 1968 as unique historical branch points. The same can be said of places. There are particular geographical locations where cultural, economic, political, and technological forces intersect and conspire to throw the world in a new direction. Venice and Florence lit the spark that led to the Renaissance, and pre–World War I Vienna served as an extraordinary cultural and scientific hotbed for modernism.

Silicon Valley, first named in 1971, has been a similar magnet for the forces that shape world history. Historians have paid attention both to William Hewlett and David Packard's garage in Professorville and to William Shockley's laboratory in Mountain View, but less attention has been paid to the extraordinary convergence of counterculture institutions in 1968 at what would soon become the northern boundary of Silicon Valley. It was not accidental that the *Whole Earth Catalog* emerged in the heart of this technology hothouse, shaped by the same forces and at exactly the same historical moment that birthed Silicon Valley.

One evening Brand brought Kesey by for a look at Engelbart's oNLine System. For an hour Dave Evans took the system through its paces, showing the writer how it would be possible to manipulate text, retrieve information, and collaborate with others. At the end of the demonstration, Kesey exhaled and said, "It's the next thing after acid."

LSD might expand your mind, but computing gave you a steering wheel. So would the *Catalog*.

O

W HEN BRAND DECIDED TO ABANDON ANY NOTION OF GOING back to the land and instead returned to Menlo Park to explore his "technology," the location of the Whole Earth Truck Store would prove to be fundamental. Despite his grand plan to become a tool supplier and educational library on wheels, another round of commune visits had provided ample market research that it wasn't a business; the communards simply had no money. The truck (a 1963 Dodge, to be precise) would be stocked and sent out several times, but it made much more sense to open an arm of the business without wheels attached.

On the face of it, the Truck Store was a hardware store, its mission to sell things that appeared in the *Catalog*. However, in addition to an eclectic array of intriguing and useful tools, many of which were hard to find, it mostly sold books, T-shirts, and other clothing, maps, and postcards.

Set just across El Camino Real from the Menlo Park business district in four thousand square feet of retail space with a volleyball court in back, the Truck Store was a storefront with a large warehouse space, where at one point Brand constructed a small bedroom, and another common room that served as a lunchroom off to one side. Raymond was able to swing a five-year lease for $450 a month, and by renting two offices, they brought it down to just $250.

Jim Wolpman had worked at a large establishment labor law firm in San Francisco. But the increasingly strident politics of the sixties led him to leave the firm, and he quickly became a people's lawyer on the Midpeninsula. He needed a place to hang a shingle, and Brand offered him space for an office at the back of the Truck Store. He also helped out by referring contacts and clients. They had met through Kesey on Perry Lane, but the two men never became close. Wolpman's law practice began by representing the victims of marijuana busts and then increasingly expanded into work defending the burgeoning antiwar movement. Brand was militantly apolitical in Wolpman's eyes, but in the *Whole Earth Catalog* he could see the influence of the Morley family hardware business. He had a ringside seat as the operation blossomed; to his mind the *Catalog* had much of the flavor of infinite hardware store, full of nooks and crannies and shelves with every possible gadget and component.

The Truck Store showcased just a fraction of that variety. By the spring of 1970 most of its goods consisted of books found in the *Catalog*, but on display could also be found Ashley stoves, a Corona corn mill, Aladdin lamps, Snugli baby carriers, and a few magazines.

Dennis Allison, a young physicist who worked several blocks away at the Stanford Research Institute, made a habit of visiting regularly during his lunch hour to peruse novel items he wouldn't find anywhere else. On one visit he found a pair of Mennonite pants that were good for gardening; another time he found a broad, sharp axlike tool, called an adze, to use in his garden. The Truck Store began as a counterculture boutique and eventually became a destination to which *Catalog* devotees could make a pilgrimage.

Serendipitously, the decision to locate the Truck Store adjacent to the Portola Institute at 558 Santa Cruz Avenue also placed Brand at the center of the emerging counterculture on the Midpeninsula. *Ramparts* magazine had been founded in offices just around the corner in 1962, before moving to San Francisco several years later. The Truck Store was also several blocks away from the already legendary Kepler's Books. Founded by peace activist and World War II conscientious objector Roy Kepler, it was part coffeehouse and gathering spot for the political and bohemian elements that were the heart of the gathering counterculture. In the evening the bookstore would play host to Jerry Garcia and his pals before the Grateful Dead was formed. (First as Mother McCree's Uptown Jug Champions and then as the Warlocks, they had begun playing at Magoo's, the local pizza parlor just up the street from the Truck Store on Santa Cruz Avenue.)

Around the corner on El Camino was East West Bookshop, an early alternative bookseller, while in the opposite direction was the storefront office of the Midpeninsula Free University, which had begun in the living room of a handful of local Marxists in the midsixties and would erupt into a powerful counterculture institution, briefly teaching as many as fifty thousand students at the height of the antiwar movement. Several blocks to the west was where the International Foundation for Advanced Study, the LSD research group, had been located until 1966, when LSD became an illegal drug.

The Truck Store was also only a few blocks from Douglas Engelbart's Augmentation Research Center laboratory at the Stanford Research Institute, and near where Bob Albrecht soon would open the People's Computer Company, which would ultimately serve as the incubator for the Homebrew Computer Club. In 1974, Raymond and the Portola Institute also helped nurture the nearby BriarPatch Food Co-op, which would thrive for the next decade. From the rich soil of this alternative community, the *Catalog* would shape the worldview of an entire generation of young Americans, key engineers and entrepreneurs in what would soon become Silicon Valley and would presage the 1970s environmental movement.

As had been the case since he'd arrived in the Bay Area, Brand's family inheritance, as well as a contribution from Jennings, was just enough to keep him from having to take a day job. Years later Brand would adopt the mantra "Live small, so you can live large." Indeed, when the Brands first arrived on Alpine Road in Menlo Park, they lived in a fifteen-foot-long box trailer that had little more than a bed and a tiny kitchen. It was a classic hippie pad—the bed was a mattress set directly on the trailer's floor. It was barely large enough for the two of them and their cat, Tao. They parked it in Diana Shugart's backyard on Alpine Road.

The *Catalog* would become an intellectual touchstone for the emerging alternative to mainstream American consumerism. In theory, at least, the back-to-the-land movement was onto something wondrous. To an increasing number of Americans, substituting a more intimate, low-footprint relationship with rural living was a brave and radical escape from an economic system that encouraged violent oppression, personal alienation, and the destruction of the environment. They would build their homes, till their fields, and sell homemade crafts to the outside world.

After he launched the Whole Earth Truck Store during the spring of 1968, Brand spent almost $6,000 on an additional Airstream trailer and a used Dodge Power Wagon equipped with extra fuel tanks and parked them on Shugart's property. Shugart would go to work at the Truck Store as an office manager as the enterprise grew.

Although Kesey was in the process of moving to Oregon, many of the Pranksters were still around, along with a number of fellow travelers. One day Jennings heard a man shouting in the street in front of Shugart's home. It turned out to be Hugh Romney, later to go by Wavy Gravy, the leader of the Hog Farm commune. He was yelling at a young woman who was on a bad trip, attempting to bring her down by threatening to staple his penis to the front of his car. Weird, Jennings thought, though no weirder than many other events she had witnessed.

Jennings assiduously avoided the drug scene, never became a Prankster, and would become the driving force and effective general manager of the *Catalog* and the Truck Store. A "hidden figure" when she was a programmer working for the navy, she was long underappreciated for her role in creating the *Catalog* as well. John Brockman visited during the Truck Store's first year and spent the day sitting with Brand painstakingly underlining passages in Norbert Wiener's *The Human Use of Human Beings*, an early book on cybernetics, while Jennings, seated across the room, worked at laying out an issue of the *Catalog*. (At the fiftieth anniversary, held in San Francisco in October of 2018, from the stage Brand would acknowledge that Jennings had been the cofounder of the *Catalog*.)

Brand and Jennings had found their way into a tight community less than a mile from Perry Lane, the bohemian focal point of the Midpeninsula, where Brand had first met Ken Kesey. Members of the Grateful Dead had lived in a rambling mansion known as the Chateau about halfway between the two neighborhoods. Dick Raymond lived just up the road a mile farther along Alpine Road in a middle-class neighborhood known as Ladera. Vic Lovell, Shugart's former boyfriend and a psychology graduate student who alerted Kesey to the LSD experiments at the Menlo Park Veterans Administration Hospital, and Richard Alpert, as well as the writers and Kesey pals Robert Stone and Ed McClanahan, had previously lived just a couple of blocks over on Homer Lane, as had several Merry Pranksters. In a cottage at the other end of the tiny neighborhood, Bill English, Doug Engelbart's chief engineer, was renting a home from Raymond. English would be closely involved in several conferences on sustainability and environmental design that grew out of the *Catalog*. His wife, Roberta, was working as a secretary for Engelbart (and as a result was probably the first secretary to ever use an interactive word processor with video display). They would both be *Catalog* volunteers, and Brand would marry them in a ceremony in Shugart's backyard.

Although Brand had underwritten the project with his family inheritance, it was operated as a nonprofit and rooted in the egalitarian values of its time and place. Wages ranged from $2 to $5 an hour; even Brand as the publisher was taking home only about $10,000 annually. The staff would grow to almost thirty by the time of the release of *The Last Whole Earth Catalog* in 1971. Volleyball in a lot behind the store became a daily event, and Engelbart's programmers and engineers would occasionally join in. (Because it was played during lunch on paid time, teams tried to stretch games as long as possible.)

The *Catalog* itself was first announced to the world in May 1968 in a Portola Institute marketing brochure, offering $8 annual subscriptions covering two issues and two supplements, and setting the single-issue price at $5. "The function of this program is improving access to tools for self-dependent self-education, individual or cooperative," the brief announcement read, adding that as a preliminary to the *Catalog*, the Truck Store would travel to intentional communities, experimental schools, and design departments during the summer to do market research.

In preparing for his new project, Brand went back to the roots of his senior year magazine-writing class at Stanford. His first stop was to visit Bill Lane, the publisher of *Sunset* magazine. He remembered a basic lesson from his class: that it would take a minimum of $1 million to launch a national magazine. But having had an early hint that new publishing technology was going to change the equation, he was undaunted.

Lane suggested a do-it-yourself approach. He counseled Brand to shoot his own photographs, but he should do so using a Hasselblad, a top-of-the-line large format camera, so it wouldn't appear that he was cutting corners on quality. Soon the decision was made not to carry advertisements. The philosophy was simple as was spelled out in the *Catalog*: "We don't carry ads anymore, if you have a product, let us see it, if we like it, you don't owe us anything."[9]

Brand also kept his hand in on multimedia projects even while he was launching the *Catalog*. In June, shortly before he and Jennings headed out on the road for the summer, Portola Institute announced *WAR:GOD*, an abstract multimedia extravaganza based on images "polarized around spirituality and violence." It was an extension of some of the ideas in World War IV that had angered David Harris. Brand used two Kodak slide projectors, a 16 mm projector, and a stereo tape recorder to create an hour-long presentation that never repeated itself because the slides would be resequenced between each performance. He took the soundtrack from almost two dozen sources, including Ramon Sender's electronic music, the Mormon Tabernacle Choir, the movie *A Fistful of Dollars*, Malcolm X, the Royal Shakespeare Company, Franklin Roosevelt, Ken Kesey, the Grace Cathedral Choir, and Adolf Hitler, among others.

Like *America Needs Indians!*, *WAR:GOD* received good reviews[10] and might have gone further, but after seeing it, Jerry Mander pointed out that most of the images and some of the audio had been "borrowed" without copyright license and that was that.

In July, Brand produced a prototype of the *Catalog*—a four-page typed "Partial Preliminary Booklist" that began with Buckminster Fuller's *No More Secondhand God* under the heading "Understanding Whole Systems" and concluded with descriptions of two magazines, the *Modern Utopian* and the *Realist*. In between, he listed favorite magazines and lumped books into three design-oriented categories: Shelter and Land Use, Industry and Craft, and Communications. By the time the first *Catalog* appeared in October, the category list would expand to include Community, Nomadics, and Learning. Brand took his inspiration for the size and format of the first *Catalog* from Steve Baer's oversize *Dome Cookbook*, which had been published the previous year, and drew together his drawings of geometric shapes, handwritten memories of building various domes, typewritten instructions, and black-and-white photos.

Baer, introduced to Brand by Steve Durkee, was an early example of the diverse assembly of technophiles Brand organized Tom Sawyer–style to help create the *Catalog*. A fellow army vet, Baer had originally been inspired by Lewis Mumford to view technology as a positive force for society, and he played a key role in the design of the dome residences at the early Colorado commune Drop City.

Brand was gradually moving away from LSD, but he still had a penchant for psychedelics. For a while an E tank of nitrous oxide, ordered weekly, was a permanent fixture at the Truck Store office. It was a convenient quick high, or "flash," as Brand liked to think of it. It was kind of a workingman's drug. You could take a hit and if you liked it, you could take another. It wasn't like acid or grass. Afterward you could drive home safely, and unlike with LSD, if you had a bad trip, you weren't stuck in hell for the next eight hours. He introduced Dick Raymond to "gas"—as he called it—and then declared that his mentor had passed the Acid Test in just forty-four seconds. Brand became highly dependent on this instant rush that quickly brought you right back to reality—until one day he just kept flashing.

In November, just after he had published the first *Catalog*, he found himself crying helplessly at the end of *Camelot*, the musical comedy-drama. Soon thereafter, he was listening to the Beatles' *Sgt. Pepper's Lonely Hearts Club Band* when suddenly the world went away, and then he found himself on the other side of a large room with Jennings shaking him, saying, "You're laughing hysterically."

Clearly, he had gone over some physiological edge with nitrous oxide, so he abruptly stopped using it. Before he stopped, however, nitrous oxide—or perhaps the combination of the explosive growth of the *Catalog*, nitrous oxide, LSD, and a crumbling marriage—would push him into a deepening depression. It would be a significant factor in his decision to place an end date on the *Catalog*: 1971—in the fall of 1969, only a year after its first publication, just as it began its exponential growth.

S ANDY TCHEREPNIN HAD GONE WEST AFTER HARVARD TO LOOK for a new life. Through friends, she spent the summer of 1968 living in Marin County with a group of mostly former Stanford students. At the end of the summer, they gave her a puppy and told her it was time to move on. They also told her that Stewart Brand was looking to hire someone for a new project in Menlo Park. Coincidentally, she knew who Brand was: one day walking across Harvard Square after class she had seen this odd fellow in a top hat, standing in the square with a billboard that read "Why haven't we seen a photograph of the whole Earth yet?"

So she hopped in her VW bug and headed to Dick Raymond's house for an interview, where she learned about the *Catalog*. Brand was looking for a typist to help with the production operation, which they planned to establish in a rambling house owned by Raymond that was set among the redwoods on Skyline Boulevard in the mountains above Stanford.

Tcherepnin had a crucial qualification—she was a good typist. After a brief conversation, Brand told her the job was hers and invited her over to the Portola Institute for a celebratory round of nitrous oxide. She became the *Whole Earth Catalog*'s first employee.

After a short time working at the Portola Institute office, the *Catalog* team set up shop in the garage of Rancho Diablo, a seventy-acre former hippie crash pad with views of the Pacific Ocean. For a while they all commuted from Menlo Park up into the hills to the garage. Later Brand and Jennings moved the two trailers up to Skyline. At one point they lived in the house as caretakers and at another point they lived in the trailers set among the redwoods on the Rancho Diablo property. Production was laid out around a wood-burning stove set in the middle of the garage; sometimes when Brand came across a book that he felt was truly terrible, he would toss it into the fire.

He kept his desk in a corner next to the garage door, while typesetting, photography, and pasteup were each arrayed in the other corners of the large, open room. They began by using beeswax to paste down each article and eventually moved on to rubber cement. Sometimes *Catalog* entries would be cut from a book or magazine and be pasted directly into the layout. In the summer when it got hot they just opened the garage doors.

The first sixty-four-page *Catalog* was printed at a small Menlo Park printing press located around the corner from the Truck Store, and then the cover, which was printed separately in color, was saddle-stitched to create the finished product. When they began, desktop publishing did not exist, but for $150 a month, Brand leased an IBM Selectric Composer, an advanced version of the company's workhorse electric typewriter that had been introduced in 1966. The composer was capable of producing camera-ready justified copy with proportional fonts and it opened the door to low-cost publishing. (The Fall 1969 *Catalog* cost only $33,000 to produce.)

The new desktop publishing tools meshed perfectly with Brand's editorial design for the *Catalog*. By purchasing an $850 halftone camera, he freed the *Catalog* from the world of print shops and graphic design houses completely. It made self-publishing not only possible but easy and adaptive in real time. He had decreed there would be no politics—although there clearly were, both between the lines and otherwise. He went about recruiting short, pithy items, mostly mentions of useful books, magazines, and journals, from an expanding list of contacts. There would be only positive reviews because the point was to provide people with what they needed, not what they didn't. The contributor's fee was ten dollars for about two hundred words plus a byline.

It was all tied together by Brand's introductory sentence: "We are as gods and might as well get used to it." (In a second printing of the first edition it was edited to read "get good at it. . . .") The notion, he later acknowledged, was borrowed from Edmund Leach, the British social anthropologist who in 1967 had given a series of lectures focused on the interconnectedness of the world and humanity's relationship to the environment. *A Runaway World?*, a book based on his lectures, begins: "Men have become like gods. Isn't it about time that we understood our divinity?"[11]

Brand's introduction contained a simple rationale, divided into "FUNCTION" and "PURPOSE." The *Catalog* was to be an evaluation and access "device"—making it possible for the user "to know better what is worth getting and where and how to do the getting." To be listed in the *Catalog*, an item must be useful as a tool; relevant to independent education; high quality or low cost; not already common knowledge; and easily available by mail.

Under "PURPOSE" the *Catalog* noted that government, big business, formal education, and the church had gone about as far as possible. Now a "realm of personal power" was ready to "find his own inspiration, shape his own environment, and share his adventure with whoever is interested."

Both the front and back covers featured a large color photo of the Earth taken by a NASA weather satellite. On the back cover, Brand added a final philosophical touch in large type: "We can't put it together. It is together."

The inside cover of the *Catalog* offered a directory dividing its contents into seven categories, ending with "Learning," telegraphing Brand's enduring passion. The interior layout had the informality of a scrapbook—items didn't line up, and they would sprawl across the pages. Some pages were so packed with a hodgepodge of images (drawings, charts, photographs) and text that they brought to mind the collage of Brand's multimedia experiments. All told there were microreviews of 135 items in the first *Catalog*, including shout-outs to the Sierra Club, Heathkit electronics kits, Kaibab boots, and Richard Brautigan's novella *Trout Fishing in America*.

You could find advice about complications when cutting the cord after birth; a short discussion of whether "sportsmen should take dope"; an endorsement of *New Scientist* magazine; a reproduction of the Ariadne astrological symbol; an article about how house shells quickly made from epoxy could solve the world's housing crisis; a discussion of "bioholography," which argued that animals take advantage of "a 'coherent' level of background ultrasound"; a cartoon of a naked woman with an elephant head; instructions (printed upside down) on how to correctly use a barometer to measure building height; sketches of the molecular structure of LSD, psilocybin, trimethoxyamphetamine, and inactive bufotenine under the headline ALL THE WAY WITH DNA; and an analysis of the causes of urban rioting—and that was just up to page twenty-four. Of course the drug descriptions came without an inventory of things to order. Three pages later one could find illustrated descriptions, with ordering information, of hacksaws; flexible files ("impregnated with super-hard aluminum abrasives"); a compact tool and knife chest; eight-and-a-half-inch shears; a "Model 355 Midgetester" for measuring voltage and resistance; an *Atlas of Landforms* from the US Geological Survey; a "G1717 Engineering Compass" that was "similar to Model G1719 Geological Compass except that it does not have a pendulum clinometer, level bubble or extension rule"; as well as information about how best to make potholes with explosives and a tribute to internal pipe wrenches. And there it was: How many people heading to northern New Mexico to start their commune had ever thought of the dilemma of poorly excavated potholes before arriving? The *Whole Earth Catalog* had. There was even a page of products from good old L.L.Bean.

But much of what filled the *Catalog* were listings for books and magazines—practical guides to glassblowing and woodcrafts and how to grow organic vegetables, but also books about dolphin neuroscience; Arthur Koestler's *Ghost in the Machine* and *The Act of Creation*; an anthology of John Cage's writings and lectures; Joseph Needham's *Science and Civilization in China, volume IV:2*; Frank Herbert's *Dune*; the USGS's *Thermal Springs of the United States and Other Countries and the World—A Summary*; Leslie M. Lecron's *Self-Hypnotism: The Technique and Its Use in Daily Living*; Matsuo Bashō's *The Narrow Road to the Deep North*; the I Ching; and scores more. *National Geographic* was in there, but so were *American Cinematographer* and *Architectural Digest*. There were guides to auto repair, home refrigeration, welding, selling honey, mushroom hunting, operating a one-man sawmill, and solar power. Endorsements could be expert or playful. ("Van Waters & Rogers is a huge supply house. I don't know anything about them except that they have a hard-bound catalog this thick of illegal-looking equipment.") Sometimes a book would get only a summary; for those of greater ambition or complexity a section run-through was occasionally offered. (The content breakdown for the *Yearbook of the Society for General Systems Research,* volume IX, included "Concession-Making in Experimental Conditions," "Some Considerations on the Notion of Invariant Field in Linguistics," and "Toward a

Unified Theory of Cognition," among numerous other such condensations for that volume alone.) Credit was given to outsiders who suggested inclusion of an item; as for his own commentary, as a longtime fan of the *Realist*, Brand adopted Paul Krassner's trademark "PK" signature style, signing his articles in the *Catalog* "SB."

From the outset, while much of the counterculture rejected computing technology for being a central component of the bureaucratic mainstream world they dismissed, Brand embraced it.

"I worked for 2 hours last night online with a fantastically sophisticated interactive computer system," he wrote, describing Engelbart's computing system to Baer. "And then dreamt most of the night among the tree of choices and clarities."

Although it was still more than seven years before the first hobbyist personal computer would appear, the *Catalog* was sprinkled with hints that the power of computing might be seized from corporations and the military. A few pages after listings of buckskin ("hair-on-calf unclipped $1.60/sq. foot") and yarn (fifty cents for a catalog and sample card) was this:

### 9100A Calculator

The best of the new table-top number crunchers is this Hewlett-Packard machine. It is programmable, versatile, and fast—more so than its competition. Portola Institute currently is using the 9100A to help kids gain early mastery of computers—it is a superb inquiry machine.

Later in the *Catalog* was an endorsement of *We Built Our Own Computers*, a British book, accompanied by a photo of the "Electric Logical Computer Exeter." And a page before the 9100A was a two-page spread on *Human Biocomputer*, which "offers the opportunity to learn and explore computers without requiring money or administrative approval." The signs were there from the beginning.

Fuller's systems thinking perspective and a cybernetic approach to information spilled out of the *Catalog*. Fuller and Norbert Wiener were highlighted up front (the first two pages of listings—the first thing a reader would encounter after the introductory material—concerned material written by Fuller), and deeper within were discussions (with ordering instructions) of Fuller's *Education Automation* and Weiner's *Cybernetics: or Control and Communication in the Animal and the Machine*. For many of the *Catalog*'s readers, it was the first time they had heard about Wiener's science of communications and control. (By the early 1970s, cybernetics would take root more deeply in Europe than in the United States, where Wiener was viewed as an arrogant and prickly outsider. Indeed, John McCarthy, the computer scientist and mathematician, had come up with the term *artificial intelligence* because, he wrote, he was trying to avoid contact and quarrels with Wiener and his devotees.[12])

Brand saw the *Catalog* not as a stand-alone document but as part of a dynamic system, and throughout its existence he added regular supplements to offer a channel for feedback in what he believed would be a self-sustaining organism.[13] Brand's insistence on this feedback loop added more currency and complexity to what thus became a sort of "living" document—*The Last Whole Earth Catalog*, published in 1971, would offer a vast menu of items sprawling over almost 500 pages.

The *Whole Earth Catalog* supplements offered a window into the communal back-to-the-land movement that had captured the spirit of the counterculture. The January 1969 issue featured this stark Prankster-friendly statement in large block letters on its cover: "BRAIN DAMAGE IS WHAT WE HAD IN MIND ALL ALONG. CHROMOSOME DAMAGE IS JUST GRAVY."

Inside was reprinted correspondence from Mark, Vicki, Bernie, Diane, Parscal, and Liz, who were writing that they were headed from Detroit to Oregon and were looking for a place to stay for an "indefinite amount of time." Incredibly excited about their upcoming adventure, they concluded by noting: "This last weekend the Doors shouted out to the audience to break on through to the other side."

In a startling handwritten scrawl, George, Liza, and Dave wrote back from their Oregon farm with an impassioned plea asking the supplicants to please stay away: "Help!!! We're all dying of leprosy!!! Bring lots of bandages, clean gauze, old rags, anything!!![14]

The production of the first two *Catalog*s was an intense round-the-clock marathon, with the burden falling almost entirely on Brand and Jennings. Despite the spreading counterculture that surrounded them, they had adopted a largely conventional marriage. They would spend long hours putting together the *Catalog* and then come home to the trailer, where Jennings would make dinner, wash the clothes, and then do the project accounting while Brand "sort of spaced out in front of the TV and got inspired."[15] The project quickly began to take its toll on their marriage.

Dick and Ann Raymond had recently been divorced, and Brand had difficulty accepting their separation. She had met a therapist, whom she would marry, and she invited Brand and Jennings to dinner one night to meet him. She was shocked when a sullen Brand set what appeared to be a Colt .45 revolver on the table and remained almost completely silent during the meal. She didn't know it at the time, but it was, in fact, not a real gun but rather a realistic toy that would attain national notoriety that year when it appeared in Lois Jennings's hand in the first chapter of Tom Wolfe's *The Electric Kool-Aid Acid Test,* which described the couple driving to meet Ken Kesey, who was getting out of jail: "And, oh yeah, there's a long-barreled Colt .45 revolver in her hand, only nobody on the street can tell it's a cap pistol as she pegs away, *kheeew, kheeew,* at the erupting marshmallow faces. . . ."[16] No matter its authenticity, the gun and Brand's brooding left a lasting impression.

While she lived in Ladera, Ann Raymond had become an informal den mother for a neighborhood group of bright youngsters, several of whom, like LeBrun, would soon go to work at the Portola Institute or the Whole Earth Truck Store. The first to arrive was Joe Bonner (he would later change his name to Dwarka), one of two sons of a Stanford University chemistry professor. He had studied art with Ann Raymond for five years, and shortly after graduating from high school, he became the *Catalog*'s first pasteup artist, living in the Brands' trailer while helping to finish production.

The IBM Selectric Composer was a revelation for Bonner. It provided a kind of immediate flexibility that hadn't previously been possible. He could simply tell the typist, "Okay, make this one so wide," and they changed the width and then printed it out again without having to reenter all the text.

The *Catalog*'s first print run was a minuscule one thousand copies. After he retrieved the first batch from the printer, Brand drove out to the beach at San Gregorio. It was the spot where he had first touched the Pacific Ocean. Sentimentally, he tossed a *Catalog* out into the waves— he considered it a symbol of launching his publication into a timeless ocean. It was a ritual he would repeat for the first several issues.

Shortly thereafter, Bill English invited him to participate in several planning sessions ahead of a demonstration that Engelbart's computer designers were preparing for an industry and academic meeting known as the Fall Joint Computer Conference in San Francisco. English had seen both *America Needs Indians!* and *WAR:GOD* and believed that his planned demonstration to the world's one thousand leading computer scientists and engineers could use Brand's multimedia expertise.

At this stage, the computer industry was dominated by a handful of mainframe computer makers, and computing itself was not directly interactive: you entered your program onto punch cards that were read into a processor that then ran your program. The output was returned on a computer printout.

That was all about to change.

Engelbart was a dreamer who envisioned a style of computing that was instantly responsive to a user who would interact with the machine through a mouse, a keyboard, and a screen. It would be almost two decades until this became the standard way that personal computers were used by the general public. But Brand was one of the few humans on the planet given an inkling of what was around the corner.

In December, at an event that would later be called the Mother of All Demos,[17] Engelbart sat onstage at the conference in front of a giant video screen and introduced this new style of computing to the world.

Even computer scientists had not seen computer mice and interactive displays before, nor had they learned about the hypertext concept for linking documents that would decades later become the basis of the World Wide Web. The computer itself was located in Menlo Park, just blocks from the Portola Institute. Brand was on hand there to operate a video camera that would relay images of the computer operators to the conference. He was an observer, but there was no video feed of the audience to see the crowd reaction to the demonstration.

When Engelbart ended his presentation, the Menlo Park team asked over the microwave relay, "Did they like it?" They waited for what seemed like a very long time, and then the answer came back: "Yes, they liked it." Engelbart had received a standing ovation.

There was a deep resonance between Engelbart's intelligence-augmentation vision and Brand's access-to-tools philosophy. Although the *Catalog* was neither fish nor fowl (he wasn't really selling things—just telling people where they could find them—and it wasn't actually a book, which confused bookstores), sales were immediately brisk. Soon, despite the fact that he had trouble persuading bookstores to carry the *Catalog*, there was a second print run of another thousand copies.

Early on, there were few reviews. One appeared in *Scientific American*, written by physicist Philip Morrison, who had met Brand when he was on his "Why haven't we seen a photograph of the whole Earth yet?" tour in Cambridge and purchased a button. Brand had sent him a copy of the *Catalog*, and Morrison described it as "good reading" from a "strangely attractive subculture, that of the dissenter and reformer who seeks to construct a philosophical, personal and economic refuge from the curious industrial society."[18] Morrison's review was important. It set a tone for the *Catalog* to be seen as "cool."

Not long afterward, the *Washington Post* columnist Nicholas von Hoffman wrote a column about the *Catalog* that was syndicated nationally, resulting in lots of attention. What really boosted subscriptions and purchases were other mentions in newspapers. At one point in 1969, "Uncle Ben Sez," an advice column in the *Detroit Free Press,* received a letter from someone asking about how to go "back to the land." "Get the *Whole Earth Catalog*," the columnist responded, adding an address to write to. That one mention generated scores of magazine sales and subscriptions.

"We're getting very friendly reviews in the underground press and extraordinary mail," Brand wrote in a letter to his mother. They had received 650 subscriptions and sold out their second printing of the first *Catalog*. They were up to seven employees, and in the summer they planned to take the operation on the road and produce a supplement in New Mexico or Colorado, "where a lot of the action is this summer."

It was soon even clearer that they were riding a rocket ship. Brand had advanced the project about $25,000 from his family inheritance to produce the first *Catalog*. After selling out the first two thousand, they did ten thousand more in the spring, followed by a second run of twenty thousand. In July of 1969, the *Catalog* operation had its first profitable month, taking in almost $16,000 in income against $8,000 in expenses. For the Fall 1969 issue, they printed sixty thousand copies and had four thousand subscribers.

For many readers, stumbling upon something in the *Catalog* would prove a transformative experience, sending their lives careening in a new direction. That was the case with Jamis MacNiven, who had been a loyal reader for several years and was working as a traveling salesman selling milking machines to small dairies in Connecticut. One day the snowfall was so heavy he and his wife literally couldn't find their home on the street where they lived. They decided that they had to move to California, where he'd gone to college. He moved back to the Bay Area and used the *Catalog* as an instruction manual to teach himself to become a general contractor. He soon talked his way into his first Berkeley home remodeling job and then was hired by a young Steve Jobs to remodel his home in Saratoga. Ultimately, MacNiven opened Buck's, a popular Silicon Valley restaurant, and he would build his own home off the grid in the Santa Cruz Mountains, not far from where the *Catalog* was produced.

Within a couple of years the *Catalog* became synonymous with the counterculture. Journalist Shana Alexander profiled Marlon Brando in *Life* magazine, reporting that she had spotted the *Catalog* lying on the floor next to his bare feet. He told her that in the publication he had finally found a statement of purpose that matched his own.

National recognition, however, meant more pressure and more work, and it quickly began to take a toll. In September, after just eleven months, Brand announced, "The CATALOG has but 20 months to live." Then he added: "The function of the skyrocket is to get as high as possible before it blows."

As work expanded, staff was added to run the Truck Store as well as produce the *Catalog*. By the fall of 1969, the team had grown to fourteen. Most of the new employees were young, some right out of high school. To them Brand was something of a distant figure who kept to himself, reading books and writing quick reviews in a small cubbyhole at the store. But as the workload of the *Catalog* relentlessly increased, Shugart noticed Brand's mood darkening. Each day, he would dig just a little bit of a coffin-shaped hole in her front yard.

At one point she asked him what he was doing. He said that after he had begun the *Whole Earth Catalog*, he asked God, "Okay, God, what's next?"

And God replied, "Sorry, Stewart, that was it."

O

IN 1968, JAMES T. BALDWIN, OFTEN KNOWN AS JAY BALDWIN OR J. Baldwin, was teaching industrial design at San Francisco State University when the departmental secretary came to him during his lunch hour and said that someone was waiting to see him.

The sandy-haired man who was waiting outside of the design workshop was wearing a white jumpsuit and sporting a small holographic disk attached to his forehead.

"I have this idea," the man said, "that with three phone calls you ought to be able to find out anything about anything that exists in the world."

"There's only one place where you can find a catalog of everything in the world," Baldwin responded. "The Yellow Pages of the Manhattan phone book."

Stewart Brand responded, "Well, I've heard you read catalogs. Why should the Manhattan phone book make a difference?"

"What's important about a product is you can get it," Baldwin replied.

A disciple of Buckminster Fuller like Brand, Baldwin was a stickler for the notion that design ideas actually work. He had made a rubber stamp that read "Bring it around and we'll take it for a drive."

Baldwin was five years older than Brand and both men had served in the US Army. They both shared a love of tools, and while Brand had left the military with an antipathy for mindless bureaucracy, Baldwin, the son of an AT&T engineer, had left with a parallel distaste for bad design.[19] Baldwin had walked away from a career as a commercial industrial designer when he realized that he believed his designs had to have human rather than merely commercial value.

He had worked for Moss Tents, an early maker of innovative camping and backpacking gear, before moving to California to attend graduate school. At UC Berkeley he was involved in the Free Speech Movement and then was fortuitously hired to teach industrial design at San Francisco State. An inveterate tinkerer and prototyper, he would become part of a small cadre of "outlaw" designers and builders who were the heart and soul of the *Catalog*.

Baldwin was one of Brand's first serious recruits. He soon became the editor of the "Nomadics" section and stayed with Brand when he reinvented the *Catalog* as the *CoEvolution Quarterly* in the 1970s. Soon after meeting Brand, Baldwin received a letter from Fuller offering him a teaching job at Southern Illinois University—the letter appeared in his office mailbox simultaneously with a pink slip from San Francisco State. He left the Bay Area, but he would remain an integral part of the informal guild of *Catalog* contributors.

It was through Baldwin that Brand met Lloyd Kahn, who would become the editor of the "Shelter" section of the *Catalog*. In 1965, Kahn quit the insurance business to become a self-taught home builder, eventually specializing in domes. He had gotten a job as the foreman on a project to build a home in Big Sur employing giant thirty-foot-long bridge timbers. It was nicely designed, but it was also remarkably heavy. That winter Buckminster Fuller had come to lecture at Esalen and spoke about building geodesic domes. Kahn was struggling with immense timbers that weighed as much as half a ton, and he was stunned to hear Fuller describe these ethereal, lightweight buildings.

He quit his job and began building domes.

The counterculture was in full bloom, and he built his own home in Big Sur complete with a terraced organic garden. When he had been a Stanford undergraduate three years before Brand, there hadn't been any discussion of Eastern religions in his Western civilization class. Now, however, the baby boomers were exploring a vast array of ideas that had been missed by Kahn's silent generation: cosmology, mind expansion, communication with dolphins, ecology, political activism, communes, and on and on and on.

As the counterculture burgeoned, he discovered he was frequently writing people letters about how to build domes. In fact, he was writing the same letter over and over again. He was thinking about publishing his own book on how to build a dome, throwing in some how-to information on organic gardening, and how to get chickens by mail from Iowa, when he met Baldwin. Baldwin told Kahn about the Whole Earth Truck Store, so he drove to Menlo Park to meet Brand. When he read the first issue of the *Catalog*, he realized that Brand was way ahead of him, and he didn't need to publish anything. He was thrilled with the *Catalog* and ordered five copies.

He also could see that Brand was a cerebral type, not a true back-to-the-lander, who needed "earthy" people like him to round out his stable of writers and researchers. In the back of the Truck Store, Brand was literally buried in books and Jennings was complaining, "Stewart, you're buying too many books!" Simultaneously, Brand recognized Kahn's value and drove down to Big Sur with his production crew in the camouflage-repainted Volkswagen bus with an army star on the front door.

Soon thereafter, Kahn sent a long letter to Brand detailing a list of interesting references resonating with the commune movement, which Brand printed in the first "Difficult but Possible" supplement to the *Catalog* in January 1969:

"I'm writing from the standpoint of having gone out in the semi-country, built a house, put in a water system, a garden, and I'm now building a shop and going to farm in the spring," he wrote. "We started with a bare hillside and had to do everything alone, and because of this, whatever pertinent information I could get ahold of was vital, because there was no one to ask."

In the early *Catalog*s and supplements Baer, Baldwin, Durkee, and Kahn would become Brand's musketeers, animating its pages with authentic do-it-yourself back-to-the-land wisdom.

There were distractions. Brand had begun to get some national visibility, and he would frequently receive invitations to speak, occasionally on subjects related to psychedelics. At the end of February, he was invited to appear on a panel at SUNY Buffalo on "Drugs and the Arts" featuring Allen Ginsberg and Ken Kesey, as part of an ambitious "New Worlds" conference on psychedelic drugs. He took the train across the country, on the way reading Paul Ehrlich's *The Population Bomb*.

After appearing on the panel, he showed *WAR:GOD* as the warm-up act for a rock concert featuring the MC5, then at the height of their "Kick out the jams, motherfuckers!" fame. Another panel featured Ralph Metzner, an LSD pioneer, and antiwar yippies Jerry Rubin, Abbie Hoffman, and Paul Krassner. Brand had met Hoffman through John Brockman, and he continued to be attracted to the "psychedelic" side of the Vietnam War protest movement. That was certainly an accurate description of Hoffman, who would become embroiled in the trial of the Chicago 8 that fall, and who took pleasure in being absolutely outrageous. One time when he and Brand went to a New York City café on the Lower East Side, as they walked in Hoffman saw some plates with unfinished food and immediately went over and began consuming the leftovers. Everyone in the restaurant was confused: What is happening? Is this theft? Or is it really good conservation of food, because otherwise that stuff is just garbage? But it's not garbage. It's still warm.

Brand loved it.

To Brand's mind, Hoffman represented what the counterculture was at its best—taking the practices that have very good reasons to exist and just messing with them a little bit to see what they're made of, what you're made of, what relations are like when you mess with them.

That evening at Leary's appearance, Hoffman and Brand made their way to the front of a large auditorium full of several thousand students and hid under a table set directly in front of the stage. Leary, his graying hair tied back in a ponytail, came onstage accompanied by his young wife, Rosemary Woodruff Leary. Before Leary could begin to speak, Hoffman called out in a loud voice: *"Hey! When's Rosemary gonna have a baby?!"* The horror film *Rosemary's Baby* had just begun playing in theaters, and Leary was visibly flustered by the taunt. He looked around and couldn't see who was baiting him. After a while he left the stage, dismissing the audience as he walked out: "We go around once in life and it's a shame to just waste it and I'm leaving now."

The crowd was confused. People were asking, "What just happened? Who insulted the guru?"

Under the table, Abbie Hoffman was cackling.

SOON AFTER HE RETURNED FROM THE DRUG CONFERENCE, BRAND, Jennings, and Joe Bonner packed up the Truck Store in his camouflage-painted VW bus full of books and headed to New Mexico. Steve Baer, the environmental designer, and his business partner Barry Hickman had been plotting a gathering of the tribes in New Mexico in the spring of 1969. The Alloy Conference took place over three days beginning on March 20. Located between the Trinity bomb test site and the Mescalero Apache reservation, the site was an abandoned tile factory near the tiny village of La Luz.

In Brand's retelling it would be the first "programmatic gathering" of the community of "outlaw designers" that he celebrated in the *Catalog*. In his words, they were:

> Persons in their late twenties or early thirties mostly. Havers of families, many of them. Outlaws, dope fiends, and fanatics naturally. Doers, primarily, with a functional grimy grasp of the world. World thinkers, dropouts from specialization. Hope freaks.

J. D. Smith, an early Truck Store employee, later referred to the gathering as "some group of baling wire hippies who can tell us how to convert our broken hairdryers into incubators."

Alloy perfectly captured the spirit of what Brand was trying to create. One hundred fifty people attended, camping among the tumbleweeds in weather that alternately baked, rained, snowed, and added a dust storm for good measure. J. Baldwin had filled his Citroën full of students and taken two days to drive from Illinois, stopping at communes in Colorado on the way. Lloyd Kahn drove from Big Sur with his wife and son in his own VW bus. When they arrived, they first built a zome, a domelike structure that had been named by Steve Durkee as a contraction of the words *dome* and *zonohedron*. They set it on the base of a hillside above the factory with cars parked below and held their meetings in the simple transparent structure where they also cooked and played music for three days.

Robert Frank, the photographer and documentary filmmaker who had attained international stature for his photographic essay *The Americans*, attended, although he would never produce anything from the event, and the film he shot is presumed lost.

Although Brand was moving away from photography, he got to know Frank well during the conference. Later that year, Frank came to California to film the documentary *Life-Raft Earth*, based on Brand's effort to draw international attention to overpopulation, and their bond became closer.

At the Alloy Conference, Brand served as an impromptu secretary, capturing sound bites and printing them largely verbatim in the next

*Whole Earth Catalog* supplement accompanying photographs he had taken, adding two pages of addresses:

> *"A temporary structure only needs to meet certain building codes."*
>
> *"We want to change ourselves to make things different."*
>
> *"You can use dope as a pure design tool."*
>
> *"You don't have to take dope to do things."*
>
> *"We are trying to develop computers that will help human beings to develop computers. The human and the computer work symbiotically."*

The idea of information sharing resonated, a good example being that Brand had gotten his inspiration for the format of the first *Catalog* from Baer's *Dome Cookbook*. The spirit of the gathering was captured succinctly by Brand's assignment of the label "hope freaks" to the participants. Baldwin and Brand in particular left with new enthusiasm and confidence that the emerging environmental movement could chart an alternative course to save the planet. Shortly after Baldwin got back to Illinois, he quit his teaching job, striking out on his own to live a nomadic life focused on creating alternative technology.

For Kahn, it had the force of something like the Monterey Pops Festival—a point in time when all the stars aligned. There were five speakers during the weekend—Baer, Durkee, Kahn, Baldwin, and Dave Evans—all sketching out an alternative ecology and society. In the view of environmental historian Andrew Kirk, it was the first moment when a tiny band of outlaw thinkers came together to attempt to articulate an environmental design philosophy that saw technology as a means to solve practical problems such as shelter and energy.[20]

Both Brand and Kahn were energized and optimistic as they drove away in their VW buses on Sunday night, when the sky was pink, lit with a rare aurora borealis pushing far south. The next day the Durkees' child was born and named Aurora.

Growing up in Illinois in the 1940s, Brand lived for the summers spent each year at Michigan's Higgins Lake at a family camp. Early on he developed a passion for wildlife and at a young age took the *Outdoor Life* magazine's pledge to protect the nation's forests, lands, air, and water.

The children of upper-class Michigan families, Stewart Brand's parents, Arthur and Julia, attended MIT and Vassar, but then settled in Rockford, Illinois, where after several years Arthur Brand would start an advertising agency.

Brand followed his older brother, Mike, to Exeter. Although he was not an academic standout, he won a coveted writing prize for an essay on the army's battle with environmentalists at Fort Sill, Oklahoma, where Mike was stationed.

After his freshman year at Stanford, Brand persuaded a cousin to hire him to work for a small-scale logging outfit in the woods northeast of Eugene. He came away convinced that there was a Great American Novel to be written about the loggers. Several years later he was relieved when he realized that Ken Kesey had written *Sometimes a Great Notion*, detailing the world that Brand had discovered.

On temporary assignment as an army photographer at the Pentagon, Lt. Brand was assigned to photograph a visit by President John F. Kennedy with a group of army nurses. When Kennedy entered the room, he headed directly toward the young photographer, but Brand backpedaled and missed an opportunity to shake the president's hand.

*18 Dec.*

*Being a report rendered to The International Foundation for Advanced study on my LSD session, 10 Dec 1962, for which I paid $ 500.00.*

*According to the notes taken at the session, I had the first goblet of LSD at 6.41 am, the second goblet at 10.00 am, and was asked to sit up at 1.07 pm, and received further LSD by injection at 2.07.*

*These times correspond roughly with the stages of the session which for convenience I call Purple Attics, Purple Helixes, Vacuum Cleaners, and Cement.*

*Purple Attics refers to the animated cartoon-like pictures my mind played with the sound of the earphone music. I recall the notion of simply perusing cobwebs through a succession of angular attics, of feeling the music was too spectacular and superficial, and of intimations that Being was large and take-able for granted but out of my then range of vision. Bodily sensations were pleasant chills and a neck-ache. I recall chuckling with feelings at things which had no obvious humor.*

*The Purple Helixes which followed the second cup of LSD (a pleasant*

After leaving the army in 1962, Brand moved back to North Beach and signed up to take part in an LSD experiment being conducted by the International Foundation for Advanced Study, a Menlo Park–based research group founded by several electronics industry engineers and Stanford professors. For Brand his first LSD experience was an intense and unpleasant one. The researchers felt that Brand was an example of someone who did not give in emotionally to the drug, and they increased his dose significantly in an effort to break through his emotional barriers.

Shortly after deciding to become a professional photographer, in early 1963 Brand received an assignment to photograph a wild horse roundup on the Warm Springs Indian Reservation in central Oregon. He spent several weeks with the three tribes that had been forced onto the reservation beginning in the 1850s. He came away with his view of the world transformed, finding a people who were more in tune with the natural environment than his own modern American society.

Setting out to explore American Indian culture in 1963, Brand took part in a number of peyote rituals. Eventually he befriended Hola Tso, a vice president in the Native American Church. Tso taught Brand the art of being a "roadman" and eventually gave him a membership card in the Church.

Brand met Jack and Jean Loeffler during an outing on the San Francisco Bay on bohemian artist Jean Varda's sailboat. Loeffler, also an army veteran, was a jazz musician who worked as a cook and a caretaker at Esalen. Loeffler and his wife helped Brand on his research journeys in the Southwest among the American Indians.

Brand's most powerful experience with LSD came while he was alone in a remote part of a Navajo reservation in Utah. In the midst of his trip he stood outside a small dwelling called a hogan, staring at Navajo Mountain, which appeared to him to be a giant, friendly deity with its arms outstretched.

The New Games movement, which Brand created in the 1970s, actually had its roots a decade earlier in an observation he made while attending a Me-Wuk Indian festival in the Sierra foothills in 1964. He noticed a group of teenage boys and girls playing an intensely physical game, not separated by gender as was the middle-class American custom. They were having an immense amount of fun and the idea stayed with him.

Brand launched his iconic "Why haven't we seen a photograph of the whole Earth yet?" campaign after spending an afternoon on his apartment roof overlooking the San Francisco skyline in early 1966. He realized that seeing the entire planet would change the frame of reference for humanity. It would take another year for NASA to produce such a photo, and it would become a powerful symbol that united rather than divided humanity.

Brand stands with Ken Kesey in front of Further, the Merry Prankster bus. Leaning out of the window in the driver's seat is Neal Cassady, the legendary Beat figure immortalized in Jack Kerouac's *On the Road*. Brand would serve as a bridge between the 1950s Beats and the 1960s hippies, but he never fit neatly into either group.

Although Brand was never an "on the bus" Prankster, he is photographed here riding the bus on its way to the Acid Test "graduation" that Ken Kesey organized to mark the end of the LSD era he had helped create.

The events leading up to the "graduation" in October 1966 were described in detail by Tom Wolfe in *The Electric Kool-Aid Acid Test*. The last of a set of wild LSD-drenched rock concerts that sparked the counterculture was a turning point for Brand as well. At the end of the evening he proposed to Lois Jennings.

In November 1966 Stewart Brand and Lois Jennings were married on the beach in Santa Cruz in a ceremony presided over by Paul Lee, a University of California philosophy professor. They were joined by Jack and Jean Loeffler and Roy Sebern, a Prankster artist who was a friend of Brand's.

The roots of personal computing were shaped by research done during the 1960s at Stanford Research Institute. Brand became a good friend of Bill English (seated), chief engineer for Doug Englebart's Augmentation Research Center and the co-inventor of the computer mouse pointing device. Later Brand would officiate at English's wedding in the backyard of a home on Homer Lane in Menlo Park.

Trained as a mathematician, Lois Jennings was the driving force behind the *Whole Earth Catalog.* She kept the books, managed the employees, and even worked at the cash register in the store. Here she is with author Gurney Norman. In 1971, his novel *Divine Right's Trip* was serialized in the pages of the *Last Whole Earth Catalog.*

Production for the *Whole Earth Catalog* took place in a garage in the Santa Cruz Mountains, west of Stanford University. A small crew would typeset and paste up camera-ready copy. Beginning in 1968, they produced two catalogs and four supplements each year until Brand closed the venture down in 1971.

Originally conceived as a way to get goods, ideas, and knowledge to his friends living rurally in communes, Brand soon located the Whole Earth Truck Store adjacent to the Portola Institute near the train station in Menlo Park. The store mostly sold books, along with an eclectic mix of tools and other often quirky gear. It was located in the heart of the late sixties and early seventies counterculture community that emerged on the San Francisco Peninsula.

The staff of the Whole Earth Truck Store worked, ate, and played together. The day was built around a lunchtime volleyball game that took place in the lot behind the store.

In January 1971, the *Last Supplement to the Whole Earth Catalog* was edited by Ken Kesey (far right) and Paul Krassner (left), the editor of *The Realist*. The issue, which was graced with a Last Supper homage by cartoonist R. Crumb, doubled as an issue of *The Realist*.

To mark the end of his publishing venture, when Brand shut down the *Whole Earth Catalog* he decided to throw a party and then surprise the guests with the announcement that he was giving away $20,000 based on a group decision about what to do with the money. He dressed in his father's monk outfit to make the announcement. The partygoers couldn't reach an agreement, and so in the early-morning hours most of the money was given to a militantly anti-money draft resister named Fred Moore. Moore would go on to cofound the Homebrew Computer Club several years later.

Brand moved to San Francisco when his marriage with Jennings ended, and then on to Marin County soon after he launched *Coevolution Quarterly*. He lived in a boathouse in Belvedere and would frequently row to work, sailing home in the afternoons.

When Jerry Brown was first elected governor of California, Brown spent a year consulting, both bringing interesting people to speak with Brown and organizing events ranging from a Space Day to a Save the Whales event. His time with Brown ended his earlier libertarianism. He came away believing that there is value in good government.

Marlon Brando became a fan and supporter of the *Catalog*
and the *Coevolution Quarterly*. Independently both men
had become supporters of American Indians in the 1960s.
Later Brando invited Brand to the island he had purchased
in the South Pacific to publish the *Quarterly*. Brand visited
but never moved his publication there.

Having convinced his publisher that he would completely edit and produce *How Buildings Learn*, Brand established his studio in a shipping container he converted. It gave him space to assemble the photographs he wanted to use as well as doing page layout. Despite being a critique of modern corporate architecture, the book was well received by the architectural community.

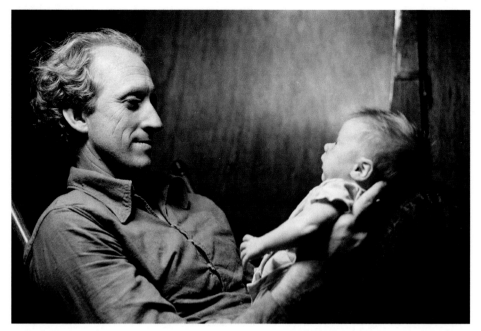

In 1977 Brand became a father after a long-distance girlfriend, Alia Johnson, visited and told him she was pregnant with their son, Noah. Brand agreed to be a somewhat involved father. Johnson would later marry physicist Robert Fuller, and Brand stayed active in Noah's upbringing.

Patty, later Ryan, Phelan came to work at the *Quarterly* and proved to be a skilled organizer of events such as the Whole Earth Jamboree shown here, celebrating the tenth anniversary of the founding of the *Whole Earth Catalog*. The couple decided to get married after they had purchased and renovated a tugboat as their new home among the Sausalito houseboats. Brand's motto became "Live small, so you can live large."

The *Mirene* began life as a "gasoline schooner" in Oregon in 1912. When Brand and Phelan found it, it was moored and rotting on Richardson Bay. Lovingly restored in several stages, it is now taken out for an annual cruise on the San Francisco Bay.

Brand married Ryan Phelan in 1983 in a Zen Buddhist ceremony at Green Gulch, a rural retreat center and farm in the hills overlooking the Pacific Ocean. To appease Phelan's mother, a Catholic prayer was read during the ceremony.

Kevin Kelly had the original idea of an event that would allow the characters in Steven Levy's *Hackers* to meet one another. Brand and Phelan helped organize the event, which was held in the fall of 1984. In response to a remark by Apple cofounder Steve Wozniak, Brand said, "Information wants to be expensive . . . and information wants to be free." Virtually everyone ignored the first half of his aphorism, and "Information wants to be free" would become the rallying cry of the dot-com era.

Although Brand had first conceived of the idea of an online community even before starting the *Whole Earth Catalog*, he founded the Whole Earth 'Lectronic Link in 1985. Part of the wave of interest in computer networks that preceded the internet, the WELL attracted a small but devoted community. Presaging the clashes over social media by several decades, six years later Brand left the WELL's board feeling that the online community he had founded was flawed.

Brand had first visited the Black Rock Desert Playa in Nevada with Jack Loeffler in the 1960s. In the 1990s he returned as a participant in the annual Burning Man gathering.

Brand looks at a prototype of the Clock of the Long Now, a project he began with computer scientist Danny Hillis as an exercise to inspire long-term thinking. Significantly, the two set out to build what will effectively be the world's slowest computer just when Silicon Valley and the world became obsessed with ever quicker internet time. Funded by Amazon founder Jeff Bezos, a full-scale version of the clock is near completion.

In the summer of 2018, after a long absence, Brand and Phelan returned to visit his boyhood summer retreat at Higgins Lake in upper Michigan.

While in the army, Brand picked up the habit of getting in a quick nap whenever the opportunity arose. The custom has stayed with him throughout his life. Here he is taking advantage of the desert sun during Burning Man.

O

THE SPRING 1969 *CATALOG* WAS PRODUCED IN THE TRUCK STORE in Menlo Park, which turned out to be a mistake resulting in countless interruptions. Afterward, Brand hitchhiked to New Mexico to "clear his head" for what would be billed as "The Great Bus Race."

Held on the summer solstice in the mountains near Santa Fe, it was a gathering of a very different kind of tribe from the small spring conclave of outlaw ecological designers who had gathered at the Alloy Conference. Kesey, who had achieved global fame from *The Electric Kool-Aid Acid Test*, was there with Further, his 1939 International Harvester bus, driven by the Pranksters while Kesey himself arrived in a white Cadillac convertible stocked with his tennis racket and lots of beer. Members of the Hog Farm commune, Paul Krassner, and hundreds of camp followers, hippies, and back-to-the-landers were in attendance.

The original plan had been for the buses to race the clock in a large meadow, but Kesey insisted that it be a real race. Brand wrote about the race in the next issue of the supplement, but later acknowledged that there had been some errors in his reportage because he "drank fluid from a Donald Duck glass" and as a result had been "too stoned to function and misrepresented some facts."

For Brand it was his last LSD trip and, in a real sense, the end of the sixties. Woodstock would take place in a few months, but despite its magnitude, the summer solstice was sundown for a chaotic decade. For Brand, too, there was a sense that a phase shift was at hand. His unease with the power of the drug, combined with an increasing number of experiences that had been either uncomfortable or just plain bad, reached a tipping point, and he stopped for good.

Stepping away from drugs, he would for several years step toward more conventional political organizing. Edward Abbey's *Desert Solitaire* had been published the previous year, and "monkey wrenching"—sabotage carried out by environmental activists—was in the air.

It was not just environmental design outlaws; the following year Earth Day would take place, and a new environmental movement was forming. There was a growing sense that a more militant approach was necessary to bring about real environmental reform. Brand fantasized about a "militant conservation organization" that would carry out guerrilla-style propaganda actions. "The instrument to play is growing Earth Consciousness," he wrote in his journal. "—And play is the means." He thought about obvious comrades in arms—Jerry Mander, David Brower, and "the Mills chick," an energetic young woman who had caught his eye, came to mind.

O

IT REMAINED A SMALL CREW EVEN AS THE *CATALOG* OPERATION EX-panded with each issue. Production was an intense, multiday affair that took place six times a year—twice for the *Catalog* and four times for the supplements. While each *Catalog* would be bigger and better than the last one, the supplements were what allowed Brand to communicate with the growing counterculture. Each one read like a messy conversation, full of opinion and improvisation. Despite the inclusion of more "outside" content, they were still a tremendous amount of work to put together.

Sandy Tcherepnin had quickly moved on from her role as a typist and was replaced by Annie Helmuth and then later by Matthew "Cappy" McClure, a Stanford history graduate. The arrival of an IBM service tech became a weekly occurrence. Beginning with the Fall 1969 *Catalog*, Fred Richardson, a skilled mechanic who had the longest hair of any-one on the Truck Store staff and affected an Amish style of dressing without buttons, joined as the production photographer, proofreader, in-dexer, and general handyman.

Jennings quickly became the store manager. "She had the adminis-trative qualities of a good First Sergeant," Brand would write in the last issue of the *Catalog*. She also had the math skills to make a good book-keeper and a sharp tongue to organize a staff composed largely of walk-ons and high school dropouts.

In addition to volleyball, the other tradition that quickly emerged was a daily communal lunch for the entire staff. Midmorning each day Jennings and Shugart would take off for the grocery store and return to prepare the midday meal.

It was almost impossible to capture how eclectic the Truck Store staff was, but when Ed McClanahan and Gurney Norman profiled the phenomenon for *Esquire*, they tried, painting this picture of a normal lunchtime scene in a room with a "bile-green ceiling" around two plastic and chrome kitchen tables covered in oilcloth.

> Let's see, there's: Freaky Pete the White Panther and Pretty Leslie the first girl to drop out of Yale and Merce the Peace Corps dropout and Tracy the dope-scene dropout and Alan the Karmic Yoga dropout (no hair on that one; he shaves his head) and Jim the dropout high school civics teacher and Diana the dropout social worker and Fred the dropout electronics technician and Dick Raymond the dropout land-use economics consultant and maybe half a dozen others, perhaps a cell of anarcho-syndicalists, an infernal machine in every pocket, revolutionaries, dope feens, Communists, communards, conscientious objectors, peace feens, love freex, vegetarians, Christers, Christians. . . . [21]

Subscriptions continued to roll in. During the summer Hal Hershey, who had joined early to do page layout and graphics design, noted that on a Monday in July they had received orders for forty-seven subscriptions on a single day. In wavy handwritten block letters, he wrote: "HELP??" in the daily journal that was kept to track business. "Dear Tribe," he wrote. "We have obviously reached a higher level of business—the problem being to regear for procedures, methods, and people to deal with the increasing influx of orders, customers, etc."

Several days later Shugart added that she had put $2,250.25 into the bank for three days of business and that the bank balance was continuing to grow no matter how hard they tried to spend the money.

Yet despite the great success of the *Catalog*, Brand was increasingly withdrawn. Shugart could sense the toll it was taking. Trained as a clinical social worker, she had never gotten the sense that Brand was emotionally interactive. Instead, while a great talker, she saw him as someone who lived in his head with the ideas he was passionate about. The *Catalog* was his baby—everyone knew that, and it placed enormous weight on his shoulders.

Joe Bonner, who had been the first layout person, was joined by his younger brother, Jay, who had just graduated from high school, for what would turn out to be just a single issue. Awash in the Bay Area's intense antiwar movement, Jay soon came to Brand with several suggestions for making the *Catalog* more connected to the political moment. To his surprise, Brand instantly rejected Bonner's ideas. He told him that he had started the *Catalog* with three criteria: no politics, no religion, and no art. Bonner was stunned because in his view the *Catalog* was full of all three. He felt as if he were being treated like a child and condescendingly shut down.

The friction would continue for the entire time that Jay Bonner worked on the *Catalog*. Another reason he was put off was because Brand was a vicious boffer, the New Age jousting sport played with Styrofoam swords called boffers. When fencing with rank beginners, Brand would attack with a banshee-like frenzy, leaving welts on his opponents.

Their discord became public in the January 1970 supplement when Brand published a long letter from Bonner attacking the *Catalog*'s "capitalist" leanings. He wrote, "From all the 128 pages of the *Whole Earth Catalog* there emerges an unmentioned political viewpoint. The whole feeling of escapism which the catalog conveys is to me unfortunate."

Bonner went on to note that his salary was just five dollars an hour, that the *Catalog* was nonprofit, and that he had yet to figure out what capitalism was. He had been radicalized by working on the *Catalog* into a personal involvement with politics, he wrote, but that guilt-based action never worked and only made the situation worse. "I'm for power to the people and responsibility to the people . . . ," he concluded.

Brand responded curtly: "Jay worked with his brother Joe doing layout on the Fall *Catalog* and was not rehired for January production, because of too many technical mistakes on his pages. Jay is 17. (I'm 31. How old are you? It matters, more than any of us like.)"

The *Catalog* and the Truck Store became a vehicle for Brand to follow his whims and chase after any kind of new idea. He reviewed a build-your-own-airplane kit and found it interesting enough that he decided to order one. Ultimately, the plane, unfinished and unflown, ended up in the barn on Bill English's property. Brand purchased a BMW motorcycle and discovered that as great an adventure tool as his new motorbike was, it was also probably more risk than it was worth. Thinking about the hazard of zipping along the freeway at seventy miles per hour, he decided to sell the bike after several misadventures.

The unrelenting pressure of running the *Catalog* soon led to signs that he was beginning to lose control. One evening he went with Jennings to see the Burt Lancaster movie *The Swimmer*, based on a John Cheever story about a man who decides to go home via swimming every pool in his suburban Connecticut town. At the beginning of the movie, Lancaster's character is treated with respect, and then at the end of the movie, as his family collapses, he is treated with contempt. The chilling portrait of a life unraveling shook Brand viscerally. That night he returned to his trailer reflecting that people could really lose it; he found himself thinking that maybe he was losing it too.

He remained close to the Engelbart team and a close friend of Dave Evans's in particular. At the end of May the two men teamed up at a weekend workshop at Esalen on the use of tools. The idea was to investigate "some of the new modes of tool use—from the satisfying do-it-yourself labor of intentional communities to the intense experience of living and working on-line with a powerful computer facility." The connection between Brand's access-to-tools orientation and Engelbart's intelligence augmentation ideas with computers was immediate; personal computers were on the horizon, simply awaiting the invention of the microprocessor in 1971, making computing affordable to hobbyists.

The relationship between the two cultures, however, was imperfect. Evans was passionate about strengthening the connections between the cybernetic and countercultural worlds, and in August he organized an

Alloy-inspired weekend conference that was named Peradam, a term from René Daumal's novel *Mount Analogue*, which was loved by Brand, Durkee, and Evans. A Peradam was supposed to be a crystal of remarkable purity that was stronger than diamond and with an index of refraction close to that of air.

Evans had persuaded Engelbart to visit the Lama commune, but as hard as he tried to bridge the gap between Engelbart and the counter-cultural community, he ended up increasing the stress on the computer scientist, who in principle was open to new ideas but who in the wake of the success of his computer demonstration in the fall of 1968 was increasingly obsessing over losing control of his group.

Evans believed it was a perfect philosophical match for the challenge confronting both the computing and counterculture communities: creating tools to attempt to scale the power of the human intellect. It was an event that should have been at the sweet spot of a range of sixties movements from human growth to psychedelics to Brand's access-to-tools and self-education crusade. The event brought together a group of half a dozen of Engelbart's researchers with Stewart Brand, Steve Baer, and Steve Durkee. Evans believed that Engelbart's "bootstrapping" vision depended on getting a whole host of people on board if he was ever to reach beyond the computer science types at the Stanford Research Institute.

Nonetheless, while Alloy had been a sparkling success, where everything came together—the people, the setting, the food, the music—in comparison Peradam felt "lame." A group came from Pacific High School, an alternative school located in the mountains behind Stanford where Lloyd Kahn had begun teaching and building domes; and the Hog Farm, a commune that was then based on a mountaintop near Los Angeles, did the cooking. A performance-art group known as the Texas Inflatables created a futuristic plastic environment to walk through.

But nothing jelled.

Engelbart couldn't cope with the idea of an eclectic vision that wasn't his. Although he was invited, he chose to stay away from the weekend retreat. He didn't like the idea. In his mind, it was just another symptom of losing control of his intelligence augmentation vision. Brand did attend, but he felt that it lacked the spark of the Alloy Conference and he chose to bury his notes from Peradam in the back pages of the January 1970 supplement, noting gently that Alloy was a hard act to follow.

O

HAVING DECREED THAT THERE WOULD BE NO POLITICS IN THE *Catalog,* Brand quickly violated his principles. Antiwar activism and labor organizing were forbidden subjects, but environmental activism was in his blood going back to *Outdoor Life*'s conservation pledge, and in 1970 poet Gary Snyder published "Four Changes," his environmental call to action in the supplement.

In the fall of 1969, *The Population Bomb* fresh in his mind, Brand began planning a public "starve-in" he called the *Hunger Show* or *Life-Raft Earth.* Ehrlich had predicted that there would be huge famines arriving imminently in the 1970s and 80s, and there was nothing that could be done about it.

After unsuccessfully approaching foundations for funding, Brand picked up the $2,300 cost of the event from his family inheritance. To demonstrate the dangers of overpopulation and the impending threat of global famine, the protesters would hold their fast in a hundred-foot-by-hundred-foot inflated plastic pillowlike enclosure. Brand's idea was to make it a game: if you broke your fast before the week was out, you would "die" and have to leave the island they had created. The plastic enclosure—unceremoniously sandwiched between a restaurant and a supermarket in Hayward across the Bay from Menlo Park—was equipped with portable toilets and an emergency "hospital tent."

Unable to get permission for a larger venue, Brand was forced to hold the event in the parking lot of a poverty program center in Hayward, clearly a disappointment from the point of view of creating a public spectacle. He described Hayward as the kind of place where you get marooned while hitchhiking.

The *Hunger Show* began on the evening of Saturday, October 11, the original idea being that the participants needed to be inside by that time and if they left, there would be no return. Robert Frank, the filmmaker, showed up and joined the fast. He was let into the plastic enclosure to film while the rest of the media was forced to set up outside and conduct their interviews over the barrier.

On Sunday, ABC-TV showed up as part of a piece they were filming on environmental activists. They pleaded to be allowed to come inside the plastic pillow to film. That created a debate within the fasting community. Many were dead set against the media, whom they viewed as the enemy. Others thought the media could be used. Still others thought they could be converted. ABC was told that if they were willing to fast, they could come inside. The TV news team didn't have the time. They filmed from outside.

Initially, about one hundred *Hunger Show* participants began the fast. The group was joined by Wavy Gravy and a busload from the Hog Farm, now famous for having served as security at Woodstock. Twenty people left the first day, some to go back to work, and some because fasting proved challenging. On Tuesday, rains drenched the parking lot, soaking sleeping bags, and numbers dwindled further. When the rains continued, the decision was made to evacuate to the Truck Store in Menlo Park. The next morning the police showed up and told them to leave, and Brand had a confrontation with an angry cop that threatened to get out of hand until Dick Raymond intervened, suggesting that a better choice might be to move once again. The rest of the troops drove up to Rancho Diablo and spent the remainder of the week in isolation, with very few of the reporters who had originally flocked to the event.

On Friday, Wavy Gravy departed with the Hog Farm bus to attend the trial of the Chicago 8. The next morning the remaining fifty-two survivors broke their fast, and diplomas were issued commemorating their achievement. Brand would stay in Ehrlich's apocalyptic orbit for several more years.

His deepening depression came at a time when his celebrity rose in concert with the visibility of the *Catalog*. Cerebral and almost a half generation older than the youth movement mainstream, he found that he was increasingly sought out as both a symbol and a spokesman. In general, attention only made him feel worse. He tried to improve his outlook with a visit to Jack Dowling, a gestalt therapist and neighbor in Menlo Park, who had introduced LSD therapy for alcoholics in the early 1960s. Their meetings were ineffective. (Dowling instructed him to shout at a chair that was supposed to represent his deceased father.)

One day he was driving up the hill to work on production at Rancho Diablo when he realized that he simply didn't want to do what he was doing. He put a bed in his reading cubbyhole at the Truck Store and would spend entire days on the mattress under the pretext of reading books to review. In his mind, each *Catalog* had to be drastically better than the one before it, and the one before that, and the next one better than that, and the next one better than that still. He was attempting to do that partly by hiring more staff, but mostly he was demanding more from himself. As a result, he was only quasi-functional at times and would end up, by his own assessment, completely dysfunctional by the time the latest *Catalog* was published. It drove a deep wedge into his relationship with Jennings, who had a no-nonsense approach to most matters and who suffered as he became more distant.

The irony was that while the burden of the *Catalog* weighed ever more heavily on Brand, it was liberating an entire generation of young Americans. Not only did millions learn how to do new things, but they were inspired by the *Catalog* to strike out in new directions.

Larry Brilliant was a young doctor who traveled freely in the world of San Francisco rock royalty. He was close to the Grateful Dead, and if not doctor to the stars, he was an insider in the West Coast music scene, which stretched from San Francisco to Los Angeles. He had first met Brand during his residency in San Francisco while Brand was organizing the *Hunger Show*.

One of Brilliant's friends was David Crosby, the singer and guitarist who had been a member of the Byrds and then in 1968 joined Stephen Stills and Graham Nash in Crosby, Stills & Nash. By 1971, Crosby was at the peak of his fame and like many in the emerging counterculture he was romanced by the ideas he found in the *Whole Earth Catalog*.

One of Crosby's passions was sailing, and he had recently purchased the *Mayan*, a seventy-four-foot wooden sailing boat that he often kept in Sausalito but was in San Francisco near the St. Francis Yacht Club when Brilliant came to visit. Sitting on board one afternoon, Brilliant told Crosby that he knew Brand, and Crosby said he was anxious to meet the *Catalog*'s creator. On a whim they jumped into Brilliant's car and headed to Menlo Park and the Truck Store.

When Crosby and Brilliant arrived, they were escorted into the back room where Brand was sitting up in bed, editing. If Brand was in a deep depression, he hid it well. Brilliant had no sense that he was in the dumps. What he remembered afterward was a "pretty fucking intellectual" conversation between the two men. Martin Luther King Jr. and Robert F. Kennedy had been assassinated just months before the first *Catalog* had been published. Several years later the country seemed on the verge of a revolution—catalyzed in part by Crosby's band, which had written "Ohio," an impassioned protest song, in response to the National Guard killings at Kent State on May 4, 1970.

"Are we going to get out of this?" Brilliant remembered Brand asking Crosby, who responded, "Nothing will get us out of it but a total change in human consciousness."

After the two men left, Brand decided that Crosby must have realized he was falling apart. "Basket case. Too many drugs. I've seen a lot of this!" he figured the rock star must have decided.

Despite the crushing darkness—which he would later describe in the *Whole Earth Epilog*: "I was dealing with a sporadic case of the Greater Whim-Whams," his midwestern attempt to be folksy and gloss over the depths he was in—he was visible on the first Earth Day in 1970, when he spoke at the Parsons School of Design in New York City. The *New Yorker* recorded his appearance in his buckskin outfit and top hat.

When asked for advice on taking action to protect the environment, he cited Kesey's advice on how to make money—he ran a good creamery, compatible with good environmental stewardship, and money followed.

In the wake of Earth Day, he was invited to testify at a House Committee on Education and Labor hearing on environmental education, held at the California Academy of Sciences in Golden Gate Park. He stunned the committee when he showed up and argued against the value of "education."

The next day the *San Francisco Examiner* headline read WHOLE EARTH MAN SHOCKS PROBERS. Dressed in his "Daniel Boone buckskin suit and battered top hat," he told the committee that their efforts would backfire: "Forget it. If someone is interested in ecology and environmental matters, he'll delve into it. If he isn't, trying to make him learn about it will only make him hate it."

Brand's remarks outraged Representative John Brademas, a liberal from Indiana: "Yours is an Adam Smith super-duper philosophy. Your rhetoric, though colorful, is really very dangerous."[22]

Brand responded that the billions spent on the space program had been worth it because it brought home the image of the Earth as a spaceship. Brademas shot back that he didn't think a space image was worth the year's $3.6 billion appropriated for NASA and that he had voted against it.

He told Brand, "I agree with your vibrations but not with your perception."[23]

Brand's comments were, in many ways, consistent with his earlier assessment of his time at Stanford: "I'm learning how to be able to learn anything, and then I can go forth and have a life. . . . I don't need the class." Telling someone they had to get to a particular place was ultimately much less effective than providing them the tools by which they might leapfrog from curiosity to curiosity until they got there—or somewhere else that was even better.

At the end of 1970, the *Catalog* editorial staff took the supplement on the road as an experiment in mobile production. Along with several members of the Ant Farm art collective, they erected an inflatable plastic dome near some hot springs in Saline Valley, a California desert seventy miles from the nearest telephone. They brought along a generator to keep the dome inflated and to power their typesetting tools; however, the wind soon blew the dome away and production was moved into the

Airstream trailer. The resulting issue was subtitled *Difficult but Possible*. The word *desirable* was likely not considered by anyone involved.

Brand's celebrity would peak with the publication of *The Last Whole Earth Catalog*. The following month, in February of 1971, he appeared, complete with buckskin suit topped this time with a hard hat, before a national television audience on the *Dick Cavett Show*.

Cavett was effusive about the *Catalog* and immediately probed Brand about why he was stopping something that was such a success. "It's an incredible publication and its very personally done by the guy who is the sort of genius behind it," he told the TV audience. "And people are very sorry, the fans of the *Whole Earth Catalog*, that this is the last *Whole Earth Catalog*, because it's such a hit, and he owes us an explanation as to why it's the last *Whole Earth Catalog*. We welcome Mr. Stewart Brand."

Brand responded with a more upbeat version of the truth. "We wanted to stop something right for once," he responded. "So many institutions sort of fade out and piddle out. It seemed like stopping it was more important than starting it."

Cavett wasn't convinced. It was a little like stopping a show while it was a great success. Eventually, Brand brought out some of the goods that made the *Catalog* unique, including boffer swords and kama sutra oil, distracting the TV host with demonstrations.

For his mother, who was a loyal Cavett viewer, the fact that the television interviewer treated Brand with respect in front of the whole world meant that he had arrived. He was now a serious adult, in her mind. In Rockford, at long last, Stewart Brand had made it.

○

AFTER TWO YEARS OF INCREASINGLY POPULAR PUBLICATIONS, in April 1971 the *Catalog*'s balance sheet was flush, it and the Truck Store both indisputable cash machines: for the nine months up to April 30, revenue had been $498,000 with profits of $276,000.

In the face of the *Catalog*'s hypergrowth, Brand decided to stop doing everything on his own. Previously they had self-published and re-lied on a growing network of distributors. This time, recognizing that the demand for the *Catalog* was going to be even more substantial, he struck a deal with Random House. The agreement paid off very nicely for all: *The Last Whole Earth Catalog* sold more than one million copies. Published in 1971, it would offer a vast menu of more than a thousand items sprawling over 447 pages. Printed and distributed by Random House, it included Gurney Norman's novel *Divine Right's Trip*, which was threaded in snippets throughout the entire *Catalog*, requiring readers to pursue it as if exploring a maze. For books such as John Muir's *How to Keep Your Volkswagen Alive* or Jeanie Darlington's *Grow Your Own: An Encounter with Organic Gardening*, a mention in the *Catalog* was instrumental in assuring bestseller status.

*Divine Right's Trip* had grown out of a meeting that Brand held shortly after he decided, in the fall of 1969, that there would be a "demise" for the *Catalog*. He assembled a dozen of his contributors with the idea of eliciting ideas for a grand finale in 1971. They went around the table, and Norman said, "I think the last *Catalog* ought to be a novel."

"That's not likely a workable idea," Brand responded.

"Well, it ought to have a novel in it," Norman responded.

"Fine, will you write it?" Brand countered.

"Yes," Norman replied.

In addition to being a financial success, the deal was instrumental in reshaping the publishing industry. In negotiating with Random House over their distribution contract, Brand got on his "environmental high horse" and insisted that the unsold books be returned in the same way that hardcover books were, rather than be pulped. This policy change would eventually help create a justification for the "trade paperback."

None of this could buy peace of mind. Brand was barely holding it together emotionally. His marriage to Jennings was disintegrating. It seemed that the world was starting to close in, and he had become agoraphobic. In the end, he kept up appearances, putting out the last *Catalog*, but he had begun to contemplate suicide.

He found his way to Pierre Mornell, a Marin County psychiatrist. In talking to Mornell, he came to believe that for many people around him psychedelics had become an all-purpose cure using drugs as a crutch. He had pulled back on his drug use, but now, looking more closely at his state of being, he saw ways in which he had used external entities to prop himself up instead of seeking true internal stability. With Mornell's counseling and guidance, he decided to shed himself of some of the things: first the *Catalog*, and later his marriage.

He had gone into counseling on his own, not as a couple with Jennings, and although they would separate reasonably amicably, without a legal battle, for decades she was wounded and would remain bitter about what she saw as Brand's selfish decision. Several years later in a biographical note in a *Catalog* update, he would agree, blaming himself for the failure of the marriage: "Mostly my doing, I reckon."

He arranged to mark his decision to kill the *Whole Earth Catalog* with the "Demise Party." Brand had gotten to know Frank Oppenheimer, the founder of the Exploratorium science museum at the Palace of Fine Arts in the San Francisco Marina District, where he had helped Oppenheimer think through some of the museum's plans as it was being developed. So he decided to throw a party with a special twist.

The *Whole Earth Catalog* rented the Exploratorium for an evening, and as a midnight surprise Brand brought along $20,000 in cash in an inch-thick stack of hundred-dollar bills with the idea that, because he had started the *Catalog* with roughly that amount, it would be fitting to put the money back out into the world and create new things that might be equally interesting, in a pay-it-forward fashion.

It was an unusual event, even by the standards set several decades later during the height of the internet boom. The invitation went out to the entire *Whole Earth* community, from employees to subscribers to reviewers to anyone who had attended the Alloy Conference or seen *Life-Raft Earth*. It was suggested that you might consider dressing as a "tool."

The Exploratorium provided optical gadgets and illusions, and there was music, dancing, food, and drink. *Whole Earth Catalog* supporters from all over the country showed up, more than a thousand people in total. A band called Golden Toad played and a nonstop, nonscoring volleyball game went on all evening.

No one told the audience what was afoot until Scott Beach, a well-known local actor, took the stage at midnight and said, "Sorry to stop the volleyball and the inhaling of nitrous oxide from balloons, but there is $20,000 that is about to be handed out to the audience." He paused and added, "Oh, I see we have your attention."

Brand had a hypothesis that, under duress, people would come up with the most amazing ideas. Afterward, he decided it didn't work out that way. He concluded that, rather, under duress people would come up with what Brand considered remarkably stupid ideas.

After Beach's announcement, Brand climbed onstage and said, "I can tell you from working around foundations for three years that they are absolutely strung out about how to use money. They don't know. If we don't know, we can't really complain about them. So we are into frontier territory here. And like on any other frontier we have got to get together and deal with our problem. It may be a creative problem, and that's our task—to find a creative way out of it."

A microphone was set up in the audience, the one-inch-thick envelope of hundred-dollar bills was handed to each speaker, and people started walking up to the mike, taking the envelope, stating what they thought should be done with the money, and then handing it to the next person. Brand was dressed in an odd monk's black robe that had belonged to his father, a gesture that was meant as a gentle homage. He stood at a blackboard and began writing down the proposals as people made them in two-to-four-word summaries. The hour kept getting later and people kept getting more and more raucous.

There were a lot of what seemed to Brand to be knee-jerk liberal ideas. One guy stood up and said, "Let's give the money back to the Indians." That prompted Jennings to go to the microphone and say, "I'm an Indian and I don't want the money."

At one point as the money, in the form of a stack of bills, was being handed from speaker to speaker, someone said, "This shouldn't be decided by one chunk. There are a lot of things that can be done with this money. Let's all decide." And then he grabbed a handful of cash and started handing it out to the crowd. Brand rushed back to the microphone and said: "Hey, I think it is more interesting to talk about what to do with $20,000 than what to do with $100. Maybe the money will flow back to the stage."

And miraculously, the money did come back—at least $15,000 of it. The rest disappeared into the night.

In the crowd was Fred Moore, an itinerant pacifist and draft resister, who had just returned from a trip to Mexico, where he had created a project called Skool Resistance in the spirit of the deschooling ideas of the radical Chilean educator Ivan Illich. Loosely affiliated with the Portola Institute, Moore, who was almost totally broke and living in the garage of a house on the Midpeninsula, had arrived that evening with two dollars in his pocket.

But after midnight, when the dispersal of the money was being debated, Moore got angry. This was just like all the bad things that money did everywhere else in the world, he decided. Earlier in the evening, he had gone up to the microphone, removed one of those dollar bills from his pocket, held it up in the air, and burned it. The point, he argued, was not about delegating the money; it was about empowering people directly by sharing information, which would help create consensus for what to do with the money. He went up to the microphone again and tried to make his point: "If we are going to build a change—in a changing new world, or whatever we want to call it, 'New Age,' then it's going to be because we are going to work together and we are going to help each other."

Although no one realized it at the time, several years later that would be the heart of his initiative to build a hobbyist computer club to share resources and information freely. From the man who now only had a single dollar to his name would come one of Silicon Valley's supreme ironies: Fred Moore, an itinerant activist who rejected material wealth as an end in itself, would light the spark of what became the "largest legal accumulation of capital in the twentieth century: the PC industry," as venture capitalist John Doerr described it.[24] Indeed, paradoxically Moore would also become the unrecognized patron saint of the open-source software movement, which in turn has become a major force in the computer industry.

That evening, however, it was well past midnight and still, no decision was reached. Someone finally stood up at the microphone and read the I Ching, which decreed, "Undertakings bring misfortune." Not a good omen. Finally, there was a vote, just on the question of saving the money versus spending it. But it ended in a 44–44 tie, solving nothing.

Not long afterward, a decision was finally made: give the money to Moore in hope that maybe it did make sense to create resources for (in Moore's words) "information, community, and educational networks."[25] At dawn, Fred Moore became the steward of the envelope.

Brand just shook his head. It had been an interesting experiment, but he never really expected to see Moore again. Maybe he'll send a postcard from Mexico, he thought as he left the Exploratorium in the morning light.

In the days that followed, Moore would come to feel increasingly trapped. To him banks were part of the problem, and so, not knowing what else to do with the money, he went home and put it in a tin can and buried it in his backyard. Word of the strange conclusion to the Demise Party spread quickly. After several newspaper accounts appeared, Moore was besieged with financial requests by both phone and mail.

And like Frodo's ring, the money wouldn't stay in the ground.

Moore was soon forcibly turned into a "people's banker" when a small group of San Francisco activists who were engaged in building a collective in a warehouse in a tattered neighborhood south of Market Street heard about the windfall. Project One encompassed a diverse set of community political projects, ranging from education to organizing to the theater to one of the first community time-sharing computer efforts, which was called Resource One and had become the final resting place for Doug Engelbart's original SDS-940 mainframe computer, on which he had developed his oNLine System—as luck would have it, ultimately paid for with a grant from a foundation created from money generated by the *Catalog*.

A few Project One representatives arrived at Moore's house and accompanied him out into the backyard, where he grudgingly dug up his tin can. It seemed as if Moore were about to break down in tears as he retrieved the can. Yet, as uncomfortable as he was, serendipity was at work.

Decades later, countless *Whole Earth Catalog* readers would describe how something they had stumbled upon in the *Catalog* had sent their life in a completely new direction. In Moore's case, his appeal at the Demise Party laid out a path for his quest to build an information network to tie all of the community and the political activists together. Working with his database on notecards, he desperately wanted his own computer as a tool to organize his network. In April of 1975, along with Gordon French, a local computer engineer, he called a meeting to create the Homebrew Computer Club. It would prove to be a crucial step toward the world of personal computing.

The demise of the *Catalog* also marked a familiar turning point for Brand personally. Just as he left San Francisco at the moment when the Haight-Ashbury district emerged, he chose to leave the Midpeninsula at the moment when politics and community were converging with technology to create a computing renaissance in the world that was to become Silicon Valley.

## Chapter 7

# CoEvolution

URING THE SUMMER AFTER HE HAD BROKEN FREE FROM THE
*Catalog*, Brand and Jennings took a cross-country road trip, exploring New England, then driving north until they reached
the wilds of Robert Frank's summer home.

There was a bit of symbolism in their destination. In getting away
from the *Catalog*, Brand would travel to the continent's eastern extreme,
almost four thousand miles away from the San Francisco Peninsula, to
Nova Scotia, Canada.

Getting to Canada had been relatively easy: head east and then
north. Other challenges he was facing were more complicated. He was
casting about for what to do with his life. Brand fantasized that perhaps
he could reconstitute himself as a "private statesman" or public intellectual like Buckminster Fuller, Ralph Nader, or David Brower. But how?

Robert Frank had been Brand's guide for understanding where photography was heading. Earlier, Frank had told him that he had stopped
being a photographer because "I noticed that when I made an exposure
my eyes closed." He confided in Brand that he believed that photography was stopping him from seeing the world.

Sometime later, Brand decided that was happening to him as well;
moreover, he realized that he had the skill but no talent for photography. That was different from writing, where early on he believed he had
both skill and talent.

And so his pilgrimage to see Frank, who had a summer home on

Cape Breton Island. The desolate island blew Brand's mind. He found
Frank on a steep, mountainous coast in an almost uninhabited wilderness. They drove to a bay that you could reach only on a dirt road, a
place above the water where huge lobsters could be easily hauled up out
of the Gulf of St. Lawrence.

Right next to Frank's house was another equally evocative piece of
property. Brand bought it on the spot.

It would be another year before he would return to build his own
home there. Ultimately, it would be in the isolation of the Canadian wilderness that he would leave Jennings in the same way he walked away
from the *Catalog*.

O

JACK AND JEAN LOEFFLER HAD COME TO NEW YORK CITY TO RAISE
money for an environmental defense fund created to protect Navajo
lands from coal mining companies. They were staying at a friend's
house on the Upper West Side in April of 1972 when Brand showed up
in a chauffeured Rolls-Royce to gather them up to accompany him to
the National Book Awards Ceremony.

In a surprise to Random House, *The Last Whole Earth Catalog* had
received the National Book Award in the Contemporary Affairs category. The publishing giant hadn't even thought to enter the *Catalog*, and
so it was intent on making it up to Brand by whisking him around town
in style.

The award had created something of a tempest in a teapot when two
of the judges, Digby Diehl of the *Los Angeles Times* and Harrison Salis-
bury of the *New York Times*, independently decided to give the award to
the *Catalog*, even though it hadn't been nominated. "In 100 years, 'The
Last Whole Earth Catalog' probably will be the only book of 1971 to be
remembered," one of the judges opined. A third judge, historian and
journalist Garry Wills, who believed that the rules specified that the
award should go to an author rather than a compilation, was so outraged that he resigned from the panel in protest.

While having a drink with his cousin Edward Hoagland, an established Manhattan author, Brand asked if the National Book Award was
actually important.

"I wouldn't mind getting one," Hoagland responded.

Later Brand arrived for the award event at Lincoln Center in his by-
then-trademark buckskin outfit. Donald Barthelme, a fellow National
Book Award recipient, advised him to speak slowly. Onstage, in front of
an audience of 1,800 people, he intoned: "If the award encourages still
more self-initiated, amateur, use-based, non–New York publishing, good
deal."

He planned to give the $1,000 prize money to an organizing effort
to hold a counterevent at the upcoming United Nations Conference on
the Human Environment, scheduled for Stockholm in June, he told the
audience.

The National Book Award confirmed Brand as a member of the cultural elite, if only by a two-judge margin. His editor, Jim Silberman, had
been amused that he was the editor of a book for which he did no editing. After the award ceremony, Silberman and several top Random
House executives took Brand to the '21' Club, where he again created a
bit of a stir. His buckskin outfit didn't include the required tie, and so
he had to quickly find one to add to the ensemble. The East Coast embrace of California counterculture could go only so far.

Spurred in part by the environmental consciousness that the *Catalog* had helped spawn, the UN conference was intended to consider "the
need for a common outlook and for common principles to inspire and
guide the peoples of the world in the preservation and enhancement of
the human environment."[1] At the time, Brand still thought of himself as
an environmental activist—albeit with an antipolitics stance.

A counterevent in Sweden, which had begun as Household Earth
and later was known as the Life Forum, had already gotten off on the
wrong foot. In March, Brand had traveled to Sweden to build bridges
with Swedish activists to lay the groundwork for what they hoped would
be an event on the same scale as Woodstock, but rather on an international stage. At one point during their meetings, Brand asserted that he

believed he could attract between one hundred thousand and four hundred thousand participants to travel to a remote camping site he had persuaded the Stockholm authorities to grant the protesters at Skarpnäck, a suburb far from the city center, where the UN conference was to be held. Afterward, however, he would ridicule himself for being so out of touch. The Life Forum organizers had no ability to recruit participants in Europe, and the cost of getting to Sweden from America ensured that ultimately less than a thousand people would attend.

Brand had begun by enlisting his usual supporting cast, including the Hog Farm commune, which was committed to shipping several of its buses to Stockholm to be part of a planned protest march. Because they had fed the crowds at Woodstock and provided security, the Hog Farm had acquired a reputation for counterculture crowd control.

Brand's marriage to Jennings had "loosened up" since the demise of the *Catalog*, and he was "waiting for the next big idea"—which hadn't yet materialized. He had taken an apartment in San Francisco with the idea of getting distance from his wife to give himself some freedom. He was in psychoanalysis with Pierre Mornell, and he had also begun actively pursuing several young women.

There was a vast gap between the Swedish leftists he met with and the group of American environmentalists who identified with Brand's counterculture roots. The Swedes were deeply suspicious of the fact that Brand was flush with cash from the success of the *Catalog*. A *Ramparts* magazine article had recently documented the CIA's infiltration of the American student movement, and the Swedes, who were "old" European socialists, were paranoid about the American counterculture. According to an account in the *Village Voice*, at one point during a Stockholm meeting, Brand passed out copies of the *Whole Earth Catalog* and "someone opened the book to a page that advertised books on growing marijuana. The socialists were alarmed and accused the Life Forum organizers of being CIA plants, intending to spread drugs during the Stockholm UN Conference."[2]

Despite his belief that the Demise Party had been a failure, it hadn't diminished Brand's ideal of philanthropy. He and Dick Raymond decided to create the Point Foundation, a new organization intended to give away the *Catalog*'s profits. The decision led to grumbling among some of the Truck Store workers, who had hoped their hard work might be rewarded in some way, but Brand largely ignored the staff, instead creating an initial board that consisted of himself and Raymond as well as environmentalist Huey Johnson from the Nature Conservancy; Michael Phillips, then associated with the Glide Foundation; Jerry Mander; and Bill English, who by then had left his role as Doug Engelbart's chief engineer and moved to a new Xerox computer science laboratory called the Xerox Palo Alto Research Center (PARC). Each director was assigned $55,000 annually to dispense as they saw fit. Brand initially assigned his funds to the Life Forum, the San Francisco Zen Center, and Steve Baer's Zomeworks, among others. Additionally, a seventh director, dubbed "Elijah"—a different person each board meeting—was invited to participate in funding decisions.

Brand's departure for Sweden was preceded by a series of traumatic encounters with Jennings, who ended up driving across the country to her parents' home in Bethesda by herself, leaving Brand with feelings of guilt, though clearly not an incapacitating level, because at the same time his mind was mostly on twin affairs. One was with Melissa Savage, a young environmentalist involved with the Sweden project; the other was with Stephanie Mills, a Planned Parenthood organizer he'd first met when she participated in the *Hunger Show*. While he had partially moved to the apartment in San Francisco, periodically he would decide that he was still in love with Jennings—this in between bouts of torturing her with admissions about his affairs. If he appeared to be acting like a scoundrel, he was not alone. He was straining at the confines of his marriage during a period of societal sexual liberation.

In April, before the Life Forum, he took off with Mills to soak in the hot springs of the Black Rock Desert. On their way east they stopped at Gary Snyder's home, but he was away, and so they visited with his wife, Masa. They then spent four days wandering in the desert and lounging in the hot springs, camping in the rain, and on occasion taking shelter in an abandoned cabin. Before they left, Mills commemorated their visit in graffiti on the cabin wall: "Two clean freaks came from hot creeks to stay in High Rock Canyon. We walked the heights and stayed two nights and loved with wild abandon."

The political chaos and organizing challenges facing his Life Forum idea were so formidable that by the time he got on the plane to fly to Sweden, he had become disenchanted with what he thought of as the "same-old-shit eco march." He was also in terrible psychological shape.

He flew over with Mills. On the long plane ride, she had been affectionate, and he had hoped they would stay together in rented apartments adjacent to a gallery the Life Forum was using near the UN conference. (They had decided not to camp with the majority of activists at Skarpnäck.) But after they arrived, Mills appeared to lose interest and dived into her project hosting salons for various international delegations. She also visibly took up with others, to Brand's consternation.

"You can't do that," he protested. "That's actually not fair!"

"Reparations," she responded. "I'm doing to you what men have been doing to women for a long time and it's about time. So just settle down and enjoy it."

Mills's disappearance plus the long plane trip and jet lag, coupled with his growing sense that his effort to organize a mass gathering was going to be viewed as a debacle, sent him into a downward spiral. He was highly visible in his trademark top hat with an eagle feather and buckskin suit. Yet news photos taken in Stockholm showed him with hollow, sleepless eyes and a bittersweet pained expression, his face reflecting what he believed was going to be an almost certain fiasco.

He led a Save the Whales march and appeared at several public events outside the official event with Maurice Strong, the Canadian oil executive who was the secretary-general of the conference. But the efforts by the American environmentalists to build a united front with the Europeans failed. A deep anti-American sentiment pervaded Stockholm, exemplified by Swedish prime minister Olof Palme, who had opened the conference by demanding that American "ecological warfare cease immediately."

During a preliminary spring meeting, Brand and his confederates had proposed supporting the Europeans with $40,000 to permit the Life Forum to share facilities with the European leftists. They were completely rebuffed. When Mary Jean Healy, a young editor at the environmental magazine *Clear Creek*, showed up in Stockholm as a fairly typical Berkeley New Leftist, she was shocked to find that she was viewed as a political conservative in the context of the intense factionalism swirling around her.

While there had been talk of chartered airplanes flying Americans in from both coasts, in the end, Brand underwrote fifty Hog Farm commune members and several buses—nominally for crowd control and feeding the expected hordes who never materialized; a group of antiwar activists committed to raising the issue of the environmental destruction that had resulted from the Vietnam War; Beat poets Michael McClure and Gary Snyder; Healy, who would publish *Open Options*, an alternative guidebook to the UN conference; and Jack Loeffler, who managed to bring a group of fifteen American Indians and Chicanos, some without passports or visas. Loeffler succeeded in getting a group of four Hopi Indians safely to Stockholm by finding some old paper that looked like parchment and making home-brewed passports, binding them in buckskin. The back of each passport said its bearer was a citizen of the Hopi Independent Nation and that "this passport is valid as long as the sun shines, the water flows, and the grass grows."

Brand felt that his efforts to make the case for a new form of personal responsibility and ecological consciousness were largely lost amid the politically charged atmosphere that surrounded the conference. Growing increasingly despondent, he began to feel dizzy and sick to his stomach. He tried to steady himself with Librium. From somewhere nitrous oxide appeared and he sampled it. He plunged into an increasingly deep depression that was exacerbated by painful calls and letters from an emotionally distraught Jennings, who was trying to hold on to their marriage. Savage, who had been his cultural guide three months earlier on their organizing trip to Stockholm, had also vanished into a new relationship, leaving Brand with a deepening sense of isolation. He recalled one of the directors of a local FM radio station in the Bay Area who had had affairs with many of his staff and then leaped from the Golden Gate Bridge. Suicide again floated through his mind.

Brand would remember Sweden as a disaster, but the reality was more nuanced. Both Walter Hickel, Nixon's former secretary of the interior, and Strong, the UN conference's secretary-general, praised him publicly, and the event ultimately did have an impact on environmental consciousness globally. For the most part, however, Stockholm proved to be a maze of the sort of political infighting that he abhorred.

Paul Ehrlich, his Stanford mentor, was at the conference as a population doomsayer. It had been four years since *The Population Bomb* had been published, and the issue of potentially mandatory population control remained a flashpoint in international forums. Barry Commoner, a rival biologist who was also in Stockholm, actively attacked Ehrlich's view. Commoner argued that there would be a demographic "transition" and that population growth would abate. At the time, Brand remained firmly in Ehrlich's camp, although years later he would change his view and acknowledge that Commoner had been correct.

One evening he accompanied Healy to hear Peter Berg, an environmental activist who had been a member of the Diggers and the San Francisco Mime Troupe, make a distinction between "one world," which he called a "bullshit transnational fuckup" idea, and "one planet," which he considered a timeless idea grounded in experience. The idea would stay with Brand.

The bright spots that eluded him were due in part to $75,000 of Point Foundation funds invested in the Life Forum. Brand had given Berkeley activist Joan McIntyre $15,000 to launch the campaign to Save the Whales, which was unveiled in Stockholm. The Hog Farm had dressed up their bus as a whale for the planned Save the Whales march that marked the coming-out party for McIntyre's Project Jonah. The Life Forum activists marched through town making whale sounds, and photos of the whale bus appeared on the front page of newspapers around the world, giving the project tremendous impact and leading to a symbolic unanimous vote against commercial whaling at the Stockholm conference.

And it had an impact on individuals as well. In Stockholm, Savage ended up sharing a hotel room with McIntyre. At the end of the conference, the two women discussed what should come next. McIntyre told the younger woman, "Well, you know, just pick a species."

Savage remained an ardent environmentalist and scientist throughout her career, and when she turned sixty, she decided to take McIntyre's advice and focus her work on saving the otters. She viewed her lifework as one of those "little vortexes" that would spin off from one of the areas that Brand dived into.

For Brand, however, his conclusion that "Woodstockholm" was a grand failure would sour him permanently on political protest—with several exceptions for local politics. When the UN conference was over, he took off with Healy to go camping near the Arctic Circle for several days. Then he flew to Washington, DC, for a tense reunion with Jennings, who told him he smelled different.

O

ON THE LONG ROAD TRIP FROM BETHESDA TO CAPE BRETON ISland with Jennings later that summer, Brand again began to fantasize about becoming a journalist. He mused about learning shorthand, assessed his writing as "bad," and began to contemplate the possibility of teaching himself to write "pretty, clear & fast." After spending a morning in a New Jersey hotel room reading the *New York Times* opinion page, he was inspired. He thought about learning languages and possibly even taking a new direction in photography.

He was starting to come out of the downward spiral that he had been in since shutting down the *Catalog* the previous year, and his head was spinning with ideas. He began discussing several projects with Marlon Brando, the early supporter of the *Catalog*, including a television adaptation of the *Catalog* and a possible sustainability project on Brando's island in the South Pacific. He was considering practicing zazen, a form of seated meditation, and would join the San Francisco Zen Center when he returned to the West Coast. He even reconsidered graduate school at UC Berkeley, perhaps returning to biology.

He jotted down a list of fifteen ideas titled "many fantasized projects for California" and inserted one in the middle—"uncouple." He'd taken a big step already in renting a basement apartment in San Francisco, but there had been excuses for that. On the trip north, he and Jennings quarreled repeatedly, and she accused him of planning to leave her. It was true, but he couldn't bring himself to force the discussion.

The planned trip to Cape Breton Island had originally been intended to complete their shared dream of building a small summer home to live in for a while each year far from civilization. However, upon arriving they both immediately had reservations. It was an overwhelming and expensive project, and they discussed giving up and fleeing back to California.

To make matters worse, the first night was spent in the rain in a tent. Trapped in a downpour, he found an upside: it was the moment when he discovered Gregory Bateson, who would supplant Buckminster Fuller as his intellectual guiding star. He had with him a copy of Bateson's *Steps to an Ecology of Mind*, published earlier that year, a tightly coupled collection of essays offering a synthesis of cybernetics, philosophy, and the social sciences, situating humanity in the natural world. Brand was spellbound.

It was not his first encounter with the anthropologist. Brand had interviewed him as part of his Stanford senior project about the Center for Advanced Studies in the Behavioral Sciences. Later, he had also seen Bateson speak at the LSD research conference at San Jose State College. But he'd never invested the time to read Bateson's work. Bateson offered a mystical and biological foundation for a cybernetic theory that placed the individual in a much broader context that stretched from the social to the cultural and ecological dimensions of the world. The ideas that Brand now encountered would lead directly to the perspective at the heart of the *CoEvolution Quarterly*, the magazine that he would found two years later.

When the weather cleared, Brand put down his book and set to work clearing the ground for an interim tepee and then shaping the long poles. After the tepee was up, he and Jennings commenced work on a twelve-foot-by-sixteen-foot shack, mostly finishing it in a month of steady hard labor. While he worked, however, Brand continued to fantasize about leaving his wife and living alone as a bachelor in his San Francisco apartment when he returned to California.

He planned to wait until they were back on the West Coast to talk about divorce, but the tension overwhelmed him. Once he fainted, coming to with his fingers clutching the grass, wondering what had happened and what might be wrong.

He vacillated for days and then finally broached the subject of living alone in San Francisco—of separating. His hand was forced when Jennings found Mills's hairbrush in the glove compartment of the truck. She was furious and he responded in kind. Jennings heard his anger as the end of the relationship that she was certain he had been building up to. She began to weep softly. "What do I do now?" she asked. "I don't know, I don't know, I don't know."

Her words echoed inside his head, leaving him feeling guilty and scared. He wasn't sure what he had done. He stood on a cliff looking out over the ocean and thought about jumping and realized he wouldn't. His mood swung and he felt a brief breeze of freedom.

There was no sudden collapse; Brand and Jennings continued to work on finishing the shack while discussing coming apart. They left for California in early August, again in heavy rain, stopping at Higgins Lake in Michigan to spend time with Brand's family. His brother Mike and his sister-in-law Gale took separate walks with each of them and afterward agreed that they were certain the relationship was ending.

They arrived in California before the end of the month. Still not formally split, they stayed briefly on the Midpeninsula with friends, and then Brand headed to his city apartment, only to find that the building had been sold and he had just three weeks to find a new home.

During the next month, he slowly ended their marriage. On occasion, he would "freak out" and run back to Jennings, but that only made things worse. "Wounded with each other we lean on each other for further wounding," he wrote in his journal. At the same time, he realized that his behavior had a terrible impact on Jennings, who wanted their marriage to survive.

He had scrambled and found an apartment on Rose Street Alley, half a block from the San Francisco Zen Center, and christened his new "pad" by inviting Mills over. That night they went to see *Fiddler on the Roof* and then ate dinner in North Beach.

During the fall he tried to establish a Zen practice but soon found that he quickly grew bored when he tried to meditate. He continued to see his psychiatrist while simultaneously dating both Mills and Savage as well as an expanding gallery of other women while spending time on the Point Foundation, figuring out how to give away the $1.5 million he'd earned from the success of *The Last Whole Earth Catalog*, all while still gripped with frequent deep bouts of depression and fainting spells.

"What I've got, these days, is a tenuous, sophisticatedly tenuous, grip on reality," he wrote. His confessed fantasy was to become a "Zen playboy," but he simultaneously realized the effect his behavior was having on Jennings, ensuring that they both remained miserable.

He had been pursuing one of his fantasies—learning to fly. Despite the failure of his earlier experiment with the BD-4 kit, he had begun taking flying lessons. He soon gave up the experiment, however, in the face of persistent panic attacks. He would find himself paralyzed by anxiety—awash in emotions that were strong enough to force him to pull his car over to the side of the road. He realized that such a state of being was not conducive to piloting an aircraft.

It began as a period of personal and intellectual uncertainty, but gradually things began to come into focus. Toward the end of September, he had dinner with Peter and Fox King, and as he drove back to his apartment he found a new sense of clarity. He would become a gentleman "gamesman," he decided, designing games and traveling the world to discover exotic ones. He would return to his original fascination with watching the California Indians playing and the "out of body" *Spacewar!* experience at the campus computer center. He would invent "New Games."

As often was the case, it was a decision with an antecedent. After the War Resisters League asked him to organize a series of events at San Francisco State, UC Berkeley, and Stanford, which he had dubbed World War IV, he felt that his events, which were based on the idea of competitive games to work out aggressions—rather than war— had been failures. However, he was still passionate about "the sheer fun of fighting." He fondly recalled his *Catalog* days as a vicious boffer.

At an Esalen event, Brand met George Leonard, the *Look* magazine writer who had published early articles that previewed the impact of the youth movement and the California counterculture. Leonard shared a passion for participatory athletics with Esalen's cofounder Michael Murphy, and his ideas resonated with Brand, who had been thinking about what he had continued to think of as "softwar"—war without real violence, the organizing principle for what would be the first New Games Tournament in October of 1973.

Despite Brand's early freelancing failures at Stanford and afterward, he finally broke into the world of professional journalism when Jann Wenner, cofounder and publisher of *Rolling Stone* magazine, gave him an assignment to write about a small cult of computer hackers who would stay up all night and play the video game *Spacewar!* in a computer laboratory hidden in the hills behind Stanford.

Brand asked Shel Kaphan, one of the young computer hackers who worked at the Truck Store, to put him in touch with the researchers at computer scientist John McCarthy's Stanford Artificial Intelligence Laboratory. SAIL had been established to build a working artificial intelligence and he had collected an eclectic group of young researchers exploring technologies like robotics, computer vision, natural language understanding, and speech recognition. Simultaneously, Bill English opened the doors to the Palo Alto Research Center. Xerox had created PARC to compete directly with IBM, and Robert Taylor, a young psychologist who had funded the development of the ARPANET while at the Pentagon, had been given the charter of rethinking the future of the office based upon computers and networks.

Being admitted into the inner sanctum of the two research centers gave Brand an even clearer awareness of the impending arrival of personal computing and the potentially pervasive impact of computer networks. Several years later he would be the first journalist to use the term *personal computer*; but more important, he was the first to relay the significance of their imminent arrival to the world.[3]

One evening in October, Brand arrived at the Stanford Artificial Intelligence Laboratory with *Rolling Stone* photographer Annie Leibovitz in tow. They had come to report on the first Intergalactic Spacewar Olympics. In doing so, he would describe the virtual world of computer hackers to a general audience for the first time, at the same time portraying the designers of both the first modern personal computer at Xerox's PARC and the recently born ARPANET community, the forerunner of the modern internet.

He used his night spent with the SAIL hackers to draw an opening vignette that made a connection between the computing culture he had discovered and the counterculture he had helped create. Women were a novelty in the computing world, and Leibovitz was treated as visiting royalty by the young researchers, who proudly showed off the SAIL robot arm and other research projects going on at the lab.

Brand's article was written in a style that was borrowed in part from Tom Wolfe and in part from Hunter S. Thompson, with a flashy you-are-there patter and mind-boggling grand observations: "These are heads, most of them," he observed of the hackers, without bothering to source his claim. "Half or more of computer science is heads." The implication, of course, was that computing was the next thing after LSD—something Ken Kesey had realized several years earlier when Dave Evans introduced him to Engelbart's NLS computing system.

Style aside, the article caught virtually all of the significant trends in computing, ranging from PARC scientist Alan Kay's Dynabook, which was the archetype of the modern laptop personal computer, to the impact of ARPANET and the fact that a new generation of computer chips, then just on the horizon, would consume far less power, making possible a generation of inexpensive consumer-oriented machines.

He also captured the original spirit that would engender the community of hobbyists from which the personal computer industry emerged beginning in 1975. Brand realized that the young computer hobbyists who were playing on the SAIL mainframe had a deep hunger to have their own computers, something that the microprocessor, which had been invented in 1971, would soon make possible. Most of the hackers weren't even sure what they would do with their computers when they obtained them—they were simply fantasy amplifiers. Brand included an appendix in the *Rolling Stone* article on how to program your own *Spacewar!* game, including a sample of computer code written by Alan Kay.

"SPACEWAR: Fanatic Life and Symbolic Death Among the Computer Bums" appeared in *Rolling Stone* at the end of 1972, stating unequivocally: "Ready or not, computers are coming to the people."[4] The term *personal computer* did not appear in the original *Rolling Stone* article. Brand would first use the term in an epilogue that he added to his first book, titled *II Cybernetic Frontiers*, published in 1974.

Comprising the *Rolling Stone* article and an extended profile of Gregory Bateson that appeared in *Harper's Magazine* in late 1973, the book was published in part as a reward from Random House for the runaway success of *The Last Whole Earth Catalog*. Brand was attempting to portray a synthesis of Bateson's "organic" cybernetics with the "machine" cybernetics of the emerging computing universe. Although it was far ahead of a world that was still largely unaware of the impending societal impact of computers and networks, Brand's prophecy was clear, and together the magazine article and book would provide the first inkling to the outside world that something transformative was afoot.

However, none of the impact that his first successful journalistic foray would eventually have was apparent to Brand that winter. He ended his psychotherapy sessions, but he wasn't out of the woods. He still suffered from precipitous ups and downs and panic attacks while frequently using Librium to get through the night, continuing to experience stomach distress while bouncing back and forth between a series of girlfriends, of whom Mills remained the most serious.

He visited Kesey's farm in Eugene at the end of the year, traveling there and back by Greyhound bus. Kesey remained a combination mentor, older brother, and competitor. But Brand always gained something from his interactions with Kesey, and this time he came away with a surge of motivation and self-confidence.

Over New Year's he and Mills drove down to Mazatlán in Mexico. After playing in the waves, he suddenly noticed a weird separation on the left side of his field of vision. He stumbled back to his VW van, put on sunglasses, took a Librium, and lay down on the seat trying not to freak out until his vision returned to normal. It would become a frequent occurrence, and years later he discovered it was a relatively common affliction dubbed "scintillating scotoma."

At the time, Mills tried to encourage him not to worry, but they were both unnerved by the episode, and it cast a pall over their adventure. He felt fragile and she tried to push on as if nothing were wrong. Later at a Mexican rancho in Baja California, reached by two days of piloting the VW over rocky dirt roads, Mills gently inquired about why he was so troubled.

"Existence," Brand answered.

"Whether or not, or how?" she countered.

"Both," he replied.

"Well, remember. You're just a speck of dust. In the sunshine."

WHEN HE WAS BACK IN SAN FRANCISCO AT HIS APARTMENT ON Rose Alley, Brand felt as if he were turning a corner. The world was changing. LBJ had died. Abortion was legalized. The Paris Peace Accords had been signed. And Jennings announced that she was filing for divorce and was involved with Keith Britton, a British engineer and explosives expert, whom she planned to marry.

A weight was lifted.

Although Brand retained a political aversion to the New Left, he made exceptions for people he considered friends, like Abbie Hoffman and Paul Krassner. Krassner had originally moved to the West Coast from New York when Brand had convinced him and Ken Kesey to edit the final *Last Supplement to the Whole Earth Catalog*. They in turn had persuaded the counterculture cartoonist R. Crumb to illustrate the cover, making it look like a very raunchy Last Supper.

When Brand moved to Rose Alley, he let Krassner move into his second bedroom on the condition that their "lifestyles" didn't conflict. Krassner later described their relationship as the "New Age Odd Couple," a contradiction in lifestyles that reached a peak the day they conducted simultaneous interviews in their separate bedrooms, Brand with Gregory Bateson, Krassner with Squeaky Fromme, the member of the Charles Manson family who several years later attempted to murder President Gerald Ford.[5]

The "odd couple" description was apt in several other ways. Krassner had a tendency toward paranoia, fueled in part by his friendship with Mae Brussell, a leftist conspiracy researcher who published frequently in the *Realist*. When Krassner told her he was planning to share an apartment with Brand, she warned him that the former *Whole Earth Catalog* publisher was a "government pig." When Krassner asked her if he would be in physical danger, she replied, "No, but he'll try to psych you out."

Krassner ignored her warning; however, there were moments when it seemed she might have been correct.[6] One day Krassner came back to his room and sensed that something was different. He puzzled for a while and then realized that it was his giant Nixon poster. The former president's eyes, which had been looking to the left, were now peering to the right. He looked at the poster carefully and discovered that the original eyeballs had been whited out and new ones drawn in.

He confronted Brand, demanding to know if he had tampered with his poster. His roommate replied that, no, he hadn't, but that Ken Kesey had been around for a while that afternoon.[7]

On another occasion, Krassner decided to read the journal that Brand was keeping, and then found his feelings were hurt when he read how he had been described.

As much as he was struggling to find balance, Brand was often the more reliable of the two. One night he was in his bedroom using an electric saw to build a bed while he listened to Krassner in the hallway brushing his teeth and complaining about his battles with the Church of Scientology. Suddenly there was a long silence and Brand peeked outside his room to see Krassner standing motionless making strange noises. As Brand watched, he toppled over onto a buffalo skin rug and continued making garbled sounds.

Brand enlisted several people from the Zen Center to help him carry Krassner to a nearby hospital. When Krassner finally came around several hours later, he admitted that in preparation for attending a final performance by The Committee improv comedy group, he had consumed his remaining stash of pure THC, suffering the predictable consequences.

Brand's divorce from Jennings at the end of November 1973 was not contentious, and they continued to stay in contact as directors of the Point Foundation. Their community property totaled about $40,000. Brand kept the Nova Scotia property while Jennings kept the $10,000 savings account and the TV.

Soon after finishing his "SPACEWAR" *Rolling Stone* article, Brand began work on a profile of Bateson. With tape recorder in hand, he attended the social scientist's courses at the University of California at Santa Cruz.

An Esalen conference led by G. Spencer Brown, who had recently written a philosophical mathematical treatise called *The Laws of Form*, enabled conversation with diverse members of the cybernetics and counterculture crowd that included Bateson, John Lilly, Ram Dass, Kurt von Meier, Alan Watts, and John Brockman.

That spring, through Bateson, Brand would dive more deeply into the world of cybernetics. He spent time with the anthropologist in Big Sur, where Bateson was building a home. When the resulting article, "Both Sides of the Necessary Paradox," appeared in *Harper's Magazine* in November, it had less of an impact than "SPACEWAR," but his dive into Bateson's world had a lasting effect on Brand's thinking.

He had come to cybernetics earlier, and although it permeated the Systems section of the *Catalog*, it was his close encounter with Bateson that would shake him loose from his devotion to the engineering-oriented world of geodesic domes and Buckminster Fuller. Bateson's assertion in *Steps to an Ecology of Mind* that the three interwoven systems of the individual, society, and ecosystem form a cybernetic system that encompasses everything would become the through line for Brand's intellectual work for decades to come.

In his introduction to *II Cybernetic Frontiers,* Brand argued that "SPACEWAR" was the better article, but he was placing it after "Both Sides of the Necessary Paradox," "because what Bateson is getting at, I'm convinced, will inform damn near everybody's lives." Both meat and machines learn, he noted, pointing out that organic and machine cybernetic systems share the quality that they both subvert "man's special pride, conscious purposefulness." In shifting his allegiance from Fuller to Bateson, Brand was able to fully exploit his training as a biologist.

Bateson's ecological perspective also pried Brand loose from the last vestiges of the Ayn Rand libertarianism that he had carried with him into the *Catalog.* In Bateson's thrall, beginning to think deeply about the concept of coevolution, he wrote in his journal, "Self-sufficiency is not to be had on any terms, ever. It is a charming, woodsy extension of the fatal American mania for privacy. 'I don't need you. I don't need anybody. I am self-sufficient!'"

In the spring of 1973 Brand was invited to the Rhode Island School of Design to speak about creating New Games. It was a warm day, and he took off his shirt while leading a group of students in a game of Slaughter, which involved up to forty players, on their knees, trying to get their team's ball into a hole dug by their opponents and vice versa. If you were pushed off the playing field, you were eliminated. (In a later permutation, you could be eliminated if an opposing player managed to remove your socks.) The shirtless Brand attracted a young architecture student named Gloria Root, who accompanied him back to his room, where she took off her shirt in return.

Root, the former Miss December 1969 *Playboy* Playmate, had the tantalizing backstory of having spent nine months in a Greek prison for smuggling hashish from Turkey, as well as having spent time with the likes of Hugh Hefner and Roman Polanski. Brand envisioned a lively ongoing romance, but when he tried to contact her in the following weeks he was ignored.

In June he returned to the East Coast to spend a couple of months in his shack on Cape Breton Island. But after he arrived, he discovered that he had not completely escaped Jennings's ghost. He had felt her presence in the small trailer where he continued to stay on occasion in the Santa Cruz Mountains, and now in Canada, memories of his stomach pains and fainting spells from the previous summer still plagued him.

He kept telling himself that he had been tired of San Francisco and the rat race of work and social entanglements. His retreat was supposed to offer a respite for simple physical tasks and planning. "Relax, Stewart, enjoy paradise," he reminded himself. That wasn't as easy as he hoped. An air freight snafu resulted in much of his gear being lost—including his meditation pillow—and he used its absence to avoid the Zen practice. He had two female guests, but when each left, he simultaneously felt unsettled and pleased. Robert Frank was around, which was good, but Brand still felt uneasy, without serenity.

In the end, however, the summer offered new clarity. But when it arrived, it turned out not to be a new idea but rather a reinterpretation of an old one. He described his moment of inspiration in the *Whole Earth Epilog* the following year, in a riff on his darker comment to Diana Shugart some years earlier: "After burning our bridges we reported before the Throne to announce: 'We're here for our next terrific idea.' The Throne said, 'That was it.'"

He had found that his new freelance existence felt lonely and precarious and he missed the artistic control the *Catalog* had afforded. *Harper's Magazine* had reached out to him with a proposal to produce a slim update to the last *Catalog*—an *Epilog*. He wrote back to the editors, pointing out that despite being out of date, the *Catalog* was continuing to sell three thousand copies each week. As he thought more about the *Harper's* offer, he realized that what he had really enjoyed was working on the supplements. They had offered more freedom than the *Catalog*. There had been a satisfying pace, and it had allowed him to follow his nose for new ideas. They had provided a way to do intellectual and artistic explorations while simultaneously creating conversations and community.

And now, Brand realized, there was a need for a new conversation that went beyond the *Catalog*. At just the moment when millions in the American student Left were fixated on dialectics and were regurgitating calls for class struggle, Bateson had alerted him to a much more powerful approach to open up new ways of looking at the world. Although the biological origins of the term *coevolution* spoke to the evolutionary partnership between predator and prey, Bateson had a broader conception that scrutinized the juxtaposition of opposites to expose "new meaning and understanding."[8] It was an idea that recalled Koestler's *Act of Creation*, but Bateson articulated it with a degree of precision and sophistication that took Brand to an entirely new level—and new clarity about what he should do next.

His model for the *Catalog* had been L.L.Bean. Now, as he roamed the vacant lands surrounding his cabin, a new idea took shape based on the *New Yorker*. It would be an antithesis to other magazines, which he viewed as "publication by committee." He had learned from his encounters with the publishing world that the best periodicals had a "mad monk" somewhere in the middle of the enterprise, someone who was devoted to the enterprise and who took responsibility for every comma.

That would be him.

At the end of the summer, he headed back to California intent on creating a new magazine, accompanied by an occasional *Catalog*. A decade later, in an interview, he would describe it as "the *New Yorker* of Sausalito." He would do the *Harper's Magazine* project as a stepping-stone toward something more permanent.

A CROSS THE GOLDEN GATE BRIDGE NORTH OF SAN FRANCISCO and just inland from the Pacific Ocean, Gerbode Valley lies in the heart of the Marin Headlands. It offers a broad, grassy plain that is home to bobcat and deer.

In October, Brand chose the site for his next act as an impresario. Once again, he assembled some of his usual cast of friends and fellow travelers to organize a New Games Tournament: Huey Johnson, who'd first alerted him to the hidden valley; George Leonard, who joined to teach aikido; and Wavy Gravy, who brought the Hog Farm to join in the festivities. To help with logistics, he enlisted Pat Farrington, a community organizer, and Andrew Fluegelman, a young attorney with a counterculture sensibility.

For Brand it wove together several threads. He had read *Homo Ludens*, the 1938 book by Dutch historian and cultural theorist Johan Huizinga, who put forward the idea that play was central to the formation of culture. But at some point, "games" had come to be seen as entirely zero-sum: one winner only. Bateson told Brand that a weakness in game theory was that there was no provision for changing the rules of the game. Thinking back to his childhood, Brand realized that as a kid he would change the rules of games all the time to make them more interesting. He decided that as they got older, children were forced into structured games by adults who in turn had been forced into such contests when they were kids.

Why not create games where the players changed the rules as needed?

Here was another Brand question that at first glance seemed simple (Why haven't we seen a photograph of the whole Earth yet?) but encapsulated a profound shift in perspective. In his mind, the first New Games Tournament would be a prototype for a new type of competition. Moreover, his "softwar" idea implied that society at some point in the future might change the rules of warfare itself to make it less violent—definitely a utopian ideal.

Several hundred people showed up each day for the three-day event to play games like the ones that Brand had already invented, such as Slaughter and Earth Ball (in which two teams pushed a giant ball, painted to look like Earth, toward each other's respective sidelines; the catch is that if one side appears close to victory, some of its members on occasion would switch to the other team and start pushing in the opposite direction) as well as boffing and classics like tug-of-war.

The idea was not only to invent new games but to change the rules of existing ones. Volleyball was played to see how many times the ball could be batted back and forth without hitting the ground, for example. Hang gliders—including one piloted by Brand—soared overhead to add a festive touch.

Afterward, he would describe the event as being only "semi-successful." Although the New Games Tournament was covered by the national press, and even *Sports Illustrated* ran a piece on the idea of alternative sports, Brand was put off by what he would call the "nicey nice" counterculture gaming types who took over the New Games movement the tournament engendered. They tended to invent games that were not very competitive and in which everyone "won." He had been interested in competitive, physical contact sports. He wasn't interested so much in ending war as in civilizing it. He would sum up his views in a short essay on "Softwar" in *II Cybernetic Frontiers*, arguing, "Let Audie Murphy, Henry Kissinger, and Howard Cosell work the same playing fields. Let Martin Buber and Pope John train the referees."

Immediately after returning from Canada, Brand had begun to reassemble a publishing staff, starting with Diana Shugart. For an office, he rented a "moldering crab shack" for $40 a month in Sausalito. Set along a working waterfront, it came with a parrot named Lolita that incessantly screamed, "Lolitaaa! I want a cracker!"

He used the Point Foundation's organizational structure to host his new publication; technically, Michael Phillips, the foundation's president, simply hired Brand as the editor in chief. (After they spent a year in the crab shack, a developer got permission to demolish the pier where it was located and so the *Quarterly* offices moved to what they described as the "longhair industrial dump" section of Sausalito in a building that retained the sign "Harvey's Lunches.")

Fluegelman was ready to abandon his legal career, and Brand hired him as the managing editor for the new publication. The working title had been *The Never Piss Against the Wind Newsletter* and Brand also considered a Batesonian *Making Circuit* before at the last moment settling on *CoEvolution Quarterly*, which captured his newfound enthusiasm for Bateson and biology.

"Ecology is a whole system, alright," Brand wrote in the *Quarterly*, "but coevolution is a whole system in *time*. The health of it is forward. Systemic self-education which feeds on certain imperfection. Ecology maintains. Coevolution learns."[9]

The *Epilog* and the *Quarterly*, which first appeared in compressed form as a 112-page supplement to *Harper's Magazine*, began with a four-point explanation about why the *Catalog* was being restarted. He had expected that someone else would fill the niche he had left, Brand wrote, but they hadn't. Second, the *Catalog* was continuing to sell briskly, even with increasingly outdated information. Third, the North American economy was "beginning to lose its mind, putting more people than ever in need of tools for independence." And fourth, he acknowledged that he hadn't been able to come up with a better idea. The *Harper's* wraparound ended with a review of a pessimistic assessment of the world's energy future, partly occasioned by the ongoing oil crisis. Below it Brand set a half-page photo of Stephanie Mills naked and silhouetted with her back to the camera walking on the desert playa and captioned "Armageddon Outa Here!"

While Bateson introduced Brand to the importance of paradox as well as to the sense of complete interconnectedness of the natural world, the biology-centered worldview of Ehrlich and Bateson pried him away from the engineering-centered influence of Fuller and Wiener that had shaped the original worldview of the *Catalog*. The magazine that emerged from these ideas opened with a special section titled "Apocalypse Juggernaut, hello," citing the head of the Arthur D. Little consulting firm who predicted food shortages, massive federal spending increases, a worsening climate, the end of Keynesian economics, and the tripling of coal prices.

A first printing quickly sold out seven thousand copies and went into a second printing. It featured an article on coevolution by Ehrlich, as well as features on subjects like energy and culture and "spiritual tyranny," and an article by future *New York Times* science writer Gina Kolata on the promise of theoretical ecology.

While Brand was launching the *Quarterly*, in the spring he immersed himself in Ehrlich's biological perspective, sitting in on his graduate ecology seminar at Stanford. The idea of coevolution was becoming a powerful concept for Brand. It offered both a new metaphor and a prism through which he would interpret the world.

Ultimately, a staff of twenty-five worked on the *Epilog* from September 1973 to September 1974, when a first printing of 350,000 copies was produced. Some of the staff had been part of the original *Whole Earth* undertaking, such as J. Baldwin, Lloyd Kahn, and J. D. Smith, but Brand remained an adept recruiter and quickly began to acquire a stable of talented young writers and editors. Echoing the original *Catalog*, the staff at the new offices shared lunch and a daily volleyball game.

The *Epilog* was not as imposing as its predecessor, *The Last Whole Earth Catalog*, but it contained some of the same idiosyncratic features. It was 318 pages and priced at four dollars, and it included a slightly ribald satire of the TV series *Kung Fu* penned by Paul Krassner, titled *Tongue Fu* and modeled on *Divine Right's Trip*, Gurney Norman's novel, which had been serialized in *The Last Whole Earth Catalog*. A new section, "Soft Technology," edited by Baldwin, focused on new kinds of alternative energy technologies.

By the middle of the summer, Brand was ready to sever his ties with San Francisco. He initially found what was essentially a garage in Tiburon, a tony Marin bedroom community across the Bay from the city, and then fairly quickly settled in a boathouse below the fashionable homes on Belvedere Cove.

Living directly on the Bay allowed him to keep both a small sailboat and a classic Whitehall rowboat tied up on his dock. He indulged his enthusiasm for sailing and began regularly commuting across Richardson Bay in the rowboat. The Whitehall was a design that had been pioneered in the nineteenth century. It was a small, elegant craft with room for a sail at the bow. It was so perfectly built that in the afternoons he would frequently lie down on the bottom of the boat and sail home downwind.

O

RUSSELL "RUSTY" SCHWEICKART HAD GONE INTO SPACE ON Apollo 9 in 1969, the first astronaut to venture outside one of the Apollo spacecraft. In December of 1972, he met Irwin William Thompson, who along with Carl Sagan and other literary and scientific figures of note had been invited by *Time* magazine to watch the final launch of NASA's Apollo program. Thompson was a gnomish philosopher and poet who had grown up in California, taught at MIT, and earlier that year, with Rockefeller family backing, had established a retreat center known as the Lindisfarne Association at the tip of Long Island.

Schweickart became friendly with Thompson, who invited him to the association's annual meeting on Long Island during the summer of 1974 to give a talk about his experience in space. Brand missed Schweickart's talk, which would become known as "No Frames, No Boundaries,"[10] but it had such an impact—half the audience was in tears when he finished speaking—that a recording was passed around samizdat-style, and in the summer reached Brand, who began playing it on a long drive to his Canadian retreat.

As he listened, he was stunned to hear Schweickart confirm from his time in space what Brand had imagined under the influence of LSD on his rooftop in North Beach. "When you go around the Earth in an hour and a half," the astronaut said, "you begin to recognize that your identity is with that whole thing. And that makes a change."

Schweickart described how it was unimaginable how many international borders and boundaries he had crossed as his spacecraft repeatedly circled the Earth, and how ludicrous it seemed that people were killing each other over these imaginary lines. He told his audience that he wished that he could take one person from each side of a conflict and say, "Look. Look at it from this perspective. Look at that. What's important?"

Brand was gobsmacked. Here was an astronaut having what amounted to a mystical experience. Schweickart had become humanity's eyeball in orbit. He made real what until then had been merely a philosophical idea.

Driving on the freeway, he pulled over and called Schweickart at his office in Washington, DC. The two men spoke for half an hour and struck an immediate bond. Not long afterward Brand appeared in Schweickart's office, and the astronaut showed him a file drawer containing dozens of beautiful prints of the Earth taken from space. Brand soon reprinted Schweickart's talk in the *CoEvolution Quarterly*.

O

THE SAME WELLSPRINGS THAT DROVE THE CREATION OF THE Bay Area counterculture in the sixties continued to run deep in the seventies. And as the political New Left began to fragment and George McGovern's idealist antiwar coalition was crushed by Richard Nixon, there was a fountain of New Age and personal growth movements. The back-to-the-land movement that had been fueled by the *Whole Earth Catalog* had ebbed, and in its place more individualistic paths to salvation and enlightenment ranging from est (Erhard Seminars Training) to Zen thrived.

Baker Roshi was a charismatic figure and a great salesman and one of the main reasons that Brand had been attracted to the study of Zen. He had quickly established an informal "Invisible College" that stretched from the Tassajara retreat center to the San Francisco Zen Center to Green Gulch, the organic farm purchased by the Zen Center in 1972, located just north of the Golden Gate Bridge near Muir Beach.

In the early seventies, Baker attracted an eclectic coterie of counterculture types ranging from Gregory Bateson, a longtime student of Zen who frequently used Zen concepts in his writings and talks, to Peter Coyote, writer Orville Schell, and Brand.

It was through the Zen Center that Brand met Jerry Brown, who had bonded with Baker at a Tassajara retreat in 1974, and began spending a significant amount of time at the center after he had been elected governor but before he took office. Like Brand, Brown had been significantly influenced by an Aldous Huxley speech; when Brown heard him speak in 1961, the writer had condemned modern education and, when questioned by the future governor, pointed Brown to the study of Zen.[11] Shortly after Brown took office, Brand sent him a letter endorsing his "dialectical approach to truth," praising the field of cybernetics for providing a similar analytical approach to problems, and offering his services. He included a copy of Norbert Wiener's *Cybernetics* with several marked passages. (He underscored that the book was a loan. He had previously sent two books to Brown, and they were returned by the governor's staff because he was unable to accept gifts.)

After Brown took up residence in Sacramento, Brand became part of an informal kitchen cabinet or "brain trust" surrounding the young governor. It included people like Baker and a handful of others from the Zen Center such as Coyote; Sim van der Ryn, who would become the state architect; and Huey Johnson, who served as California's natural resources secretary.

Brand was so wide-eyed and enthusiastic about the new administration that he took on a temporary full-time position for several months in the capital, pronouncing Sacramento "Oz—munchkins, witches, wizards, the motley band with Dorothy, and all. There's dazzling magic," adding, "As Jimmy Breslin said of politics, it's all done with mirrors and smoke. What isn't?"

He would make weekly trips to Sacramento, and he acquired a three-piece suit for his new role, an effect that completely failed, according to Brown's biographer Orville Schell, who noted, "The concept was good, but this particular piece of haberdashery hung on him like a shroud over an unveiled statue."[12]

Brand's role was to invite a wide-ranging group of thinkers to a speaker's series for Brown and his staff called What's *Actually* Happening. Among those he roped in were Buckminster Fuller, Ray Bradbury, Carl Sagan, Jacques Cousteau, E. F. Schumacher, Ivan Illich, Gregory Bateson, Ken Kesey, Amory Lovins, and Herman Kahn. (Brand decided that the intelligent conservatives he had invited to meet with Brown—Kahn, Milton Friedman, and William F. Buckley Jr.—were more open-minded and better company than the intelligent liberals.) On occasion he would record the interviews and reprint them in the *Quarterly*, giving his magazine added influence.

For more than a year during Brown's first term in office, Brand held the title of special assistant and was paid $2,000 monthly. He also acted as an impresario for a Save the Whales concert and Space Day, put on by the new administration.

Before he worked for Brown, much of Brand's learning had come from reading books, but in Sacramento, he discovered that those in government were hardworking and that they didn't care that much about politics. He decided that his libertarian friends had almost exclusively never served in government and had no idea how it functioned. At the *Whole Earth Catalog*, they'd had little time for government. But after working for Brown, Brand decided that good government had value.

He stayed for only part of Brown's term as governor, deciding that it was too challenging to perform his role as assistant without moving to Sacramento and giving up the *Quarterly*, which he was unwilling to do. Later he would take particular pride in having persuaded the governor to direct one of his agencies to produce the *California Water Atlas*. Brand became the chair of the planning group and the first edition of the *Atlas* was published in 1979.[13]

It was not just his political worldview that was altered. Before he departed, he was forced to confront some of his most closely held beliefs, in the process reframing the way he saw the world. In the run-up to Space Day—intended to celebrate the launch of the space shuttle and the California aerospace industry that created it—Brand visited the Rockwell International space shuttle factory in Los Angeles. Steeped in a counterculture of "small is beautiful," it turned out to be an overwhelming experience. He stood with Schweickart and several Rockwell "suit wearers" and looked up at this enormous structure teeming with busy, capable men and women who were all doing things he couldn't begin to comprehend.

He found it mind-boggling and revolting at the same time. He fled to the men's room, feeling completely alienated by the technological gigantism and yet simultaneously compelled by what he realized was a tour de force. "Technology, kiddo," he wrote in his journal several days later. "This is today what the great sailing ships were in their day. Get with the program or stick with your spinning wheel."

The visit would have the same impact on him that his visit to the Warm Springs Indian Reservation had had, radically reshaping his sense of man's relationship with nature. Brand had always been an optimist about technology, but this was different. He had to choose sides—a decision that would set him apart from the counterculture movement he had been instrumental in creating.

He decided to get on board.

Space Day was underwritten by eleven aerospace companies and the United California Bank. Showing that he didn't want to completely be labeled a shill for the space industry, Brand invited R. Crumb to "cover" the event for the *Quarterly*. The cartoonist took away a different message from his host's: "Finally it was over. . . . I staggered out the door, my mind reeling from the eight hours of lies, double-talk, persuasion, and downright madness I had just sat through."

The highlight of the resulting cartoon was a panel portraying an oblivious Stewart Brand flirting with poet Michael McClure's blond daughter.

O

THE FIRST ISSUE OF THE *QUARTERLY* WAS PRODUCED BY A STAFF of about a dozen people and continued to reflect the *Catalog*'s back-to-the-land roots, with discussions of sources for obtaining wheat and reports from several communes.

Brand continued to work at being provocative. He had been frustrated by quarrels with his *Harper's Magazine* editors, and he flashed his editorial independence by including a query to readers on the best sources for mail-order pornography, accompanied by a stark photograph of a naked woman with her legs spread apart. He had met Huey Newton, the militant leader of the Oakland-based Black Panther Party, and he turned the third issue over to the Panthers to edit. Newton had fled the country for Cuba in the face of a murder charge, and so Elaine Brown, who ran the Panthers in Newton's absence, led the project. Later Brand interviewed Newton, who was then in exile, for the *Quarterly*.

The *Quarterly* production cycle had been established as a series of six-week-long pushes and equally long breaks where the staff would temporarily disperse. It was largely collegial with lots of activity and idiosyncratic personalities. As an editor Brand put his earlier army philosophy to work—that the best managers were the ones who were smart and lazy, leaving it to their employees to take their own initiatives. As a management approach, it mostly worked well. However, there were on occasion skeptics, such as the production employee who sent Brand a letter describing what it was like to work for him:

> The other day in the production area (trying to maintain my reputation as a wit) I said that sometimes I pictured you, the publisher and boss, you Stewart were joyously pursuing your dream in the driver's seat of an old VW microbus—the John Muir kind—careening happily up this curving mountain road with both hands tight on the wheel and a big toothy SB smile (well, that's the way I pictured it as I said it) and the way I pictured it, we were in the back, the windows blocked off with Cost Plus madras bedspreads, bouncing from the ceiling to the wall to the floor—not knowing where we were going and very very carsick.

He had evolved into the sharply idiosyncratic editor he had imagined would be necessary to create a publication with a unique voice, but he relied heavily on the network of friends and intellectual fellow travelers he had created at the *Catalog*. Steve Durkee supplied cover art and illustrations; J. Baldwin came back into the fold as a section editor; and

Steve Baer wrote an early article attacking the idea of converting to the metric system. Brand was opposed to the metric system but for reasons different from Baer's, arguing that he was against metric measurement because it was too efficient—slowing down to do conversions forced you to think. (Eventually, Brown would appoint Brand to the California Metric Conversion Council, in a move that was attacked as an effort to undermine the process.)

The Summer 1975 issue was the first to have significant impact, thanks to a cover story on the Gaia hypothesis, proposed by British chemist James Lovelock and microbiologist Lynn Margulis, for the first time presenting their concept to a nontechnical audience.

The Gaia hypothesis argued that the Earth's biosphere was a complex and naturally self-regulating system; as Brand wrote in his introduction, the innovative theory treated "the anomalous Earth atmosphere as an artifact of life and comprehends the planet itself as a single life"—which was quite up his alley.

The issue included another innovation: "Listen, it is very likely true that little computers are on the way to revolutionizing human behavior," Brand announced. "Eager not to be left out, we herewith become the first general magazine with a computer science department." The section on personal computers was compiled by the now twenty-two-year-old Marc LeBrun, one of the group of enthusiasts who had that spring launched the Homebrew Computer Club, which in turn would light the entrepreneurial fire leading to the emergence of the personal computer industry in Silicon Valley. LeBrun called attention to a 128-page "underground" publication, *Computer Lib/Dream Machines*, published the previous year by Ted Nelson, a former sociology graduate student and coinventor of the concept of hypertext—Doug Engelbart had simultaneously come up with the same idea—that would eventually lead directly to the creation of the World Wide Web. (Like a range of publications, *Computer Lib* was modeled on the format of the *Catalog*.)

There were also walk-ons and discoveries based on manuscripts that came in over the transom. Anne Herbert, a young writer from Ohio, was one such discovery. A talented poet and writer of fiction, Herbert would soon move to Sausalito to join the *Quarterly* as an editor. She would be remembered for the phrases "Practice random kindness and senseless acts of beauty" and "Libraries will get you through times of no money better than money will get you through times of no libraries."

Brand also failed on occasion to reel in the big fish he invited to write for the *Quarterly*. An entreaty to Woody Allen went unanswered. However, Norman Mailer had the courtesy of responding: "Dear Stewart Brand," he wrote in 1975. "I'm IRSed to the tune of 100 times 10 times 100, and so writing for the metaphysical essence of turds, once known as cabbage. Alas, I can't afford your rates. Wish I could. Cheers, Norman Mailer."

It would be Gerard K. O'Neill's twenty-five-page cover story in the *Quarterly*'s Fall 1975 issue that established the publication, its voice, and its unique identity. O'Neill was a physicist who had done original work in the field of particle accelerators, and Brand had first stumbled across him at a World Future Society conference that summer and been wowed.

"O'Neill's Space Colonies," the cover announced. "Practical, Desirable, Profitable, Ready in 15 years." Within, O'Neill proposed a vision of a vast industrial-scale space-based platform for the exploration and colonization of the universe, a concept that horrified many of the *Whole Earth Catalog*'s devotees. The issue, which also featured an article by E. F. Schumacher, of *Small Is Beautiful* fame, and a transcription of a conversation between Governor Brown and Bateson, would drive a wedge between Brand and some of his closest environmental allies and cement his reputation as a "techno-utopian."

O'Neill laid out his argument that there was no good reason why an off-the-planet colony shouldn't be an established fact by the year 2000. Brand added a full-throated defense of the idea in his introduction: "Space Colonies show promise of being able to solve, in order, the Energy Crisis, the Food Crisis, the Arms Race, and the Population Problem. . . . Space is part of the wildness in which lies 'the preservation of the world.'"

The ensuing debate stretched over multiple issues of the *Quarterly* and brought out a series of big names on both sides of the issue, including Lynn Margulis, Wendell Berry, Paul and Anne Ehrlich, E. F. Schumacher, Lewis Mumford, Buckminster Fuller, Ernest Callenbach, David Brower, Carl Sagan, Gary Snyder, and Russell Schweickart.

None of the attacks were more bitter than Berry's, whose writings on agriculture and the environment had been showcased in the *Catalog.* Calling Brand's enthusiasm a "warmed-over Marine Corps recruitment advertisement," he wrote, "As you represent it, a space colony will be nothing less than a magic machine that will transmute little problems into big solutions. Like utopians before, you envision a clean break with all human precedent: history, heredity, character."

The controversy echoed through the pages of the *Quarterly* and in *Space Colonies*, a 1977 book on the dispute that Brand published. He reported that positive mail had far outweighed negative responses. Moreover, he added that almost all of the 122 unsolicited positive responses came from college students—a fact he considered to be a sign that the younger generation might avoid the cynicism and passivity of his own.

For Brand the idea of colonizing space was mostly a thought experiment that was in line with his faith in the ability of technology and tools to improve the quality of human life. It showed a consistency that stretched back to the opening sentence of the *Catalog*: "We are as Gods and might as well get good at it."

However, his thinking on the ideal of self-sufficiency that had helped inspire the back-to-the-land communal wave had changed. Writing in the *Quarterly*, he declared that "'self-sufficiency' is an idea which has done more harm than good. On close conceptual examination it is flawed at the root. More importantly, it works badly in practice."

As with many things, it was his encounter with Bateson that had caused him to rethink his philosophy of independence. "It is a damned lie," he wrote of self-sufficiency. "There is no dissectible self. Ever since there were two organisms life has been a matter of coevolution."[14]

O

THE *QUARTERLY* NEVER EXPERIENCED THE ROCKET GROWTH OF the *Catalog*. After four years of publication, subscriptions were hovering around 20,000 and newsstand purchases added another 35,000 in sales—still not enough to break even. It was a precarious existence. Although Random House had published *The Last Whole Earth Catalog*, Brand had decided to switch to Penguin to partner for the publication and distribution of the *Epilog*. They had printed 450,000 of the *Whole Earth Epilog* in 1974, and it turned out to be far too optimistic, putting the Point Foundation, their holding company, in debt to Penguin.

Fortunately, a benefactor emerged at a crucial moment.

Marlon Brando had long been a fan of the *Catalog* and particularly of J. Baldwin and his writing on nomadic life and ecological design. Shortly after the publication of the first *Catalog*, he had raved about the publication in an interview in *Life*, which had proven to be one of the viral moments that pushed dramatic sales.

Brando had stumbled upon the tiny atoll of Tetiaroa, located north of Tahiti, in 1960, when he was scouting locations for *Mutiny on the Bounty*. While filming, he fell in love with his Tahitian costar, whom he married. He would ultimately purchase most of the atoll. He called Brand at one point and said that he wanted him to run the island for him as a sustainable retreat. The idea intrigued Brand, and Brando invited Baldwin and Brand to visit him in Los Angeles at his home on Mulholland Drive. The actor picked them up at the airport and proved to be the world's most distracted driver, routinely stopping at green lights and running through reds, scaring the wits out of his two passengers.

Ultimately, nothing came of the proposal, though Brand and Baldwin did get to visit Tetiaroa. However, the actor remained friendly and one day in late 1975 he called with a very convenient question: "Would you like $20,000? It's the end of the year."

"We'd be absolutely thrilled!" Brand responded. "And would you like to come by and visit?" When Brando visited in Sausalito, he proposed that they move operations to Tetiaroa and produce the *Quarterly* from there. Isabella Kirkland, who was working for the *Quarterly* as an artist and graphics designer, listened as Brand and Brando decided that they would have to use bicycle-powered generators to produce enough energy to run the electronic typesetter and waxer to do layout. She was excited about the idea of moving to Tetiaroa, but having the actor in the *Quarterly* office seemed bizarre. He overflowed the small chair he was sitting on in their cramped rooms. Although the supplement had proved that it was "difficult but possible" to produce the magazine remotely, it was clear to the staff that Tahiti would not be "difficult but possible"—it would be way too much of a stretch. The idea evaporated after Brando was gone.

○

WHILE PLAYING HIS ROLE IN JERRY BROWN'S KITCHEN CABInet and growing into his role as a magazine editor, Brand had no shortage of female companionship—but nothing stuck.

His "great love" continued to elude him. For several years he had a long-distance relationship with Alia Johnson. She had been Donna Rasnake when she first wrote Brand asking him for an interview with *Reason* magazine when he came to Washington, DC, for the World Future Society conference.

She soon became Johnson (Johnson was her birth name, and Alia was taken from a character in the science fiction novel *Dune* who was described as a "virgin witch"), and was living on a commune run by George von Hilsheimer, a former columnist for Krassner's *Realist*, who was collecting silver coins. She and Brand began their long-distance relationship at the commune deep in the Virginia countryside.

Brand had continued to return to Cape Breton Island each summer, with a never-finished vision of building a larger home to replace the shack that he had built with Jennings. At the end of the summer of 1976, he visited for several weeks with Johnson and Schweickart and his wife. Although their brief romance was ending amicably, it was one of his best vacations.

He was back in California the next spring when Johnson showed up at his boathouse unexpectedly and told him she was pregnant with their child. She informed him that she wanted to keep the baby and was asking his consent; she would raise the child without him. They spent an intense night talking about it, and in the morning Brand told her "fine."

Several months later she returned to introduce him to his son, Noah Johnson (as a teen he would change his last name to Brand), who was born in Denver in May of 1977—eight pounds, with red hair and blue eyes. Brand greeted the occasion from a distance, noting in his journal: "Mother ecstatic. Father interested, evidently relieved, still puzzled." (Johnson eventually married Robert Fuller, a physicist who had been president of Oberlin College.)

In November, Brand wrote his mother to let her know that he had become a father. "A nifty kid," he told her. "At present I see Noah every couple of weeks and I provide some things for him—a camera, a stroller, etc. I can go dote anytime, or they come here. As for the future we're taking it as it comes. I may or may not acquire a family of my own."

O

PATTY PHELAN HAD BEEN A SOCIAL ENTREPRENEUR SINCE SHE was a teenager. The youngest of seven children in a well-to-do Pasadena family, in high school she turned a small tutoring program into a year-round mentorship and summer camp for kids who were growing up in poverty.

She had puzzled her parents with her passion for helping the underprivileged, and she entered UC Berkeley intent on getting a teaching credential. By the time she graduated, she had shifted her focus to publishing. While she was a student she had put together a community services directory, loosely modeled on the *Whole Earth Catalog*, called *Savvy*. It was intended to be a compendium of all the services offered in the East Bay for people who were low-income or were challenged with

problems like housing or with navigating the other social services offered by the local government.

She had learned how to create her directory by reading an article
in the *CoEvolution Quarterly* that was a step-by-step account of how the
*Whole Earth Catalog* had been created. A fresh college graduate, she decided that a job at the *Quarterly* would be a good stepping-stone.

Learning quickly that Brand was difficult to reach, she simply rang
the *Quarterly* office and in her sweetest girlfriend voice said, "Hi! Is Stewart there?" She was put through to him and immediately made her pitch
to come to work. Brand was intrigued, and after her interview, during
which she slapped a copy of *Savvy* down on Brand's desk as her résumé,
he hired her on the spot, even though they had no open positions. Her
title was "projects." Initially, her salary was half paid by the Comprehensive Employment and Training Act, a Nixon administration federal program that subsidized public services jobs. Her hiring "demolishes the
last of our capitalist pretensions," Brand noted in the Gossip section of
the *Quarterly*.

He was immediately attracted to Phelan, who at twenty-five years
old was fourteen years younger than Brand. But he was also chastened
because he had made the mistake of chasing a *Quarterly* staff member
several years earlier, creating all of the problems attendant with frater-
nizing with employees, so he kept his feelings to himself.

Early on, however, he tapped Phelan to help him in organizing a
tenth-anniversary event to celebrate the founding of the *Catalog*. The
idea was to celebrate to "help us get used to the idea of having been in
business for ten years."

The initial plan was to go back to the Gerbode preserve, where the
New Games Tournament had been held a half decade earlier. As Brand
got deeper into the project, however, he began to have second thoughts.
What had seemed to be a grand celebration began to feel stale, more
like looking backward than at a new idea.

Brand's original vision had been an eco-festival with the usual cast
of characters plus some. The jamboree was intended as an "educational
and cultural" experience, with speakers, exhibits, music, and a re-creation

of the original New Games Tournament. What he failed to realize was that in the intervening five years the deserted valley where he had held that tournament was now part of the Golden Gate National Recreation Area.

Initially, Brand was able to obtain the enthusiastic support of the National Park Service plus a commitment from the National Guard. However, once word of the event leaked out, he ran into stiff resistance from environmental groups, homeowners, and Sausalito city officials. There was a great deal of pessimism about his intentions. In the end, the park's superintendent gave Brand a permit for a far more constrained event.[15]

Approximately twelve thousand people attended the celebration, which had hits and misses. Brand had unsuccessfully attempted to recruit celebrities such as James Watson, the codiscoverer of DNA, and Marlon Brando. Watson pleaded that he was too busy, and Brando said he would show but didn't. Joan Baez wasn't there, but her sister Mimi Fariña sang. Wavy Gravy entertained the crowd, and poets like Alan Ginsburg and Peter Orlovsky read from their work. Almost sixty five-minute "lightning" talks were interspersed through the event.

Brand, outfitted in his jaunty black top hat with an eagle feather and looking like he needed a shave, closed the event saying that he was most enthusiastic about the idea of being able to take a bath. He also talked about the importance of second acts and suggested that everyone should have some hobbies that might open up into new lines of work.

One afternoon while they were scouting locations for the event, Brand and Phelan lay on a hillside in the late afternoon sun. He had taken off his shirt, and he suddenly realized that she had taken off hers as well. Phelan, who was living with someone else at the time, had fallen for Brand head over heels, and it would soon be apparent to him that he had finally found his great love. Their friendship went quickly from dating to something much deeper.

Indeed, their bond was sealed soon after they began dating. One day after a ballet class Phelan came running back to her apartment in Sausalito, rushing to get ready to go out to the movies. She ran the water for a bath, but before she could climb into the tub, she found herself doubled over in pain. She had no idea what was going on or whether it would pass, but she struggled to put her clothes on. She was panicking when Brand knocked on the door. She crawled out of the bathroom.

Greeting him still doubled over, she moaned, "I'm so sorry. I'm sure it's really nothing, but I just really can't even put on my pants." She kept protesting that her severe pain would pass, and he kept telling her, "I'm going to take you to the hospital."

She was mortified when he finally picked her up and put her in her car and drove her to Marin General.

He took her to the emergency room at about seven p.m. Finally, after an examination, Phelan was admitted to the hospital at midnight. The doctors were concerned that she was bleeding internally, and they wanted to keep her for observation. Her doctor said that they suspected she had a burst appendix. Brand went home deeply worried.

It turned out to be a burst ovarian cyst. When she came out of surgery, Brand was there with a small pillow he had fetched from her apartment.

That was it. "This guy's a keeper," she decided.

# Chapter 8

# Anonymity

ART KLEINER WAS A STRUGGLING GRADUATE STUDENT AT the University of California at Berkeley School of Journalism when he decided to send Stewart Brand an unusual proposal. A transplanted New Yorker, Kleiner hadn't had much contact with the *Whole Earth Catalog*, but he was a devotee of magazines and was aspiring to a career in writing and editing. He was particularly struck by the fact that the *CoEvolution Quarterly* and *Mad* magazine were the two publications that seemed to be making a go of it without running advertising. He was taking a course on graphic design and supporting himself by working as a typesetter, so he submitted an article on the history of magazines that was completely camera-ready. The visuals appealed to Brand, so he invited Kleiner over. "You can come in and paste it up on our equipment," Brand told him.

In addition to being graphically adventuresome, the 1979 article was prescient in several ways. In his concluding paragraph, Kleiner wrote:

Computers will eventually break down the structure of the print media. Once magazines, newspapers, and books start to come in over the home terminal, or the terminal down at the corner computer center, then the boundaries between them won't be necessary; they'll merge into a steady flow of information, stories, pictures, design, photographs. You might never have to stop reading.[1]

Brand had found his guide back into the digital world he had first stumbled upon almost two decades earlier in the Stanford computer center, which was now encroaching on the world. Kleiner proposed another article that Brand accepted, this one focusing on the emerging world of online services.

Kleiner had happened into the online realm one day while he was driving his car and listening to someone talking about how people were sending messages to each other via computer. Early on, he found his way to a networking experiment known as the Electronic Information Exchange System, or EIES, a National Science Foundation–funded computer conferencing system being developed at the New Jersey Institute of Technology. The system operators gave him a free account in return for his work as a support person.

EIES was a granddaddy of the coming world of virtual communities, and Kleiner wrote a second article for the *Quarterly* that served as a guided tour to the new electronic services, titled "Better Than the Next Best Thing to Being There." Not long afterward, Brand called up Kleiner and asked him to come over to Sausalito to meet. Kleiner assumed he was going to discuss writing another article, but Brand had something else in mind.

Despite the appearance to Kleiner that the *Quarterly* was surviving on subscriptions alone, it wasn't. It was a money-losing proposition that was persisting because the occasional *Whole Earth Catalog* was highly profitable. Shortly before Brand had reached out to Kleiner, John Brockman, now Brand's literary agent, had sold another version of the *Catalog*. It was early January 1980, and the plan was to have *The Next Whole Earth Catalog* in the market for Christmas. "I would like you to be the research editor," Brand told him.

It was the most remarkable offer Kleiner, who was still mired in his journalism school thesis, could imagine. He joined to work on the *Catalog* and would spend the next five years as part of the *Quarterly* staff, first putting together a massive, completely reworked *Catalog*, a 608-page rendition that would be considered by many the definitive statement of the genre.

Kleiner became a fixture at the *Quarterly* offices during an era when the publication grappled with the impact he had foreseen of new communication technologies on the print world. He moved first into a cooperative apartment at the north end of Sausalito with Anne Herbert and Lori Gallagher, who was the *Quarterly*'s office manager. They called their home Vague Acres, and used it as a crash pad while they basically lived, ate, and even slept at the office on Gate 5 Road, which curved to parallel the waterfront. Later, Kleiner moved to San Francisco and would bicycle to work across the Golden Gate Bridge, frequently sleeping overnight in the magazine's offices and then riding home the next day.

By 1981, the counterculture that Brand had helped create was not so countercultural. The Vietnam War was over. The most famous Hollywood star was not a long-haired rebel but a newly elected, right-wing president. Psychedelics were being replaced by cocaine. Vegetarian cuisine could be found in supermarket freezers, and the most prominent celebrity antiwar protester had shifted her focus to profiting from her aerobics videos.

Despite the shift, the arrival of the newest version of the *Catalog* was greeted with great fanfare. Brand himself was treated like a rock star, the subject of innumerable newspaper and magazine profiles charting his course from Stanford to the army to North Beach to the Pranksters and onward. "Ah, well, the wings of man come hard, and Stewart is anchored by sensible gravities at his core," intoned Calvin Fentress in his *Washington Post* profile. "Earthbound thus, he might as well get good at it, knowing it's always a kind of sleepwalking anyway. He will do what he can. Stay open. Stay foolish. Keep the transmitter on."[2]

In the *Christian Science Monitor*, Stewart McBride added: "Brand's embracing of Milton Friedman's neoconservative economics and his enthusiasm for Gerard O'Neill's space colonies have recently gotten him into hot water with the organic, 'small is beautiful,' appropriate technology constituency. But Brand seems too preoccupied with the search for tomorrow's frontiers to be looking over his shoulder or slowed down by his critics."[3] Among those critics was Sue Halpern in the *Nation*: "The *Whole Earth Catalog*, once considered the hip bible, has become a slick testament to the acculturation of the counterculture."[4]

O

FOR BRAND IT WAS A PERIOD OF INCREASING FREEDOM TO EXplore various passions. He bought part ownership in what would be his last sailboat, the *Bakea*, and spent time in the Northwest hoping to make it seaworthy enough to bring it to the Bay Area. After making a series of repairs, he began to sail it down the coast with two skilled friends. They soon found that the boat had a built-in list to port, and so they parked it in view of the boat designer's home in Friday Harbor to shame him into fixing the problem. The designer ignored them, so Brand gave up and had the boat shipped to Sausalito, where it was anchored out among the fleet of aging vessels that dotted Richardson Bay.

He was aware of his own aging too. At forty years old, Brand was a good decade older than most of the *Quarterly* staff, and he was regularly confronted with a younger generation that was both jarring and intriguing.

Isabella Kirkland was one of the twentysomethings on the staff who played volleyball, ate, and partied together. She had been at the *Quarterly* for more than a year when one day Brand came out of his office and asked her, "What is it you do here exactly?" He was puzzled because she did a little bit of everything. She ran subscriptions, proofread, helped with marketing, and could do letterset if needed.

At the time, Kirkland was living in San Francisco's seedy South of Market neighborhood while attending the San Francisco Art Institute. She was also heavily into the burgeoning Bay Area punk rock scene. Her hair was pink; she wore combat boots; and her skirt ended substantially above her thigh. She started telling him about the concerts she'd attended, and Brand asked if she would take him.

Soon thereafter he accompanied Kirkland and another *Quarterly* staffer to hear the Dead Kennedys at Temple Beautiful, an abandoned synagogue that had been adopted by the punk scene. Brand had helped create the original San Francisco acid rock scene in the sixties. This was better, he decided—the music rawer, the crowd closer to the performers, a real sense of camaraderie in the audience, and none of the "fake" love of the flower generation that he had come to disdain.

But unlike then, now he was simply a cultural tourist—the scene was already trending, and his inclination had always been to depart if he felt his role was merely receptive and not creative. He didn't go back.

Several nights after the concert, the punks were still on his mind, however. He dreamed he was a member of a country-punk band and that his instrument was a chain saw: "At the correct musical moment I would fire it up BRRAAAAAP!!! and have at the stage itself."

O

ECOLOGISTS SPEAK ABOUT AN EDGE EFFECT THAT OCCURS WHEN-ever two biomes meet. An unusually rich habitat is generated at the edge. And that was what Brand found on the Sausalito waterfront, a tidal basin both physically and metaphorically, with a habitat that sustained artists and quasi outlaws. He had stumbled into this world, first in 1960 as a young Stanford graduate, and again after he returned from the army in 1962, when it had been home to artists and poets and professors. It was still a rich mix of people with a villagelike closeness, everyone from welfare recipients to wealthy architects and TV people.[5] It also provided a bit of longed-for continuity. Steinbeck's *Cannery Row* had remained his touchstone, and the Sausalito waterfront was similar to the world inhabited by its protagonist, Ed Ricketts.

When Brand left San Francisco and began to both live and work on Richardson Bay in 1977, he was still attracted by the romance of what Buckminster Fuller described as "outlaw areas." Fuller had noted that human habitation of the Earth was concentrated on a tiny portion of the Earth's surface and that for most of human history the oceans had been outlaw areas beyond the reach of the law. He had pointed out that most technology developments had occurred in the outlaw areas because of the toughness of nature, citing space as an obvious future arena for invention. Fuller's idea was that technology innovation happened first and most effectively in the areas that were most challenging for human life, forcing creativity.

Brand had first noted Fuller's idea in the June 1970 *Whole Earth Catalog* supplement, where he had celebrated the notion of an innovative outlaw culture. Fuller's ideas resonated with Brand's sixties counterculture world populated with twentysomethings whose lives consisted of a range of illegal activities.

Fuller had convinced him that there would always be outlaw areas, indeed that a society of laws required them. The question was whether you preferred a highly "coevolved" outlaw area like the Waldo Point scene "with the benefit of 20 years of mutual accommodations between the transients, permanent residents and surrounding town and county," or would take your chances on whatever might spring up elsewhere.

All of this was on Brand's mind when, in the summer of 1980, with almost no warning, eviction notices were given, and bulldozers started destroying structures at the Napa Street Pier.

Brand, who continued to describe himself as a conservative (though one who refused to read the *Wall Street Journal* because he was horrified by its editorial pages), was increasingly skeptical about the value of activism on the national stage, and in some ways, Fuller's thinking played into this: focusing on the edges, which in this case meant local. Thus, willingly, Brand became a local activist.

In the late 1970s, the city of Sausalito had changed zoning laws to make living aboard a boat illegal. The change went through quietly and almost nobody realized that the houseboat community along the Napa Street Pier had become an outlaw community.

That quickly changed after the city took ownership of the underwater "street" that followed the path of the Napa Street Pier and quietly issued eviction permits. Grange Debris Box & Wrecking Co. showed up

early one morning with axes and bulldozers to be confronted by an
angry crowd of waterfront residents held back by local police.

In the wake of the destruction of Bob's Boatyard, Brand convened
a meeting of neighbors where slides of the destruction of the waterfront
community were shown. There was a lot of bitterness and a series of
first-person accounts of disruption and loss. The meeting turned out an
eclectic group with a diverse set of talents. Brand asked the protesters,
"Well, what are we going to do about this?"

"I'm willing to work with the council," he continued. "It seems to me
that if they can change the zoning that easily, they can change it right
back."

They put together a group to organize the opposition to the zoning
changes and when one of the women at the meeting asked, "Why isn't
there an art zone?" they decided to give that name to their newly formed
group. When Art Zone applied for funding from the San Francisco
Foundation to support their organizing effort, Brand wrote in a cover
letter for the grant application: "If there were no Sausalito waterfront
there would be no *Whole Earth Catalog* or *CoEvolution Quarterly*."

For years the group worked to focus attention on the waterfront and
to convince Sausalito residents of the cultural value of the waterfront
community. In the end, they were able to radically change the city's orientation toward the waterfront, and in doing so a group of dedicated
waterfront residents, including Brand, built a vibrant community movement. A member of the activist group eventually became mayor of Sausalito and later a Marin County supervisor.

In the long run, however, it was not an unqualified victory. The developers persisted and gentrification won; as is often the case, the outlaws
and the libertines ultimately gave way to upper-middle-class hipsters.

SHORTLY AFTER SHE BECAME INVOLVED WITH BRAND, PATTY PHELAN
left her job at the *Quarterly* to work at Planetree, a patient-centered
medicine nonprofit that had been founded by Angie Thieriot, the wife

of *San Francisco Chronicle* publisher Richard Thieriot. Phelan and Brand had been a couple for more than two years when they decided it was time to move in together. Neither his boathouse flat on Belvedere Cove nor her apartment seemed appropriate and so they began scouring the Sausalito waterfront for a place to live.

Commissioned in 1912, the *Mirene* had been a working tug in the Oregon logging industry for decades before it showed up in the Bay Area in 1975. By 1982 it had been left to die, and when they first rowed out to look it over, it was in such questionable condition that parts of the boat were rotten to the touch. The engine was gone, and the tug had been stripped of almost everything, including its rudder.

They purchased the *Mirene* for just $8,000 and then spent the next two years and $180,000 converting her into a houseboat berthed at the South 40 Dock. A decade later they would put the boat back in dry dock and add a new engine as part of making the boat seaworthy again.

The South 40 Dock became the best example of what Brand's friend and neighbor Peter Calthorpe would describe as the "new urbanism." Calthorpe was inspired in part by living in the houseboat community to realize that "walkability"—the notion that all of the resources a resident might need, from shopping to work and entertainment, were within easy walking distance—was the key to creating a livable community. That insight would be at the heart of Brand's break with the back-to-the-land movement. He had realized early on that he was an urban person at heart—thus his flight from commune life in favor of the Bay Area. Now he came to agree with Calthorpe, who in 1985 wrote that in terms of impact on the environment, dense urban cities were the most benign forms of human settlement. It would later become one of the key ideas in Brand's own *Whole Earth Discipline*—a book that evoked cries of betrayal from many in the environmental movement when it appeared in 2009—that cities were the "greenest" of human communities, with the least impact on the environment.

The plan to move onto the *Mirene* also led to Phelan's decision to marry Brand. In July of 1983, just as they were about to move aboard,

Joe Kane, a San Francisco freelancer, penned a laudatory profile for *Esquire* around the thesis that Brand was not the hippie he was often portrayed as but was rather a tough competitor whose personality was perfectly suited for the coming digital decade. That was not what caught Phelan's mother's eye, however. The story included a full-page photo of Brand and Phelan posed in front of the *Mirene*, just as they were about to move into it, and her mother, a devout Catholic, was not at all happy with the notion of her daughter living in sin.

Phelan had had no intention of marrying Brand, but in the wake of her father's death, seeing the pain she was causing her mother, she softened her views. In October of 1983 they were wed by a Zen priest at Green Gulch, the San Francisco Zen Center farm near the Pacific Ocean. In their vows they didn't promise to be forever faithful to each other—rather their common bond was to protect all sentient beings. To appease Phelan's mother, a Catholic prayer was read during the ceremony.

Their choice of Green Gulch was also a statement about their support for the Zen Center, which earlier that year had been rocked by a scandal involving Richard Baker, the abbot. He was publicly revealed to have been sleeping with a number of his students, including Anna Hawken, the wife of businessman and environmentalist Paul Hawken, who was Baker's close friend as well as Brand and Phelan's.

Over Phelan's initial objection as well as pressure from the Zen Center, Brand chose to run an article about the scandal written by Katy Butler, a *San Francisco Chronicle* reporter, in the *Quarterly*; it effectively ended his decades-long friendship with Baker. Their relationship had become more complex before the scandal tore it apart. At the time, Brand was a member of the Zen Center's Outside Financial Advisory Board, and he and Phelan would regularly attend Sunday talks given by Baker at Green Gulch, where Hawken and his wife were living in a small house they had built. However, Brand had grown increasingly wary of Baker's charisma and how he employed it.

Hawken, also someone Brand had known since the early sixties and a major funder of the Zen Center, threatened legal action after he discovered his wife's affair with Baker, touching off the crisis.

The Zen Center scandal fractured a web of close friends and forced members of a tightly knit community to choose sides. Phelan was caught between her husband-to-be and a new group of people with whom she had become close. "Stewart, he is your friend," she pleaded with him, speaking of Baker. She believed that friendship came first.

Brand was unmoved. "I can't sit back," he told her. "This is going to happen in other communities." The pressure wasn't just from Phelan: the Zen Center itself wrote Brand, asking him to allow their community to "heal" privately.

Butler's article in the *Quarterly* opened with a scene describing a confrontation between the Zen Center's board of directors and Baker, who was accused of having a sexual relationship with a student who was also the wife of a man described as his best friend. Several other female students had come forward with similar accounts. The article did more than describe his sexual behavior. Baker had abused his power to live well and travel widely. He had used the Zen Center's funds to buy himself a new BMW. He was forced to step down from his position as abbot and ultimately left the Zen Center.

Brand wrote a personal introduction to the article. He described encountering Baker at a philanthropy conference early that year where he said, "In most religious groups, the rules gradually relax as you get more senior. At Zen Center it's the reverse; the rules are stricter the more senior you are." Brand noted that he had nodded approvingly at the time and responded: "Western philosophers teach only with their talk. Eastern philosophers do it with their whole lives."

He described his sense of responsibility for permitting Baker's distorted style of leadership, adding, "I gradually became peripherally involved in the community's efforts to understand what had happened and to reshape things so it wouldn't happen again. This article is perhaps part of that effort."

In response, Baker wrote an angry letter to Brand's mother attacking him. She told Brand that she had ignored it.

O

L IFE ABOARD THE *MIRENE* FUNDAMENTALLY TRANSFORMED BRAND. His decade as a bachelor had made up for the frustrations he had felt in his marriage to Jennings, but now he felt centered and comfortable and there was a consensus among their friends that Patty Phelan had made him a nicer person. At the same time, after editing and publishing the *Quarterly* for more than a half decade, he again felt bored and ready for something new.

Increasingly delegating more of the *Quarterly*'s editing to a rotating group of staffers, Brand traveled, spoke publicly, and, to stave off the ennui, launched new projects, some of which failed and several of which succeeded spectacularly. He tried to push back against the "self-help" orientation that had seized America in the seventies. As part of that effort, he offered a series of classes focused on helping others. Like many of Brand's schemes, Uncommon Courtesy: School of Compassionate Skills had bubbled for many years as an idea before he launched it. At the end of 1971, he had written a long essay on the subject of "What is most worth doing now?" to the directors of the Point Foundation. In the essay, Brand imagined the creation of something he described as a "Peripheral Intelligence Agency," which would employ a group of "Free Agents" whose role would be to use their resourcefulness to do the maximum amount of good with the minimum expenditure. The Point should explore the humanitarian uses of "applied laziness," by which he meant that the amount of work wasn't the issue, rather the degree of impact was. A decade later he wrote, "Most of the real good done is accomplished by amateurs in their spare time with their left hand and the corner of their eye."

In 1982 he hired a friend of Phelan's to serve as director of the Uncommon Courtesy school, in which students would spend about $50 to take classes "aimed at teaching people skills necessary to become better and more efficient helpers or 'do-gooders.'"

He planned other classes, on firefighting skills, home care, local politics, and creative philanthropy, all with the intent of making "doing good do better."[6]

He enlisted his friends, persuading Paul Hawken to teach several classes, and got some national press attention, including positive articles in *Esquire* and the *Christian Science Monitor*. In early 1983 Brand joined with Hawken and Michael Phillips to teach Business as Service, a two-day bus tour of some of the Bay Area's best for-profits and nonprofits. The excursion featured lunch at the Zen Center's Greens Restaurant, an evening lecture, and visits to the Fort Mason "art park," to a collection of businesses run by ex-convicts and ex-addicts called Delancey Street, to the East Bay to see Nolo Press, and to Hawken's own Smith & Hawken. In the *Quarterly*, Brand praised the garden supply business, asserting that the higher the "integrity" of a business, the more its customers would support it. The tour filled a fifty-seat bus and was similarly at capacity when the trio repeated the event several months later.

In the fall, Brand ran a weekend seminar titled Creative Philanthropy along with Hawken and two local philanthropists, targeting those who may have recently inherited wealth and were trying to do something useful.[7]

Distracted by other projects, however, he soon lost interest in Uncommon Courtesy. Managed in an offhanded way, the School of Compassionate Skills lasted until early 1984, when it fell victim to Point Foundation belt-tightening. Brand missed the irony when summarizing the demise: "All this was good people doing work for good service to the customers, but so structured it threatened our fundamental service."[8]

AT THE END OF 1972, BRAND HAD DEFIANTLY ANNOUNCED IN HIS *Rolling Stone* article on *Spacewar!*: "Ready or not, computers are coming to the people."[9] It would be another decade, however, before he would acquire his own computer. When it arrived, the machine was a Kaypro II, one of the last successful entries in the hobbyist period of the personal computer age. It was what was referred to as a "luggable" com-

puter, about the size and weight of a heavy sewing machine. It was modeled after the earlier Osborne 1, which had revolutionized the neophyte
industry by bundling a complete suite of business software for less than
$2,000—word processing, database, spreadsheet, and, most significantly, a
communications program that connected the machine to a device known
as a modem. The modem dialed via the telephone network and connected
to a growing array of information and bulletin board services, or BBSs,
which provided small-scale platforms for personal computer users to share
information and software, usually one user at a time.

From the perspective of today's internet-connected world, Kaypro's
applications were almost laughably low-tech. But for Brand, not only
was his Kaypro a powerful writing tool, it also unlocked the door to an
emerging world of online services.

Cerebral and democratic, it was a perfect fit for Brand. In the spring
of 1983, he wrote in the *Quarterly* that he had been waiting half his life
for word processing software even as he acknowledged a bumpy road
early on, describing it as a bit like learning to drive in a Model T that
would have a flat tire every five miles.[10]

Brand came upon the computer by agreeing to become a faculty
member at the Western Behavioral Sciences Institute in La Jolla. In
1981, cofounder Richard Farson, a psychologist, had begun a series of
programs pioneering the use of teleconferencing to build communities
of scholars in areas such as education, leadership, and the development
of social policies. The institute also set up distance-learning programs
in which the faculty would teach classes using the EIES teleconferencing
software that Kleiner had discovered some years earlier.

The yearlong classes were conducted in a hybrid manner: they involved online exchanges alongside a pair of weeklong, face-to-face meetings with Brand's students in La Jolla. The students were an eclectic mix
of corporate chief executives and managers, some of whom were in the
midst of midlife crises, trying to figure out what their next step would be.

One of Brand's courses focused on something he described as Benign Genres. In his mind, the *Catalog* had been a genre, New Games was
a genre, in fact, his whole approach to organizing public events was a

genre. With his students, he tried to spell out how he did it, which forced him to further look for patterns in his own wandering path.

Brand was paid a modest monthly fee for moderating the online courses and given free access to the EIES service. He quickly fell in love with the idea of living online. At that point, the computer network world and most of the personal computer world was limited to text. (Indeed, the marketing phrase used by Infocom, an early publisher of text-based interactive games, was "the best graphics are in your head.")

O

**B**RAND HAD LEFT SILICON VALLEY JUST AS IT WAS BEING BORN IN the early 1970s. The region was now exploding as computing proliferated and each new generation of technology reached a broader portion of the population. At the end of 1982, *Time* had named the personal computer Man of the Year.

The first glimmers of the media revolution were being brought back to the hidebound New York publishing world by Brand's friend and literary agent John Brockman. Brockman had touched off a personal computer publishing business frenzy by selling a ten-book series for IBM PC owners to Simon & Schuster for $600,000. The deal proved to be a disappointment when the market for how-to computing manuals was quickly oversubscribed, but its magnitude alerted an industry that remained notoriously conservative about using new technologies that there might be a market for books about personal computing.

Brockman had come to San Francisco in the spring of 1983 to attend the West Coast Computer Faire, an annual computer hobbyist exhibition. (It had been at the first Faire, held in 1977, that a twenty-two-year-old Steve Jobs and his twenty-six-year-old partner, Steve Wozniak, had introduced the Apple II.) Brockman's literary agency was focused on scientists and technology writers. Now he embarked on a strategy of trying to find and represent star programmers, selling their work to the new software publishing industry. He had recently added an image of a floppy disk to the logo on his company stationery.

Brockman was also just beginning to perfect the idea of a book auction in which he would persuade writers to prepare simple proposals without the traditional accompaniment of sample material, and then share them with a small group of publishers who could bid on the titles. At Brockman's urging, Brand quickly cobbled together a twelve-page proposal for a *Whole Earth Software Catalog*, modeled after the original *Catalog*. The idea was for the *Catalog* itself to be updated annually, with some of the advance diverted through the Point Foundation to support the quarterly publication of the *Whole Earth Software Review*, which would attempt to keep up with the cascade of new products released as the next full edition was being worked on. Brockman and his wife, Katinka Matson, then circulated it to eight of New York's largest publishing houses.

Brockman could be brazen, and he demanded contract terms that the publishing industry would normally never grant, such as an unusually high royalty rate and complete editorial control. Doubleday bit, offering $1.3 million, an unheard-of amount for a trade paperback book that created general astonishment and outrage among the rival publishers.

Doubleday told Brockman that Brand had one day to accept or the offer would be withdrawn. He had initially told Brockman that the deal was "too much money, it might screw us up." Then he said he needed to speak to his financial adviser. At midnight he called Brockman back and said he hadn't been able to find him, adding that he was going to think about it overnight and throw the I Ching. Brockman couldn't tell whether he was joking or not. In the morning Brand called back and said he would accept.

O

KEVIN KELLY WAS A HIGH SCHOOLER IN A SUBURBAN TOWN IN New Jersey when his mother alerted him to the *Whole Earth Catalog*, having read a story about it in *Time* and figured it might be something he would be interested in. It would take Kelly another year to actually find a copy in a bookstore in Woodstock. Like many others of his generation who had found the *Whole Earth Catalog*, it changed his life.

Steve Jobs later described the *Catalog* to Stanford graduates in 2005 as "Google before Google." For Kelly, the *Catalog* was a roster of possibilities, a portal that suddenly opened up into endless hallways, offering access to an unending procession of doorways, each a world of its own. Kelly was already a catalog person, and now he felt as if he had found the catalog of catalogs. He felt as if it were whispering to him.

When he got to college at the University of Rhode Island, he began trying to contribute to both of Brand's publications, without any luck. Still, he took the implicit message inherent in the *Catalog* to heart: you should invent your own life. It was part hippie ideal, but underlying it was a utilitarian set of values demonstrating that tools exist to give the confidence needed to strike out on your own.

So he did, dropping out after his first year. Kelly became an inveterate traveler, spending the better part of a decade in Asia and wandering through more than a dozen countries as an independent photographer and travel writer. While his early contributions had been uniformly rejected, he now had better luck. Brand published the first article he submitted to the *Quarterly*, a piece on how to explore the Himalayas on very little money. (He had spent several years in the region during the 1970s.) Not only that, but the article ran with one of Kelly's photos on the cover. It was one of the happiest moments of his life. Kelly felt as if he had arrived as a professional journalist. He decided that his dream job would be to edit the *Whole Earth Catalog*.

Several years later he would visit in Sausalito, but Brand wasn't around. There was no immediate way to work at the *Catalog*, and so he decided to dabble in science, taking a job in a microbiology lab at the University of Georgia. He continued to write travel articles and started a small mail-order business selling travel books. Simultaneously, like many young writers at the time, he found his way to nascent online networks by using the Apple II computer in his university laboratory. As Kleiner had, Kelly dived into the expanding world of virtual communities, online chat rooms, and computer bulletin boards. (Among his articles from this period was "The Birth of a Network Nation," for the *New Age Journal*, where he described the online world as a new country.)

Kelly had continued to stay in touch with Brand, occasionally proposing a project or even a new magazine, without a great deal of traction. Then he learned about a new software industry conference called Soft-Con being held in New Orleans. He sprang for a long train ride from Georgia to New Orleans in the hope of finding Brand at the conference.

Successfully tracking him down on the exhibition floor, Kelly pitched Brand the idea of producing a *Whole Earth* "greatest hits," what he described as the *Essential Whole Earth Catalog*. Brand was evasive, as he already had two consuming projects on his hands, but he agreed to stay in touch, and eventually would give the green light on the *Essential* version Kelly had proposed.

One day not long afterward, Kelly received an email from Brand, asking if he would like a job editing the *Quarterly*, replacing Kleiner, who was taking over the job of editing the *Software Review*. The first thing Kelly did was write a letter to Anne Herbert, one of the *Quarterly* editors he trusted, asking if she thought he would fit in. Although there was no question that he saw himself as a *Whole Earth Catalog* kind of person, several years earlier he had also undergone a conversion to Christianity while traveling in Jerusalem. His wasn't a born-again, Bible-thumping Southern Baptist–style Christian—he was more in the Thomas Merton mystic tradition—but he was still concerned that he might be an alien enough object to ask if he was going to be a problem.

Herbert didn't think so, but when Kelly showed up in Sausalito looking like an Amish farmer, and Brand learned that he had hired a converted Christian, he had second thoughts. How would this young acolyte's Christianity mix with the *Quarterly*'s scientific orientation? he wondered.

When he learned that Kelly was a deacon in the Faith Presbyterian Evangelical Church, he inquired, "Fundamentalist?"

"Not a comfortable word," Kelly responded. "Try Bible-preaching."

When he moved to Sausalito, the new editor's first purchase was an aging Toyota that he acquired for twenty-five dollars and used to get to church on Sundays.[11]

Kelly rented a room close to the *Mirene*, several berths over, belowdecks in Peter Calthorpe's houseboat. It was just a concrete shell, almost like a dungeon, but he didn't mind. Calthorpe felt that the young editor, who would emerge each morning and chat briefly on his way to work, was a bit like a gnome.

Brand had little time to fret over his new hire. He had the money from Doubleday and a limited time to build a *Whole Earth Catalog*–style book from scratch. He found office space for the new project on the second floor of a nondescript commercial building across the street from the *Quarterly* offices. Space was quickly filled with a new corps of staffers—editors, design and production artists, staff assistants, research librarians, staff researchers, assistant researchers, "domain" editors, contributing editors, line editors, proofreaders, office managers, clerks, and camera operators. The list was endless, and most of them were paid at a rate significantly above the *Quarterly*'s ten-dollar-an-hour communal wage.

The way he ran the operation was that Andrea Sharp, the *Whole Earth* business manager, would come to him and say, "Stewart, we're running out of money." That would set off a mad scramble for new revenue-producing ideas. A new version of the *Catalog* or a book idea or perhaps a new conference or service.

There was a brief period when he discovered spreadsheets, and for a while he dived in deeply, fantasizing about all of the incredible successes that would finally generate money to conjure with, leading to cool new projects. "Wow," he imagined, "I can control the future with these spreadsheets, fantastic!"

Sadly, it was fantasy. What a spreadsheet is supposed to tell you is the trouble you're in, because it reflects real numbers. But he didn't use them that way. Instead, he created his own numbers, which frequently had little bearing on reality. While he was launching the *Software Catalog*, the cost of running the operation was roughly $100,000 a month, creating a precarious high-wire act. It worked only until the book advance ran out.

Kleiner would later decide that agreeing to the Doubleday offer was Brand's worst business decision. On the surface, however, the idea made some sense. A generation of *Catalog* readers was now older and a bit wealthier and increasingly coming in contact with the new digital world. He was gambling that the *Whole Earth Software Catalog* would provide the same kind of counsel that it had offered in the preceding analog era. Moreover, Brand believed it seemed logical that the counterculture generation he had led would follow him into cyberspace.

"They have no immunity at all to personal computers," he told a *Christian Science Monitor* reporter.[12]

It turned out that they did.

O

THERE WAS ACTUALLY A PROFOUND GENERATIONAL SPLIT OVER the arrival of personal computers and computer networks. It could be seen in the gulf that emerged along Gate 5 Road as defined by the higher wages offered on one side of the road. Everyone would rally together for the daily volleyball games, but the two cultures were heading in different directions.

Matthew McClure could see the divide clearly. One of the early typesetters for the *Catalog*, he had departed with the caravan that left Haight-Ashbury for the Farm, the Tennessee commune started by Stephen Gaskin, where he had become the editor of their publishing business. McClure had a background similar to Brand's—he had gone to prep schools and Stanford—and when he learned that a new *Catalog* effort was underway, he came out and interviewed and was hired to help run production. Their similar backgrounds made for an easy rapport between the two men. McClure became an effective lieutenant in the new publishing effort. Brand was the boss, but with a hands-off management style, and McClure had enough of a technical background to make him a valuable employee. He did have a bit of culture shock in coming back to the West Coast from his commune. The new editors of the *Software Catalog* were big lunchtime drinkers, and when he joined them, he found that he would on occasion come back from the meal with several drinks under his belt, making for a challenging afternoon.

Joe Kane was an adventurous young writer who had come from several years of working at *Rolling Stone* to initially work as a *Quarterly* copy editor. Before going to New York, Kane had grown up in Silicon Valley, where in junior high school he had the distinction of punching a young Steve Jobs in the nose—and breaking it—when he found Jobs and another boy harassing his brother in the schoolyard.

Kane had arrived at the *Quarterly* in the early 1980s and had taken over a regular column in the *San Francisco Chronicle* that Brand had recently begun before quickly deciding it wasn't his cup of tea. Brand could be a compelling writer, but he was never able to master the art form of the weekly eight-hundred-word essay.

Despite Kane's Silicon Valley roots, he had little interest in the personal computer cultural explosion that was beginning to transform workplaces everywhere. He lived in San Francisco and commuted to Sausalito each day on a Miyata 12-speed bicycle.

Early on, he experienced a John Henry moment that confirmed his skepticism about the emerging networked world. Kleiner and Brand and several others were struggling to configure a Radio Shack Model 100 portable computer to connect it by a modem so they could electronically file weekly *Whole Earth* columns that Kane would soon inherit at the *Chronicle* offices.

"I bet you I'm quicker," Kane taunted them, and he grabbed a paper copy of the column, hopped on his bike, and rode across the Golden Gate Bridge and along the waterfront to reach the *Chronicle*'s offices at Fifth and Mission.

He won the race.

Brockman's deal for the *Whole Earth Software Catalog* in 1983 helped cement Brand's futurist credentials, which had already been dovetailing with the digital changes that were afoot. He was, after all, the man who had coined the term *personal computer*, a badge to pontificate on what was over the horizon. Paid speaking engagements (often high-paying affairs for corporate clients, including trips to events in exotic locations) began to roll in.

At one point, Prime Computer, a high-flying minicomputer maker based along the Massachusetts Route 128 tech corridor that was about to be swept away by the PC onslaught, invited him to the British Virgin Islands to address a corporate sales retreat. He came as the bearer of bad tidings. "What are you going to do," he asked them, "when PCs act like minicomputers?" He donned his biologist's cap and told them that lines were blurring between their species, something that spelled opportunity to his ecologist's point of view. However, not for dinosaurs like Prime Computer.

The media reception for the *Whole Earth Software Catalog* was decidedly uneven. There was the predictable round of articles chronicling the fact that the nation's hippie icon was going digital, and the mainstream press was largely enthusiastic, playing on Brand's celebrity status.

There were also skeptics.

Computer industry columnist John Dvorak pointed to the sixties counterculture flavor of the *Whole Earth Software Review* when it arrived in March of 1984: "There is a leftover element of the long-forgotten late 1960s that permeates the magazine," he wrote.

It was Erik Sandberg-Diment, a computer columnist for the *New York Times*, who posed the most telling critique. "Far too many mediocre books about computers are hogging space on bookstore shelves this season, for publishers are still jumping on the silicon bandwagon," he warned. The digitized version of the *Whole Earth Catalog* was a disappointment: compared to its original "imaginative" namesake, "the digitized version is a rather drab collection of predictable reviews."

Rather than the black-and-white low-production newsprint values of the original *Catalog*, for the *Software Catalog* Brand decided to opt for glossy and expensive paper stock, with bold red headlines and color photographs. This made budgeting even more challenging, because the new book and the *Review* would have to compete with existing computer magazines, of which there were many. (By 1984 there were well over two hundred monthly and weekly business and hobbyist publications focusing on just one subject, according to Brand's estimate.[13] In true *Whole Earth* tradition, a number of these publications were endorsed in the new *Catalog*.) Brand's original policy of running only positive reviews was a quaint affectation in a newsstand niche where readers wanted strong opinions. The economics became even more difficult because he had also decided to adhere to the original *Catalog* values of not accepting advertisements. He thought to himself at one point, "There has never been a wronger time to be in this particular genre of periodicals." But by then he had already jumped into the pool.

"I can report on my own continuing dismay, delight and intensely educational confusion as a shopper as well as a would-be cataloger," Brand wrote in the first issue of the *Whole Earth Software Review*, describing his maiden visit to the vast COMDEX Trade Show in Las Vegas the previous November.[14] The new *Catalog* went a good distance toward dispelling some of that confusion. Like previous *Catalog*s, it culled and distilled, it advised on process in addition to tangible objects. There were parallels throughout, in specifics (*The Last Whole Earth Catalog*'s recommendation of Wolfgang Langewiesche's *Stick and Rudder: An Explanation of the Art of Flying*; the new version recommended Microsoft's *Flight Simulator*) and expansiveness. The *Software Catalog* was divided into ten main sections: Playing, Writing, Analyzing, Organizing, Accounting, Managing, Drawing, Programming, Learning, and Etc. (which included, among other things, entries on music, weaving, artificial intelligence, nutrition, exercise, astronomy, meditation, postal services, gambling, and—forever dear to Brand's heart—slide show control).

But the fact that Brand took the stance of being a gawker outside in a hothouse technology culture that was unique and as insular as the sixties counterculture had been two decades earlier didn't help. As Dvorak intimated, the entire enterprise conveyed a *Stranger in a Strange Land* sense of eyes-wide-open naivete.

Although the second *Software Catalog* appeared at the end of 1985, it all came to a head a year earlier. In a lengthy profile[15] on Brand and the *Software Catalog, San Jose Mercury* reporter Michael Malone—who had first seen Brand in his "high Indian" phase, when as a young boy growing up in Silicon Valley he went with his father to see Brand perform his *America Needs Indians!* multimedia slide show—revealed that the *Whole Earth Software Catalog Review* was being folded together with the *CoEvolution Quarterly* after just three issues.

It had quickly become clear that it wasn't a viable business. *Catalog* sales in the United States were unimpressive and subscriptions to the *Software Review* could be measured in the hundreds per quarter. (Sales of the UK edition were almost nonexistent.) It was left to Paul Hawken, the Point Foundation's financial officer, to break the news to the *Review* editors that their publication would be folded into the *CoEvolution Quarterly* to create the newly combined publication that Art Kleiner had named the *Whole Earth Review*. Kevin Kelly, who was editing the *Quarterly*, was put in charge of editing the two publications together into a strange mixture that pleased neither audience.

The coup de grâce, however, was delivered by Richard Dalton, who had served as editor of the first two issues of the *Review* and then handed the job off to Kleiner. In a gossip column called "Inside," in what would be the final issue, Dalton wrote: "Shocking news: most people don't care doodly squat about computing."

B RAND HAD CELEBRATED THE NOTION OF THE "OUTLAW AREA" in the January 1970 supplement to the *Catalog.* "One thing we need is better outlaws," he wrote.[16] He had argued against reasonable laws, created by reasonable men, suggesting that they proscribed innovation. Now a generation of mostly young men were finding adventure and mystery by breaking into the then largely corporate and military world of the first computer networks. They had already created a vibrant dark version of the hacker culture that Brand had been the first to alert the world to more than a decade earlier. Like virtually all of his generation, however, he chose to ignore the warnings and plunged ahead into the beckoning world that science fiction writer William Gibson described as a "fluid neon origami trick."[17]

And there was much to be excited about. Among those invited to present at the first Technology, Entertainment and Design (TED) Conference were Brand and MIT professor and architect Nicholas Negroponte. Their encounter would reframe the next decade of Brand's life.

The event, held in a hotel conference center in Monterey in February of 1984, was a little-noticed affair attended by a relatively intimate group (compared to later conferences) of 250 artists, writers, musicians, corporate execs, and scientists united by their "faith in the computer."[18]

IBM mathematician Benoit Mandelbrot and *Megatrends* futurist John Naisbitt both spoke, but it was Negroponte who stole the show. Dressed in a dapper gray suit and tie and with a rich mane of longish hair, he showed off in his "TED Talk" (the term had yet to become a marque—and, in some quarters, a pigeonhole) a variety of futuristic technologies for interacting with computers, including touch screen manipulation, which would not become commonplace until a quarter century later with the introduction of the iPhone.

Negroponte was an uncompromising digital utopian. "The icebox will realize when it's out of milk. It will tell the answering machine, and it will tell you. . . . Toasters will know about toast," he proclaimed.[19]

Phelan was smitten. "He's incredible!" she exclaimed to Brand. He agreed.

When they met later during the conference, the two men instantly formed a mutual admiration society. Brand had written a positive review of Negroponte's book on the future of architectural design, *Architecture Machines*, and the MIT professor not only expressed his admiration for the *Whole Earth Catalog*, but he recited the review verbatim.

Negroponte was about to launch an interdisciplinary center at MIT to explore the new digital world, to be named the Media Lab, and he suggested that Brand spend a quarter to see how the Media Lab evolved. He was even willing to pay him to visit. Brand tucked Negroponte's offer into his back pocket.

O

BRAND KEPT HIS HAND IN A GROWING ARRAY OF OUTSIDE ACTIV-
ities, including several nonprofit board seats and his lecturing position at the Western Behavioral Sciences Institute, where he went at the
end of 1984 for his twice-annual face-to-face stint with his students. It
was there that Steve Jobs played an indirect role in plunging Brand even
more deeply into the emerging digital universe.

Phelan had accompanied Brand to the La Jolla campus, and one day
they were about to sneak off during lunch to a secluded skinny-dipping
spot known as Black's Beach when they were cornered by a roly-poly
man on a mission. In 1969, Brand had met Larry Brilliant while Brilliant was training to be a doctor in San Francisco and Brand was organizing the *Hunger Show*. Brilliant had later befriended Steve Jobs when
the young computer entrepreneur was on a spiritual pilgrimage in India
while Brilliant was there working on the international effort to eradicate
smallpox.

Brilliant had discovered computer conferencing while he was doing
medical relief work in Nepal, and Jobs had helped him acquire a 300-
baud modem to contact the outside world via satellite. In 1983, Jobs visited Brilliant at the headquarters of Seva, a medical foundation the
doctor had established to help cure blindness. Brilliant had created a
simple communications software program called Seva Talk, which they
were using to run the organization that was spread all over the world.

Brilliant asked Jobs for money to support Seva, and the brash computer entrepreneur responded, "Build your own fucking product! Build
your own fucking company! Make your own fucking money! Fund your
own fucking non-profit! Give that to Seva!" And then he paused and
smiled and said, "And I'll help you."

Jobs gave Brilliant the first $5,000 to create a computer software
company called Network Technologies International, or NETI. Brilliant took the company public on a Canadian stock exchange, raising

$6 million, despite not having publicly announced any business for the conferencing system they had built. Quickly, however, they announced deals with AT&T and General Electric.

Despite his momentum, Brilliant was completely at sea as an entrepreneur and a technologist. An associate professor of medicine in Michigan who had leaped into the grubby world of business, he felt as if he had no idea what he was doing. He went out to California to participate in a three-day meeting about EIES and computer conferencing and ended up crying in frustration in the office of the dean of the School of Public Health at UC San Diego.

At the time, he was flush with capital, and he approached Brand with the idea of putting the *Whole Earth Catalog* online in some interactive fashion. They quickly came to an agreement that called for NETI to invest $250,000, including a refrigerator-size Digital Equipment VAX minicomputer, in a new joint venture. Brand would supply the labor from his staff.

Brand had no interest in digitally reinventing the *Catalog,* so he listened to what Brilliant had in mind and then pretty much ignored it. From his experience with EIES at the Western Behavioral Sciences Institute, he understood that the online world was a deeply different kind of medium from a book or a magazine. Moreover, he had learned from his original *Catalog* experience that the supplements and the feedback loop they created with his readers was what he wanted to re-create in the digital world.

Art Kleiner was sent off to Michigan to work out the details of the joint venture, and Brand began to brainstorm. Passionate about building institutions, he fantasized about the details of this virtual world he was creating from whole cloth. First, he took a piece of stationery and made a list of possible names, and one of them felt just right: the WELL, or Whole Earth 'Lectronic Link.[20]

He then made a series of design decisions, even before he hired Matthew McClure as the effort's first employee. The WELL would be relatively low cost, billing would be entirely by credit card, users would be the owners of the content they created ("you own your own words"), and, in a stroke of marketing brilliance, he gave free accounts to a group of technology journalists, ensuring that the WELL would gain an out-of-scale reputation in the emerging digital universe.

Unlike Brand's earlier efforts, however, the WELL was neither first nor unique. The world was already full of giant, albeit more costly, on-line service providers. Compuserve had begun in 1969, and the Source had been founded in 1978. In the Bay Area, Community Memory had been created in 1973 to act as a local bulletin board service with the idea of placing terminals in public spaces. Not only was there also a thriving world of thousands of BBSs, but in 1983 a system called FidoNet had begun linking them together into a nationwide electronic mail service.

Moreover, the "cyberculture" for which the WELL would become emblematic had actually been thriving nationally since 1980 in the form of Usenet, an anarchic set of computer conferences or "newsgroups," which circulated on the nation's computers that ran in academic and corporate computing settings. (Concepts such as FAQ, flame, sock puppet, and spam all emerged from Usenet.)[21]

Indeed, the culture that blossomed on the WELL was something of a throwback. Despite its state-of-the-art minicomputer, so-called WELL beings seemed to believe that the medium was as much a step backward toward the nineteenth-century literary salon as a step forward into the future.

"People used to write letters all the time," Howard Rheingold, a *Whole Earth Review* editor and a key member of the WELL community, told the *New York Times*. "Now we do again. It's not for everybody. It's for people who enjoy communicating through the written word."[22]

Moreover, the community that sustained the WELL was not entirely online; rather, it became a creative hybrid, with lots of face-to-face socializing in which Brand frequently participated. In fact, the WELL may have ultimately owed much of its early success to its discovery by one of the nation's most eclectic countercultures: Deadheads. Devotees of the rock group Grateful Dead, the Deadheads, who included modern-day hippies, professionals, and computer enthusiasts, were a floating community that would gather at the band's dozens of concerts each year, and those who were online stayed in touch between concerts via the WELL.

Brand had learned an early and important lesson from his time communicating with the Western Behavioral Sciences Institute students via EIES. At one point, they had created an anonymous online conference, and he watched as it had instantly become pathological. Anonymity immediately triggered the basest of human behaviors, he discovered. People masqueraded as others, behaved viciously, and then the victims would act the same way to exact some revenge. As a result, he outlawed anonymity on the WELL. Each user was tied to their identifiable account and linked to their credit card.

However, while outlawing complete anonymity, the design of the WELL permitted pseudonymity: a user was permitted to have several aliases that were not his or her real name. Years later, Brand would come to see this as a major design flaw that would ultimately drive him away from the community he had created. It would also give him an early and unique window into the online threat of fake identities and unverifiable and untraceable assertions that would come to haunt the entire world.

Nevertheless, even though he grew troubled early on and relatively quickly moved on from the electronic community he had created, Brand would remain publicly optimistic. Although the first issue of the combined *Whole Earth Review* was titled "Computers as Poison," Brand was outwardly sanguine: "Computers suppress our animal presence," he told an *SF Focus* interviewer. "When you communicate through a computer, you communicate like an angel."[23]

Just a few years later he would realize how naive he had been.

SHORTLY AFTER HE ARRIVED IN SAUSALITO, KEVIN KELLY READ Steven Levy's *Hackers: Heroes of the Computer Revolution*, a book that portrayed three generations of "white hat" computer hackers (the good guys) ranging from the young programmers at MIT's AI Lab decades earlier, through the Homebrew Computer Club, to the then new world of video game design. (The term *hacker* had only recently entered the national lexicon. It had been just a year since the movie *WarGames* had introduced the American public to the notion of network-accessible computers and bright young computer wizards.) Levy identified what he described as "the hacker ethic," epitomized by the MIT students who, beginning in the 1950s, had "hacked" projects for the simple joy of designing computers and software as an end in itself. According to Levy, this ethic was a set of ideals premised on access to computers; free access to all information; distrust of authority; meritocracy; the belief that computers could create art and beauty; and the faith that computing could improve your life.[24]

Kelly went to Brand and said, "You know, the funny thing about this is, as far as I can tell, none of these guys have ever met each other, these three generations of hackers. Let's bring them together." Brand loved the idea and he persuaded Kelly and Phelan to help organize an event to be called the Hackers Conference.

They began assembling a group of elite computer programmers and hardware designers from around the Bay Area, crowding them into the *Mirene* for weekly evening meetings. High-profile computer designers such as Lee Felsenstein, who had designed the Osborne 1, and Andy Hertzfeld, one of the key designers of the Macintosh, trooped up to Sausalito to help with the organizing.

The gathering took place over a November 1984 weekend at a funky retreat center at Fort Cronkhite, a former army base set in the Marin Headlands near Gerbode Valley. Ultimately, from the roughly 400 they'd invited, the *Mirene* team attracted about 150 computer hackers and a sprinkling of technology journalists. They were people like Ted Kaehler, who was a *Whole Earth Catalog/CoEvolution Quarterly* devotee. Growing up in Palo Alto as the son of a Lockheed Missiles & Space Company engineer in the 1960s, he had belonged to his high school science club. Kaehler had attended Stanford University, fallen in love with computing, and then gone off to study computer science in graduate school at Carnegie Mellon, where he thrived under the tutelage of pioneering AI researchers like Raj Reddy and George White, both of whom would end up at Xerox's PARC in the early 1970s. Kaehler had followed his professors and helped develop the Smalltalk programming environment, which became the model for both the Apple Macintosh and Microsoft Windows.

Kaehler had read *Hackers*, and when he heard that Kevin Kelly, Brand, and Phelan were planning a meeting of all the hackers, he called Kelly and asked if he could invite his own group of hacker friends. Kelly gave him the green light, and as a result, several PARC researchers were in the mix at the first Hackers Conference.

Kaehler threw himself into the conference. He moderated two of the panel discussions and he would stand behind the cameramen who filmed the first hackers meeting, giving them cues on whom to focus on. While walking to dinner the first night, however, he ran straight into a side of Brand he was unprepared for. As they headed down a hill near the conference center, Brand asked Kaehler how many of the people walking ahead of them he knew. "Why, none of them!" Kaehler responded. He was familiar with the Xerox crowd, but the hackers' community, which included game developers and "graybeards" from MIT and Bell Labs, was new to him. In response, Brand appeared to Kaehler to grow aloof. He didn't speak to Kaehler again during the conference, and Kaehler felt slighted, deciding that he must not be a member of the hacker elite.

While Brand may simply have been intimidated by all of the Silicon Valley star power, there was also no disputing his attraction to people who were in the limelight, whether they were well-known intellectuals, CEOs, or governors. It was what Kesey had noticed when he had accused Brand of "cleaving" to power. When Brand had read the quote in a generally flattering *Washington Post* profile,[25] the words had stung.

The Hackers Conference would become an annual event, drawing together a digital subculture that was passionate about the machines and programs they designed. The first gathering was marked by several heated debates that would continue long after everyone had gone home. While Apple cofounder Steve Wozniak argued that the term *hacker* represented the child in everyone, a well-known programmer named Brian Harvey warned there was a dark side as well, and the general public would soon come to see the word as synonymous with computer outlaws who broke into computers for sport and profit.

Another important debate took place over the economic value of software. Richard Stallman, a young MIT programmer who had developed freely shareable versions of several key programming languages, stated that his goal was to make all software free. That angered other attendees such as Apple Macintosh designer Bill Atkinson, who responded, "Hackers want me to give [this] Quickdraw code away, but there is this thing called IBM and I want Apple to be around in 20 years."[26] Others, such as former *Quarterly* editor Andrew Fluegelman and former Microsoft programmer Bob Wallace, who had coined the terms *freeware* and *shareware*, respectively, tried to walk a middle ground, giving away their software and asking for donations.

It was Brand, however, who framed the moment most accurately, and in doing so helped spark the open source software movement that would come to transform the computing industry.

Wozniak was arguing that when private companies kept software away from the public, "that's a hiding of information, and that is wrong."[27] At that point, Brand, who had been silent during the session, stood up and said:

> It seems like there's a couple of interesting paradoxes we're working with here. That's why I'm especially interested in what Bob Wallace has done with PC-WRITE and what Andrew Fluegelman did before that with PC-TALK. On the one hand, information wants to be expensive, because it's so valuable. The right information in the right place just changes your life. On the other hand, information wants to be free, because the cost of getting it out is getting lower and lower all the time. So you have these two fighting against each other.[28]

He saw clearly what he had absorbed in his apprenticeship with Bateson. It was a classic example of the dilemma that two conflicting messages present—a double bind. In a phrase, he captured the moment that

became a flashpoint for a set of multibillion-dollar information industries and a rallying cry for an emerging digital culture, which, while not
Marxist, passionately believed that locking up digital property was theft.

The nuance was missed, however. Only the second half of what he
said, the aphorism "Information wants to be free," quickly came to frame
the digital era. Entire industries—newspaper publishers in particular—
took the notion of free digital information as gospel and tried to build
digital versions of their businesses based on giving away their product
and eventually—largely unsuccessfully—trying to make it back based on
advertising revenue. Even more transformational, it became the cultural
and business foundation of a wave of Silicon Valley companies that would
become among the world's most profitable and powerful based on the
same idea.

And like almost all of Brand's ideas, there was a backstory. He had
actually laid out the idea in a talk he gave that summer at the Western
Behavioral Sciences Institution, shortly after he and Kevin Kelly received a copy of the galleys to Levy's book. At the end of July, one of his
students said to Brand after his lecture: "Your point about information
wanting to [be] free turned out to be more controversial than I thought it
would be—particularly when coupled with the 'hacker's ethic' idea. Several people told me they thought you were selling 'stealing.'"

Ultimately, a range of software activists would coalesce around the
term *open source software*, fundamentally changing the way an industry
competed and simultaneously unleashing dramatic collaborative forces
leading to free public services such as Wikipedia. It was also adopted as
a rallying cry by a generation of disruptive Silicon Valley start-ups—
sometimes in ways that were contrary to the openness implied by Brand's
declaration. "Information wants to be free" would come to define an an-
tiregulatory stance taken by new internet industries that argued successfully that their "innovations" were of such benefit to society that they
should not be fenced in as more traditional industries were.

Brand's original description of the paradox was lost. All that anyone remembered was "Information wants to be free." Just as he had defined the counterculture and the environmental movement of the 1970s,

Brand had captured the essence and the contradictions of the emerging digital era.

O

S OON AFTER THE WELL LAUNCHED IN EARLY 1985, BRAND BEGAN to feel that he was once again burning out. One night he sat with Doug Carlston, the cofounder of the software company Broderbund and a member of the Point Foundation, and poured his heart out. He was tired of writing the kind of stuff that was appearing in the *Quarterly* and he was tired of being known as the guy who put out the *Catalog* a decade earlier. He needed to find a way to entirely separate himself from the institution he had created.

He arranged what he thought would be a summer-long paid sabbatical. In the end, however, he would never return to actively managing the *Quarterly* or its successor, the *Whole Earth Review*, except in a largely symbolic role. To mark the transition, he moved his stuff out of his office on Gate 5 Road, and with Huey Johnson's assistance, Brand and Phelan were able to engineer a house swap, trading their sixty-four-foot tugboat in Sausalito for a ten-thousand-acre farm adjacent to the Masai Mara game reserve in southern Kenya.

They clearly got the better end of the deal. Their African guest was six feet six inches tall and soon decamped for a rental in Belvedere. In contrast, Brand and Phelan had an extraordinary time spending several months with two servants, a household pet cheetah, and another couple who lived on the farm who became their friends on safari.

Phelan had persuaded Brand to travel to Africa in part because of their shared fascination with Beryl Markham's memoir *West with the Night*. The book had long been out of print before it was picked up and republished by a boutique Berkeley publishing house in 1983. Markham, who had been a bush pilot, adventurer, and racehorse trainer and who was the first person to solo nonstop across the Atlantic from east to west, was now down on her luck at the end of her life and living near a racetrack in Nairobi.

While taken by the book, Brand was still not as enthusiastic about their coming trip as Phelan, who was ready to try to hunt for Markham. That changed, however, when he met the writer Peter Matthiessen at a houseboat party on their dock shortly before they were scheduled to take off.

A legendary novelist, wilderness writer, Zen proponent, and former CIA officer, Matthiessen was a member of a long line of Western adventurers ranging from Teddy Roosevelt to Ernest Hemingway.

Brand was in awe of Matthiessen, but when Phelan confided that they were getting ready to visit Africa, Brand demurred, saying something like "I've seen the postcards."

"Don't even think that stupidly," Matthiessen shouted. "Until you have smelled and seen the dawn in Africa you have no idea what you're talking about."

On their way to Africa, they first stopped at MIT to visit Negroponte, who renewed his invitation for Brand to be a paid "visiting scientist" at the new MIT Media Lab. Negroponte was on the cusp of opening the lab in a $45 million building designed by I. M. Pei. With Jerome Wiesner, electrical engineer and former MIT president, Negroponte had spent seven years fundraising for the lab, which would be housed in the futuristic metal and glass structure. The intent was to create a research center at the intersection of engineering, social science, and the arts.

From Negroponte, Brand learned that the new Media Lab was a descendant of the Research Laboratory of Electronics, or RLE, an earlier multidisciplinary laboratory that Wiesner, a veteran of MIT's legendary World War II–era Radiation Laboratory, had led. MIT's Artificial Intelligence Laboratory had also grown out of the RLE, which in turn had spawned the hacker culture that Brand discovered at the Stanford Artificial Intelligence Laboratory when MIT AI Lab cofounder John McCarthy headed west. Negroponte's Architecture Machine Group, an early effort at an advanced computer system to prototype human-computer interaction, was also a descendent of the RLE.

The Media Lab's intellectual tradition was reminiscent of what

Brand had first come in contact with as a college student when he encountered cybernetics through Wiener and Warren McCulloch and information theory from the writings of Claude Shannon and Robert Fano.

At MIT, visitor and host discussed the idea of coauthoring a book, but Brand had already learned enough about Negroponte, who had a large ego and a forceful personality, to realize that would be problematic. Instead, he negotiated a contract that gave him access to the Media Lab and in return agreed to pay special attention to Negroponte's research.

He then traveled to England, where he and Phelan stayed with Peter Schwartz, whom Brand had known casually when Schwartz worked at the Stanford Research Institute (SRI) as a futurist and who was a board member of the Portola Institute.

Schwartz, who was fond of saying that he actually was a "rocket scientist," had studied aeronautical engineering as an undergraduate at the Rensselaer Polytechnic Institute. He had been hired at SRI as part of a futures group led by Willis Harman, one of the founders of the Menlo Park–based LSD group where Brand had taken part in the 1962 drug experiment. (Schwartz had shown up at SRI just in time to be given the office that had previously been occupied by the mysterious Al Hubbard.) At SRI, Schwartz had been early to foresee the impact of new digital technologies (he had consulted with the screenwriters for *WarGames*, alerting them to the emergence of the new subculture of computer hackers), with a special interest in scenario planning.

The scenario-planning method was defined by twin intellectual heritages: "thinking the unthinkable" nuclear war-gaming by Herman Kahn at the RAND Corporation in the 1950s, and the ideas of Pierre Wack, a planner at Royal Dutch Shell who had brought from his intellectual apprenticeship with the French mystic and philosopher George Ivanovitch Gurdjieff a focus on narratives rather than quantitative charts and data.

Although Schwartz had been an antiwar activist and a member of the Students for a Democratic Society in college, in 1982 he joined Royal Dutch Shell to lead the firm's scenario-planning group. At the time, Shell was an unusual company in the energy industry. It was not as buttoned-down as a company like Exxon, and just the fact that it had hired a Californian like Schwartz was indicative of the fact that it was more open to ideas, even to environmentally friendly ones.

Shell's head strategic planner was Arie de Geus, who had previously met Brand when he visited Northern California while on the hunt for fresh ideas. De Geus had spoken about doing some research on how organizations learn in order to improve Shell's planning process. Brand told Schwartz that he had been deliberating about how complex systems learn, and Schwartz responded that de Geus was pondering the same thing.

From London, Brand and Phelan headed to Italy, where Richard Thieriot, the *San Francisco Chronicle* publisher, and his wife, Angie, had rented a spectacular villa in Venice. Brand was bowled over by the tiny city-state that had survived on its wits for eight centuries. Wandering in the city by foot and boat, he explored Venice's heritage as a mercantile empire. Inside St. Mark's Basilica, the Italian-Byzantine cathedral at the eastern end of the Piazza San Marco, he peered into the magnificent golden domes overhead and was shocked. Until that moment he had been seeing Europe, Europe, Europe, Rome, Rome, Rome, Italy, Italy, Italy. Now he looked up and saw Constantinople. The striking mosaics showed deep Asian influence. "Did I just step out of Europe into Asia just going through these doors? How can that be? The fuck is going on?" He was overwhelmed.

For an afternoon he remained in the basilica and watched as the tourists poured through, looking up agape but mostly appearing to be numb. He was stunned to see that no one smiled. He lingered, unable to remove his grin, and the golden images beamed down, offering an infinite, mineral perspective.

Only gradually, by reading deeply, did he come to understand that, finite in width, yet infinite in historical depth, Venice fit perfectly as a frame for his conversation with Schwartz about complex systems that evolved and learned.

His first thought was to move to Venice and burrow in to write an extended "systems" history of the city. That quickly morphed into the idea of a fictional trilogy whose protagonist would be a physicist who would solve mysteries in a modern Venetian setting, affording the couple the opportunity to live in the city for several years.

By the time he got to Kenya, Brand was deep into fantasizing about his new life as a novelist.

O

ARRIVING IN AFRICA IN EARLY JULY, BRAND AND PHELAN SETtled into a fairy-tale existence. The ranch was owned by David Hopcroft, a fourth-generation white Kenyan with a Cornell doctorate in wildlife ecology. Phelan was as happy as she had ever been as they walked, ran, and toured on the lands with Phillip Tilley, a hunter who lived with his wife on the ranch.

Phelan fell completely in love with the large animals that roamed freely across the Masai Mara, particularly the cats. Brand and Phelan had private safaris, driving hundreds of miles and camping in the bush. They would sit all day in their safari vehicles and watch sleeping lions. Brand decided that if he ever wrote about Africa, he would call it "Sleeping with Lions."

They also tracked down Markham and visited her several times shortly before her death in August of 1986. She had heard only vague word of the American success of *West with the Night*, and if there had been profits, they were not visible to Brand and Phelan.

Meanwhile, Venice was threatening to become an all-consuming project. Brand thought about writing his book without an advance, possibly selling his sailboat and his house in Nova Scotia to underwrite his work. While out on an early morning run one day, Phelan pointed out that Venice was a perfect counterpoint to the space colonies that Brand had become fascinated by a decade earlier. Human designers planned space colonies in a matter of years, while Venice, an island ecology in its own right, had been designed organically over centuries.

His Venetian reverie was soon punctured by a letter from Andrea Sharp, who wrote to inform him about the precarious financial situation back home in Sausalito. As gently as possible, she noted that Kevin Kelly was working without a salary and the decision had been made to turn Brand's paid sabbatical into unpaid leave. "I did want you to know that there aren't additional funds from us going into your account right now," she wrote, adding that the WELL's revenue was around $2,000 for June.

The letter touched off an identity crisis. Brand had left America thinking of himself as an integral part of the organization he had created. Now he felt at sea. A continent away and out of contact, he felt like the Point board of directors had decided to discard him. It seemed they believed that he was distracted and used up and no longer a revenue-producing part of the operation.

He had established himself as a respected editor and a publisher, and now, at forty-eight years old, he faced the task of starting over. He would need to earn a living, and his fantasies of living a Bohemian literary life in Europe would have to wait.

It was a branch point. He had run a nonprofit business more or less continuously for the past seventeen years. Now he felt as if he were tired of being in charge, meeting payroll, being present from ten in the morning to six at night Monday through Friday at a minimum, being stuck with some people he had little interest in knowing.

It was time for a change.

O

BACK IN SAUSALITO, HE MOVED HIS OFFICE TO THE *MARY HART-line*. It was a dry-docked fishing boat, a rickety and a soon-to-be disintegrating affair that he called *High Tide*. Hidden behind a fence just across a courtyard from the *Whole Earth* offices, it remained a comfortable two-hundred-yard commute.

Kevin Kelly was now running the *Whole Earth Review*, and the WELL was slowly gaining new users—it had passed six hundred during the summer—and Brand had two new projects with which to reinvent himself.

While he was in Africa, the letter had arrived from Negroponte confirming his appointment at the MIT Media Lab in the winter of 1986. On the way back from Africa he once again visited Schwartz in London and met with his scenario-planning partner at Shell, Napier Collyns. The two men suggested that Brand spend time at Shell as a consultant with the idea that the company also sponsor him in arranging a set of conferences on organizational learning.

The offer struck an immediate chord. Soon after his arrival at Stanford, Brand had celebrated self-reliance, believing that independent learning was more effective from the bottom up. It was something he had come to early on and had been catalyzed in his *America Needs Indians!* project, the *Whole Earth* enterprise, and his fascination with software and personal computing. Brand wrote Robert Rodale, the publisher of the Rodale Press, tentatively looking for a half-time "patron" grant. Rodale was an admirer of the *Catalog* and Brand appealed to his "incessant curiosity." "What if something provocative can be learned about how systems learn?" he asked.

Brand had been fascinated by this question for decades, as had been made demonstrably clear in his prototype of the *Catalog*, which began with the heading "Understanding Whole Systems." Here, too, was a way of picking up where Bateson's last book, *Mind and Nature*, which drew parallels between personal, cultural, and evolutionary learning, left off.

Brand planned to spend the first three months of 1986 with some of the "top people" using unconventional learning theories—namely, Alan Kay and AI pioneers Marvin Minsky and Seymour Papert—at MIT. He would continue to stay involved in *Whole Earth*, but his sole activity would be "investigating how Systems Learn," to create a lexicon of effective learning strategies that would hopefully lead to a rudimentary taxonomy and ultimately a general theory of learning.

He proposed the idea of a Rodale scholar at large, but his appeal fell on deaf ears. In the end, he would make do with the consulting fees he was earning as well as his mother's continued largesse.

Brand had written that one would have to be a "moron" to want to spend January in Cambridge, but Kay's offer of his apartment just off Harvard Square was impossible to refuse. So it was that on a chilly and cloudy January morning he found himself walking through the Square to the T-stop feeling like a delighted child: "this is me . . . a smart scholar studying how to save the world."

He fell completely in love with the hyperintellectual Cambridge scene—the campus, the magazine kiosks, the bookstores, and the endless supply of intellectuals at the two universities. In his first month at MIT, he gave a series of lectures on his experience publishing the *Catalog*, keeping the students engaged by frequently changing the subject and veering off into odd anecdotes, then trying to swerve back, surprising his audience by converging on his argument from an unexpected angle.

After his lectures in January, he plunged into two Media Lab research projects. One was an effort focused on the future of electronic publishing, which was close to Negroponte's heart, and the second was Kay's Vivarium project, a computer learning simulation that Apple Computer was funding.

Vivarium began as an "ecology-in-a-computer" simulation when Kay was at Atari. (Ann Marion, the researcher who came up with the Vivarium concept, told him her goal was to create puppets that were capable of pulling their own strings.) Kay had attracted a remarkable brain trust of advisers, including Minsky; Frank Thomas, the Walt Disney animator who created Bambi; Jim Henson, creator of the Muppets; Tim Gallwey, author of *The Inner Game of Tennis*; and Richard Wurman, founder of Access Press and the TED Conference. When Brand joined, he and Wurman were given the assignment of putting together a "cave of knowledge," a repository of information for the kids who were the Vivarium designers. During his stay, however, Brand would remain largely an observer, taping interviews and doing the groundwork for the book he decided to write about the lab.

Negroponte had instilled a "demo or die" culture in his new Media Lab, and although the lab would be criticized for its ratio of hyperbole to tangible outcome, what the critics failed to realize was that its "product" was a generation of designers who would shape the thinking of technology companies around the world. Media Lab alumni would populate Silicon Valley companies, and as Hollywood consultants, they would have a dramatic impact on American popular culture.

Brand for the most part embraced Negroponte's vision of the future. A key for him came in reading Ithiel de Sola Pool's 1983 *Technologies of Freedom: On Free Speech in an Electronic Age*, which foresaw the transition from the print to the electronic age and explored the challenges of governing cyberspace. He relied on Pool's perspective in interpreting what he saw at the Media Lab.

Just as Brand had found his way to the early wellspring of personal computing in the late 1960s and early 1970s at SAIL, PARC, and Engelbart's Augmentation Research Center, now he found himself surrounded by the technologists who were building the digital world that would become both the World Wide Web and the "Internet of Things." One key idea that was deeply embedded in the Media Lab was the notion of personalized media. What Negroponte called the Daily Me, the concept that each person's newsfeed would be personalized, was seen as an inevitable future with little understanding of the darker reality that would become known as filter bubbles several decades later.

One of the people Brand met during his three-month stay was Danny Hillis, a cerebral young supercomputer designer who was a protégé of Minsky's. Several years earlier Hillis had founded Thinking Machines, a radical supercomputer company based on the idea of linking tens of thousands of small microprocessor chips to work on problems simultaneously. It would be a forerunner of today's internet-centric computing—now described as cloud computing.

Hillis had shown up one day for lunch at the Media Lab. At the time, his company was just delivering its first computers and the lab had taken delivery of an early prototype. Hillis invited Brand over to his office just off Kendall Square on the edge of the MIT campus. The two men quickly found common ground.

"One of the things I'm really interested in is the connection between the kind of learning an individual does and the kind of learning an ecology does as it evolves," Hillis told him.

"Yes please. That's my main interest. Say more about that," Brand responded.[29] It was the start of a close friendship and partnership that would lead in a range of different intellectual directions.

Fortuitously for Brand, Hillis's journey would take him to California. Unfortunately for Hillis, that relocation was born of the collapse of Thinking Machines. For a while one of its supercomputers was ranked first on the list of the world's fastest, but the company mishandled negotiations to be acquired by a larger computer company and subsequently went into bankruptcy. Afterward, Hillis moved to the West Coast to work as one of Disney's Imagineers at its blue-sky research laboratory. Before long, he would also buy a houseboat in Sausalito and become a part-time neighbor of Brand and Phelan's.

Brockman had been able to obtain a good advance for Brand's 1987 book, *The Media Lab: Inventing the Future at MIT*, and so after spending three months in Cambridge, Brand headed back to Sausalito to write for several months before traveling to London to take on the role of Royal Dutch Shell business consultant.

More than a year later, when it appeared in September of 1987, *Media Lab* made the bestseller lists in the Bay Area but not nationally. However, Brand's timing was impeccable. The nation had still not woken up to the arrival of computer networks, but personal computers were now firmly embedded in America's culture, and there was a growing awareness that the PC and computing in general were changing the world. Significantly, the book cemented his credentials as a futurist. He would find a steady demand for speaking at events that would bring as much as $5,000 per lecture.

Several reviewers pointed out that he often seemed starstruck by technology, perhaps unwilling to step back and consider what might go wrong. The *New York Times* reviewer noted, "When he assesses potential problems arising from the new technology, Mr. Brand worries that it is easy to be paranoid; others might worry that it is also easy to be complacent," adding, "At times, Mr. Brand seems overwhelmed by imminent utopia."[30]

There was truth to that.

The book cover itself, with the aid of the Media Lab researchers, created a minor sensation. The front cover image was an Escheresque outline of the faces of two people in conversation, separated by a holographic image showing a three-dimensional sprinkling of letters. It gave the book a glitzy sense that "the future has arrived." Although it was a commercial success, the book did not come without the inevitable turmoil that accompanied working with a personality as big as Negroponte's. When Brand showed the Media Lab director his manuscript, Negroponte hated the portion of the book that focused on a visit that Peter Schwartz, the Dutch Royal Shell scenario planner, had made to the Media Lab and threatened to withdraw his cooperation. Significantly, this included the images that the lab had promised to provide for the book.

For a while Brand was traumatized by Negroponte's tantrum. It seemed like a year's worth of labor, largely spent without weekends off, was about to go up in smoke. He turned to Paul Hawken for counsel, and his friend encouraged him not to freak out. In the end, MIT physicist Phillip Morrison, who had also become a good friend, persuaded him to stand his ground.

"You don't need his photographs," he told Brand. "Do what you want with the book. It's your book, not his book."

Brand stood his ground and prevailed. Negroponte eventually relented and supplied photographs. Later, after the book played a key role in cementing the mystique of the Media Lab, he acknowledged that "sometimes it's nice to be wrong."

At various points in the book, Brand noted that society had significant decisions to make. Yet when it came to darker prospects, he kept mum. One night in February, at home in Kay's apartment, he and Phelan talked about the consequences of the ultraconvenience that the Media Lab was intent on bringing to modern life. She suggested it would corrupt people in the same way they had seen Western civilization corrupting traditional tribal life in Africa. Worse, "people become a slave to their phone." Maybe "Phone Slave" is the correct term for what the lab is making, he mused. That thought, two decades before the introduction of the iPhone, did not make it into the pages of his book.

After returning to the West Coast in the spring, he sat in the *Mirene* one morning, reading about the Chernobyl nuclear disaster. "I wonder what kind of calamities are awaiting us in communications technology?" he wrote in his journal. "Financial freak-outs I suppose, massive false-story propagation, I dunno what else. Back to The Machine Stops," a reference to E. M. Forster's short story in which civilization collapses when technology fails.

Those dark premonitions, however, were largely kept to himself.

# Chapter 9

# Learning

FROM A HELICOPTER, THE BRENT OILFIELD PLATFORM DID not appear massive. It came into view as a tiny dot lost in a vast sea. Once delivered from Aberdeen, Scotland, to the Cormorant Alpha platform in the East Shetland Basin of the North Sea, Stewart Brand had a persistent feeling of slight motion, kind of dodgy and fluttery, a jiggle and then a roll. Despite his sailing experience, it took him awhile to get his sea legs. This was clearly a much bigger houseboat.

In the third month of his Royal Dutch Shell consultancy, Brand was sent north in September to see firsthand how field operations worked. On his flight from London to Aberdeen, he had peered down at the Scottish countryside. When he wasn't watching what appeared to him like an unending green golf course sliding under the plane's wing, he drifted into a reverie, thinking about his quest to find a general theory of learning. If he was lucky, the trip would provide a new way of understanding how organizations could change.

He had arrived in London in July, juggling his new role as a business consultant and finishing his *Media Lab* book. Now finding himself on a city-scale high-tech structure set above a rolling ocean in four hundred feet of water, he had none of the stomach-turning alienation that had overcome him when he'd first seen the space shuttle in Southern California a decade earlier.

This, too, was the diametric opposite of the do-it-yourself human-scale technology he had extolled in the *Whole Earth Catalog,* but unlike a spaceship, this was more—well, more down-to-earth, literally. It was comprehensible to imagine the platform as an extension of the *Catalog* in a way he could not comprehend the shuttle. And, frankly, it was one of the coolest things he had ever seen. He was thrilled to be around something that was so inaccessible and at the same time significant. The clang and the bang and the visibility of machinery—it felt designed for old school experts, much like the early NASA.

The irony of the idea of one of the world's iconic environmentalists going to work in the heart of the fossil fuel industry did not perturb him. Visiting a North Sea oil platform was his opportunity to see how the world's infrastructure worked. His charter was to help a giant corporation learn how to "learn," and he needed to see it from the inside. Both he and Peter Schwartz had no concern that they were perhaps being co-opted; rather they saw it as a rare opportunity to carry on a crusade for environmental sustainability from the inside. And right now, it felt like one of the highlights of his life.

Touring Cormorant Alpha, capable of storing one million barrels of oil, Brand wasn't focused on its impact on the environment. Rather he plunged into the company's effort—impeded by its unions—to restructure the oil platform workforce of two hundred into an organization that simultaneously performed both production and maintenance. He took copious notes while trailing after the Shell engineering manager who was hosting him.

The tour over, he was quickly back on his helicopter. He put on his headphones and they lifted off, everything relatively quiet and calm except for a terrifying, ear-shattering *beep!* before every announcement from the pilot.

O

THREE MONTHS EARLIER, WHEN HE'D ARRIVED IN LONDON, Brand had no certainty that he would have what it would take to become a successful consultant, much less a corporate guru. It hadn't helped that Schwartz had picked Brand's arrival week to disappear on holiday for three weeks, leaving him without any guide or tutor in understanding the ins and outs of his giant benefactor. It felt a little bit, he thought to himself, as if he had just set out doing fancy somersaults on a high wire and then noticed he had no net.

On his first day, dressed in the same respectable three-piece suit he had worn on his Sacramento tour of duty for Jerry Brown, he sat in his office in Shell Centre with nothing on his schedule and no particular agenda, asking himself, "Why am I here?"

Eventually, the phone rang, but the two calls he received were merely greetings without assignments. He decided he would lie low with his door closed until lunch, get some reading done. He considered working from home on Tuesday.

Things brightened, however, when he met Arie de Geus.

A tall and charismatic figure, de Geus welcomed Brand by telling him, "It's great to have a buccaneer in our building. We like people who have an entrepreneurial approach to things, and we don't see many of them here. In any case, you're welcome here, and we look forward to working with you."

After several weeks, Brand began to acclimatize to Shell's secretary-heavy corporate culture. It seemed like every attempted contact with the company's executive staff was met with one arrogant secretary after another asking him, "And who are you with?"

Brand had picked an interesting moment to embed himself at the heart of the global fossil fuel giant. At the end of 1985, Saudi Arabia had ceased price supports and begun increasing production to take market share. By the time Brand arrived in London, the price of a barrel of oil had fallen 58 percent and the oil and gas industry had laid off more than a hundred thousand workers around the world.[1]

A Dutch accountant with a deep interest in philosophy and psychology who had joined Shell in 1951, de Geus was an adherent of Kahn's scenario approach to strategic planning. He also introduced Brand to his "planning as learning" philosophy and some simple ideas to implement it, such as regular job rotations for executives.

When Schwartz returned, he provided a list of key Shell executives to meet and debrief, and Brand picked his way through the corporate hierarchy, introducing himself widely within the organization. Cathleen Gross, Schwartz's girlfriend, had found Brand and Phelan their quarters, and the two couples became good friends, having frequent weekend adventures around London. On occasion, Brand would carpool into Shell Centre with Schwartz, who lived close by.

When he wasn't engaged in such outreach, he tried to find time to finish the *Media Lab* book. After completing two chapters, he decided that he so hated what he had written, it was hard to even reread. He had no choice but to soldier on, writing in a "Dear Paul" conversational style, framing his writing in the form of discursive notes to his friend Paul Hawken and with the intent of fixing things in later drafts.

At Shell, Brand had been commissioned to organize a series of international learning conferences to bring together a small group of company executives and outside thinkers. The group would meet twice each year and the events were to be modeled after the legendary Macy Conferences, a set of interdisciplinary scholarly conferences that had begun in the early 1940s.

In August, Kelly wrote to apprise Brand of the status of the various Whole Earth projects. They had finished the *Essential Whole Earth Catalog*, scheduled to be published in September, and he told Brand that they would need to sell at least 136,000 copies if they were to see revenue beyond their advance.

The WELL, which was still growing at a modest pace, was "slowly silting up and becoming paralyzed by gridlock," Kelly acknowledged. There had already been an acquisition offer by a New England internet service provider, but it turned out that the tiny company wasn't able to come up with any funding. The entire staff was still working for ten dollars an hour with "no overtime pay, no paid benefits, no paid lunchtime, no paid volleyball, no paid vacation." Despite the financial squeeze, Kelly had decided to give everyone a dollar-per-hour raise, adding, "We 'can't afford it,' but we never will. Raises first, income will come, or we'll find other work."

After returning from his trip to the North Sea at the end of September, Brand hosted Negroponte on a visit to Shell Centre to describe his ideas about the digital future. It did not go well. Usually charismatic, Negroponte was uncharacteristically flat, and in return, the executives had been polite but noncommittal, which led the MIT futurist to write off his hosts as clueless corporate suits.

Brand was caught in the middle, frustrated by Negroponte's haughty stance toward Shell and his somewhat dismissive attitude toward him as well. Why, Brand wondered, was he knocking himself out working nights and weekends on a book that would probably help Negroponte more than himself? For a while he considered just dropping the entire project or instead quickly "hacking" the book out, taking the advance, and using it on "real work."

O

DANICA REMY WAS SITTING AT HOME READING THE *PACIFIC SUN*, a Marin County local paper, when she stumbled across an intriguing classified ad hidden in the paper's back pages: "Hi, I'm looking for someone who lives in Southern Marin, knows computers, wants to travel the world and meet interesting people and who will work from home. Ask for Ann."

From the phone number, she immediately suspected that Stewart Brand must have placed the ad, and she was certain that she was the perfect person for the job. With her mother, who had a dual PhD in statistics and social welfare, she had built an early personal computer kit. The *Whole Earth Catalog* had been part of her childhood, and Remy had also worked for Peter Coyote, introducing him to personal computers.

Remy had dropped out of college to become a ski bum and had supported herself by teaching filmmakers how to use personal computers—everything from scriptwriting to production planning. She sat down and wrote Brand a fan letter, then picked up the phone and called Coyote and told him, "Call Stewart and tell him he should hire me."

Several days later on a Friday at 5:30 p.m., Brand called her and said, "I read your letter and Peter called me and said I should talk to you."

"Great, I would be thrilled to come in and see you," she replied.

"How about tomorrow morning?" he responded.

She was planning a blowout of a party that evening with thirty-five people coming over for daiquiris.

"That's fine, when should I show up?" she said.

The first thing the next morning she staggered in to meet him on the *Mirene*. Seventy-five people had applied for the job and he had narrowed it down to Remy and one other candidate. That evening he called her back and told her the job of running the Learning Conferences was hers, with the first of six planned conferences scheduled in just four months, in May of 1987.

While at Shell the previous summer Brand had set the goal of the conferences: to expand the breadth and depth of what was thought of as learning. Initially, they would be intimate, with a gradually widening focus. "Let them start quite small, quite selective," he wrote, "with the idea, the first ones are a germ structure for the later ones, and that the conference sequence itself is one of learning growth and self-creating structure."

In addition to Shell, both Volvo and AT&T had signed on as sponsors, and from 1987 to 1989 Remy and Brand organized six conferences that brought together a small group of scientists and engineers to meet with the corporate executives. The meetings were organized so that the corporate sponsors sat outside of the circle as observers to the discussion.

Each of the conferences took place in a different location, ranging from the first meeting held at Biosphere 2—an elaborate sealed environment intended to explore what it would be like to live in outer space, which was being constructed outside of Tucson—to Hillis's Thinking Machines Corporation in Cambridge. Others were held in places like Sweden, Big Sur, and Costa Rica. Each of the conferences was organized around several "learning journeys" selected by Brand and Remy— essentially brief field trips to visit interesting places and organizations.

There were also activities like paintball sessions and improv theater for the participants. From the airfare to the accommodations, everything was first class.

The overarching theme remained developing a deeper understanding of how companies learn. In between the conferences, the members would stay in touch via a private online conference that was hosted on the WELL. Also, Brand would regularly send books that he thought shed light on questions of organizational learning along with his comments— an impromptu book club for the participants.

Sixteen invitees showed up for the first meeting, among them de Geus; Bo Ekman of Volvo; Hillis, Minsky, and Papert from MIT; Peter Schwartz, who had recently left Shell to move back to California, and his wife, Cathleen; Mary Catherine Bateson (Gregory's daughter), an anthropologist; and naturalist and future *Whole Earth Review* editor Peter Warshall. Several others, including Alan Kay, were invited but couldn't attend.

Brand introduced the event by describing it as a "fishing expedition." They were not striving to build a "learning-bomb," at least not at the outset. "We're trolling for insights," searching within one another's current work and interests.

By now, early in 1987, he had submitted the *Media Lab* manuscript. The breakthrough had been when he decided to confront Negroponte, emailing him a note that said bluntly, "I need encouragement." With cooperation, everything finally clicked.

Later that year the success of his book would result in a new stream of offers to expound publicly as an invited speaker on the coming digital "revolution" that was increasingly captivating the world. While he was physically present just across the courtyard from the *Quarterly,* and still a member of the Point Foundation board, it had been almost two decades of editing and managing, and he realized it was time to move on. Now he had the freedom to think about what was next and to bond with his son, Noah, who was then ten years old. During the summer the two of them drove together from the Detroit airport to Higgins Lake, the first of several summers spent together swimming in the warm, clear

waters Brand had adored as a child. He was touched that his son was
taking to the world where he had grown up, but he still found it a place
that he was glad he had escaped from.

On long hikes in the Marin Headlands, journal in hand, he would
daydream about how to reinvent himself. Phelan told him that his minor
"angst" was simply the "postpartum" feeling that was normal in the
wake of his book project. She was right, he acknowledged, but he found
it difficult to shake his continuing irritation with Negroponte.

What can I do that lasts? he wondered.

He was two years from fifty, and he decided that he would have
thirty more years of effective work, albeit "waning" in the last ten (or
even fifteen) of those years. His passion for understanding learning was
at the front of his mind, followed by an interest in communication technology or "ecologies," as he thought of them. He jotted down a simple list
of possibilities—discover additional "Benign Genres," invent useful technologies, make art that reverberates, live in an exemplary way, stop or
flag something bad, search for valuable principles, invent good games,
start conversations, create a new type of public library. Underlying it all
was the recursive curiosity that had been behind all of Brand's enter-
prises: What questions should we ask so that they would keep asking
themselves usefully?

The idea of writing a book with Paul Hawken on "how businesses
learn" crossed his mind, but most of all he thought more frequently
about pursuing something that he had observed while visiting the Media
Lab. Despite being a visual jewel, the new Media Laboratory building
had proved to be an imperfect workspace. In contrast, Brand had
learned from people like Hillis and Minsky that one of the most loved
buildings on the MIT campus was the one known as Building 20, so innocuous that it merely had a number, not a name. However, Building
20 was attractive precisely because it could be continually and easily re-
fashioned for new uses. Built as a temporary structure during World
War II, with its wooden beams, planked floors, leaky windows, and
ground-level concrete floor, elegance was out of the question. Yet its
warehouselike nature and the fact that MIT never removed its "tempo-

rary" status meant its inhabitants felt no compunctions about moving walls, rewiring as needed, refashioning office spaces, and more or less doing what they wanted to the space to make it more workable at a given moment.

Out of that came Brand's celebration of "low road" architecture. "We shape our buildings; then they shape us," he jotted in his journal in May, quoting Winston Churchill. Then he added his own coda: "Then we shape them." This was the opposite of Pei's Media Lab building, an architectural approach Brand described as "form follows funding."

From these thoughts would come (after many years of research and writing) *How Buildings Learn*, a book that would be widely praised both as a critique of the architectural profession and for a deeper theory of how different aspects of cultures and societies evolve at different rates, which Brand later named pace layering.

At the same time as Brand was contemplating his future, Phelan was starting *Phelan's Equestrian Catalog* (no shock that a Brand would choose that format), intended to be sort of a cross between Smith & Hawken and Patagonia, the outdoor equipment maker. Phelan's catalog did well and won design prizes, but ultimately a recession and a dysfunctional board of directors undercut her hard work, and after five years she sold the business for "a song."

For Brand, the Learning Conference had been an opportunity to move closer to a corporate world he had been remote from in the sixties and seventies. And while no general interdisciplinary theory of learning emerged from the gatherings, they did serve to launch the Global Business Network, which Brand would cofound with Schwartz and several others, as well as provide a springboard for Schwartz's book *The Art of the Long View*.

The planning for GBN began at the first Learning Conference meeting, where Schwartz had discussed how much they should pay themselves—he thought a $250,000 salary each was about right. Brand's reaction was "Well that's an awful lot," and Schwartz responded, "Well, I have an expensive wife."

Brand, who did not have an expensive wife, considered it for a few moments and then decided that "getting that kind of money sounded like a swell idea."

Later, when Schwartz actually offered him a position, which would have meant commuting to work and spending each day in an office cubicle coordinating people and projects, he felt repulsed.

○

THE JAPANESE PUBLICATION OF *THE MEDIA LAB* TOOK BRAND AND Negroponte to Tokyo in April of 1988. In an unusual arrangement, Brand's contract with the Japanese publisher had given Negroponte the lion's share of the royalties, 65 percent versus 35 percent. (Negroponte had brokered the deal and Brand had acquiesced because Negroponte, with his deep connections, "owned" Japan as far as Brand was concerned.)

It was a quick trip, but it renewed his fascination with Japanese culture and architecture. On the book tour, he stayed at the Akasaka Prince Hotel, a hypermodern aluminum and glass fortress adjacent to the traditional Kitashirakawa Palace. After checking in, he took off on an hourlong stroll through the neighborhood feeling alternately reverential and nervous. He was in awe of Japan's material culture; it seemed as if everything fit together perfectly, refined Japanese craftsmanship evident even in humble modern buildings. Japan, it seemed, had succeeded in combining Venice's two different dominances—trade and crafts—simultaneously.

Although he had left the *Whole Earth Software Catalog* behind, he remained immersed in the computing world. Apple had struck a deal with the Point Foundation to develop a CD-ROM version of the *Whole Earth Catalog*. Apple financed the development of the project, which began in earnest in July 1987, based on its HyperCard technology, an early effort to implement the hypertext ideas of Engelbart and Ted Nelson, an iconoclastic computer philosopher who was Brand's neighbor on the South 40 Dock—ideas that would eventually lead to the World Wide Web. The digitized *Catalog* was delivered to personal computer users via then brand-new CD-ROM optical disk technology. The computer game publisher Broderbund had agreed to publish the digital version of the *Catalog*, but it would prove to be a conspicuous failure—largely because CD-ROM players were still selling for $500, ensuring that the market for digital content remained minuscule.

Brand's office was just down the street from Autodesk, a software company whose principal product was AutoCAD, the most popular computer drafting program for PCs. Flush with revenue and with an eccentric founder, John Walker, the company was funding forays into new technologies, including Nelson's hypertext and virtual reality. Brand was a frequent visitor, and in early 1989 he listened to a talk given by Lawrence Livermore National Laboratory's Russell Brand (no relation), a gaunt and tousle-haired computer hacker and artificial intelligence researcher; a few months earlier he had been one of an elite cadre of computer systems operators who had fought off the first internet worm, launched by Robert Tappan Morris, the son of the chief scientist of the National Security Agency's National Computer Security Center.

It was the moment when the nation was first alerted simultaneously to the power and potential threat of computer networks. Meant to be a harmless "Kilroy was here" bit of electronic graffiti, because of a small programming error the Morris worm tore through the nation's then brand-new internet, initially raising fears of a foreign invasion or a cyberattack—although that word had not yet been coined.

Russell Brand warned that the Morris worm was just the tip of the iceberg. After the talk, John Walker, AutoCAD's CEO, suggested that the world was in a brief golden period between the first warning and a real catastrophe. But Stewart Brand decided that while the legal system moved glacially, technology was moving at light speed and that the "crackers" (as he referred to bad guys, to distinguish them from "white hat" hackers) wouldn't easily have the upper hand.

He was far from the only one to belittle the threat. The warning signs about the dark underside of the new online world were everywhere, but there was also a gold rush mentality that, like actual gold rushes in the past, fixated on the ore and not the hazards of extraction.

"There are fortunes to be made here," one Autodesk technologist enthused to Brand on one of his visits. "It's totally virgin territory."

Soon afterward, Brand listened to an Autodesk talk by a young computer hacker named Jaron Lanier. Lanier, whom Brand had met during the first Learning Conference at Biosphere 2, was a computer scientist who was pushing the boundaries of what the technology could do. (Several years earlier, in a review of one of Lanier's games in the *Whole Earth Software Catalog*, Art Kleiner said, "If this were still the psychedelic era, every game would be like *Moondust*.") His inventions would make him one of the pioneers of virtual reality technology, designed to create exotic simulated computer worlds experienced through computerized headsets.

It had been three decades since Brand had watched two young computer hackers playing *Spacewar!* in the Stanford computing center where he had his first inkling of what William Gibson later dubbed "cyberspace." Now Lanier was tutoring him on what that new world would mean when it arrived.

For Lanier, who had grown up on the *Whole Earth Catalog*, virtual reality was going to be the biggest thing since mankind had landed on the moon. Riffing on the opening line of Brand's original *Catalog*, he argued, "We're actually frustrated by our inability to be gods and change the world. That's one fundamental compromise of human life and virtual reality sort of fixes it."

Brand was impressed by the demonstrations, but he also picked up niggling concerns that the enthusiasts were getting ahead of themselves. "Look! We can pick up books," one newcomer to the virtual reality worlds exclaimed. To which Ted Nelson responded, "That's nice, but can you read them?"

SITTING IN PETER SCHWARTZ'S BERKELEY LIVING ROOM, JOHN Brockman discovered a Stewart Brand he had never seen before.

Despite the commercial failure of the *Whole Earth Software Catalog*, Brockman's literary agency was booming. Beyond the book deals, Brockman had long played the role of impresario, and that continued after he became a book agent. In 1981 he created a salon dubbed the "Reality Club," where he hosted informal talks attended mostly by New York City–based writers, scientists, and intellectuals.

On occasion, he would take the Reality Club lectures on the road, and in the fall of 1988, with Fritjof Capra, a client and a physicist who had written *The Tao of Physics*, he organized a talk by Susan Griffin, a prominent feminist scholar, at Schwartz's home.

Brand was there along with the inner circle of the Global Business Network, which had recently been launched with a small group of corporate clients, each of whom paid an annual $35,000 membership fee.

The idea underlying GBN was to bring together a diverse group of out-of-the-box thinkers to work with corporate and government clients who would join the "network" as paying clients. Schwartz had tapped Jay Ogilvy—who previously ran the Values and Lifestyles program, a marketing research group at SRI that had separated from Stanford in the face of the antiwar movement in 1970 (it was renamed SRI International in 1977)—to join the GBN effort along with Brand, Napier Collyns, and Lawrence Wilkinson, a television producer and media executive.

One evening Schwartz invited his four friends to his home, and they sat around a pool table in his basement while he outlined his ideas on a flip chart. He had come up with the name the day before and his idea was that they would have three interlocking businesses around consulting, a network of experts, and producing media to convey business ideas. The first two ideas would work well, while the third one completely failed in the face of the deluge of free content that came with the internet.

While GBN's public reputation would be as a consulting firm that facilitated scenario planning, it rose to prominence based on several big ideas: that globalization would change everything, computer networks would change everything, and the internet would change everything.

As the internet gradually wove its way into every nook and cranny of modern life, GBN prospered as a forum for clients who were grappling with the changes brought by the "networked" economy. Ultimately, the consulting group would build a network of "remarkable people," roughly one hundred big-name scholars, poets, consultants, musicians, technologists, activists, and science fiction writers who could be brought together to help clients speculate about alternative futures.

Between 1988 and 2000, the firm would grow rapidly and become synonymous with the emerging Silicon Valley perspective that the world's problems could be tackled with technology-centric solutions. Schwartz prided himself in saying that he didn't make predictions but rather facilitated the development of scenarios that made it possible for corporate and government decision makers to think more broadly while they were planning. Nevertheless, he became known for the idea of a sustained economic "long boom" resulting from the convergence of four powerful technology trends—computing, communications, biotechnology, and nanotechnology—that would supposedly last until 2020. Originally appearing in 1997 as a *Wired* article[2] written with Peter Leyden, Schwartz's idea of a long boom took book form in 1999, just months before the dot-com bubble burst in early 2000.

The collapse hit particularly close to home. Discounting the likelihood of a variety of "black swan" events, the GBN executives minimized not only the probability of global investment bubbles and leveraged debt obligations but also the risks to the stability of their own businesses. Although it had grown rapidly, during the downturn that followed, GBN itself was forced into the arms of Monitor, a rival consulting firm.

Before that and for a number of years afterward, Stewart Brand was able to enjoy a part-time relationship with GBN. He was paid a half-time salary, roughly $125,000 annually, serving as an elder statesman for the consulting practice and curating a book club, patterned after the one he had created for the Learning Conference, which shipped two books monthly to GBN Worldview members.

He joined frequently in the scenario-planning sessions but was able to avoid the daily schlep to the GBN office in Emeryville. He was never closely involved in day-to-day business activities but would be deployed as needed in various meetings and presentations. It was an ideal situation, affording the freedom to pursue books and other projects that caught his interest.

Brockman had known Brand as he had risen to become the icon of the counterculture. On that early evening in 1988, however, he felt as if he were looking at a changed man. Griffin was an anticorporate radical and Capra was a deep ecologist. The GBN consultants, on the other hand, were completely at home with their clients tossing around corporate nicknames like Big Blue and Manny Hanny. The GBNers spoke in an inside-baseball patois, and Brockman watched as Brand chuckled warmly, breathing it all in. Brockman felt bewildered by the familial warmth Brand showed toward what he thought Brand in the past had seen as a faceless and antihuman corporate world. In Brockman's eyes, Brand still affected an off-grid vibe; his custom-sheathed knife was omnipresent on his hip, certainly not typical dress wear among the GBN crowd, but instead of a signifier, it suddenly seemed like a barnacle.

As GBN grew, Brand's transformation became more evident to many of those who had known him in the days of tepees and Garnerville. Though now a successful movie actor, writer, and documentary film narrator, Peter Coyote remained a committed activist who was deeply suspicious of both technology and corporate power. Not long after he was added to the list of GBN's "remarkable" network of intellectuals, he asked to be removed. Although the network members were each given a $2,500 annual sinecure, Coyote decided he was simply being used as window dressing to sell corporate clients expensive high-concept ideas.

Brand took another view of their dispute. He had concluded that most of what the sixties counterculture advocated was, and remained, wrong. In his journal he recounted what he thought had failed: "Drugs, communes, spiritual practice, New Left politics, solar water heaters, domes, small farms, free schools, free sex, on and on. Right about a few things too, and for us for proposing them, but my bad."

Brand was headed yet again in a new direction, on an arc that would culminate seven years later with a *Fortune* magazine profile titled "The Electric Kool-Aid Management Consultant."

Occasionally it created situations in which he felt he was a strange bedfellow. When Nils Gilman, a young historian with dot-com entrepreneurial experience, joined GBN, his first scenario-planning exercise took place in San Francisco with a small group of CIA officials led by a man who introduced himself only as George—which Gilman assumed was not his real name. Both Brand and Schwartz were part of the planning exercise, and Brand introduced himself by saying that he had organized the Trips Festival in the 1960s. Then George said that he was one of the longest surviving agency employees involved in drug interdiction programs and that he was burned out, suffering what he called "policy fatigue." Brand quickly corrected the record to point out that while he had had a lot of experience with recreational drugs, LSD had been legal in January of 1966, and that he hadn't taken any since 1969, in any case. With impish glee, Schwartz chimed in that he could not make a similar claim.

T HE FOURTH MEETING OF THE GBN NETWORK WAS HELD AT AU- todesk in September of 1989. The theme for the all-day meeting was "People in the Nineties." Significantly, cyberspace did not take center stage; it was relegated to an add-on "learning journey" in the tradition of the field trips that Brand had offered as part of each of the Learning Conferences.

Opening the morning in a large Autodesk conference room, Schwartz began the day by saying, "We put remarkable people together in intellectually stimulating circumstances, present them with a few provocative views of what's new, and see what kind of insights emerge. So far it seems to be working well."[3]

He then introduced Brand, who, having quickly gone from student to tutor, described the virtual reality demonstrations that would be available to explore during the day. "Keep in mind that we're seeing a technology-in-progress," he warned. "One point to keep in mind is how rapidly a severe shift in how we use communication technology can happen. You will be flying a Wright Brothers contraption. Think about what a 747 version might be like."[4]

For the rest of the day, however, it was striking how completely cyberspace remained off the table as a significant factor in reshaping global culture and politics. Instead, the discussion stayed decidedly predigital. The Worldview members were treated to an exploration of global politics and culture led by GBN network members like writer and activist David Harris (who years before had criticized Brand's Slaughter game) and Brand's old acquaintances Paul Hawken and Orville Schell. At the end of the day, Brand and Rusty Schweickart led the GBN members on a decidedly physical-world tour of the Sausalito houseboat community.

Brand had not been able to shake the idea of a book on the new world of digital information. In 1971, Abbie Hoffman had written *Steal This Book*, an irreverent countercultural attack on the capitalist system. Hoffman died in April of 1989, and that fall Brand toyed with the idea of naming his project about exploring the world of digital information *Copy This Book* in Hoffman's honor. In the end, he chose *Outlaws, Musicians, Lovers & Spies: The Future of Control.*

JOHN BROCKMAN HAD PERFECTED THE ART OF BIG MONEY BOOK auctions, and he circulated Brand's proposal to twenty-two New York publishing houses. He got offers from just two, ultimately obtaining a $100,000 advance. Though it sounded like a healthy sum, the contract, per publishing industry norm, paid half of the $100,000 up-front and the second half upon publication. After taking out Brockman's agent fee, taxes, and basic expenses, it meant that the one or two years of writing and researching might pay less than $25,000.

Nevertheless, it was an alluring project. "Information wants to be free" continued to grow in influence as a rallying cry for the new digerati. Moreover, everywhere he turned, from virtual reality to computer hackers in both black and white hats and the emerging internet, digitized information was becoming a defining force. He accepted the offer.

Then immediately he decided that he had made a giant mistake. He had always done best when he was out on the edge, pursuing something that nobody else saw. Now he realized that while he had been early to see the advent of a digital world, he was no longer alone. Lots of people he knew were working on their own books on the subject.

Worse, his old phobias had returned. Once again he found himself afraid of heights and spooked while driving over bridges. It was a sure sign that something was amiss. He struggled through a difficult trip to New York, where he made a daylong presentation at a conference on multimedia in Albany. He was beset by travel fatigue and, worse, subject fatigue: some of the allure that computing had held for him was waning.

The next day, as he headed to a GBN presentation in New York City on the subject of global finance, he felt the same disinterest plus the anxiety that still plagued him when he traveled. On the train into the city, he contemplated his future. He had heard rumblings that he was being considered by the MacArthur Fellows Program (known for the genius grant). When the awards were made and he was not one of the winners, he had been crestfallen. It set him thinking: "Well, if I did get a McArthur [sic], what would I do with the money? I don't want to do another Brockman-inspired technology book. I want to do the book about buildings that got me really worked up." He called Brockman and told him he was returning the advance and going to work on the buildings book.

In the end, his original proposal would appear in the *Whole Earth Review* as an article in the summer of 1990, where he confessed at the outset that he had "chickened out" on the project, adding that even if it had been published, it would have had little impact. He added that he was handing his ideas off to Kevin Kelly, who would publish *Out of Control: The New Biology of Machines, Social Systems, and the Economic World* in 1994—the same year that Brand would finally publish the book that was closer to his heart.

After walking away from *Outlaws, Musicians, Lovers & Spies*, Brand found that his work-avoidance neurosis had evaporated, and he charged ahead. It reminded him of when he had dropped out of army ranger training and quickly moved on to get his Army Airborne School badge in 1960. It put him in a state of euphoria, and he penned the first of a series of proposed opening lines for *How Buildings Learn*: "In the real life of buildings permanent is temporary and temporary is permanent."

○

NOT LONG AFTER LAUNCHING THE GLOBAL BUSINESS NETWORK, Lawrence Wilkinson had lunch with Brand in Sausalito and learned about his new book project focused on the difference between "low road" and "high road architecture."

Brand was already framing his research in terms of his "questions that ask questions" learning metaphor. He had outlined his book in a letter to Sim van der Ryn in the summer of 1987. His early working title was *How Buildings Learn or Fail to Learn: The Strategies of Bad Design*. The title, he wrote, reflected his larger agenda to explore learning in complex systems, and the book was part of his larger, more theoretical quest.

When the idea first came to him, Brand realized immediately that he had stumbled onto a concept—buildings over time—that he would have to himself. Indeed, Hillis convinced Brand that while others could write books on the environment or the emerging digital world, if he didn't pursue his buildings idea, no one else would.

Still, for a while, he vacillated on committing to the project. He contemplated shrinking *How Buildings Learn* to an essay and moving on. Other potential projects constantly distracted him. "Information wants to be free"—not what he had originally said—had rapidly become a meme, and he thought about a book on the economics of information. Simultaneously he mused about another book idea, *The Restoration*, focusing on how to fix the world's damaged places coupled with a checklist of what the environmental movement had gotten right and wrong. His fictional Venice trilogy continued to haunt him.

Hillis had suggested that rather than try to shortcut the project, he might instead supplement it with something artistic—something like his quest for a photograph of the whole Earth, or perhaps a photographic exhibit in the tradition of the 1964 Museum of Modern Art photography exhibition "Architecture Without Architects."

Instead, Brand decided to teach the subject in the form of a seminar at UC Berkeley, titled How Buildings Learn. He had just eight students and he quickly learned that he was not a compelling teacher, though his failings were compensated for by world-class guests. Christopher Alexander, Peter Calthorpe, and Sim van der Ryn all gave lectures. When he set out, he thought that the seminar might evolve into a position at the UC Berkeley College of Environmental Design, but that didn't materialize, which at least left him with more time to write.

"Do you want to see where I'm going to do this?" he asked Wilkinson at the end of their lunch. The two men walked over to Gate 5 Road, where Wilkinson expected to see some cubbyhole office aboard the tugboat *Mirene*. Instead, they stopped before they reached the dock at a long white shipping container set up on cinder blocks. Brand's landlord had once had big ideas for containers and had acquired a large number with the notion of some kind of avant-garde construction project, but after giving up on the project, he had instead lined them up as a public storage facility.

Brand had another idea. Peeking inside the eight-by-eight-by-forty-foot shipping container, Wilkinson discovered that it was completely retrofitted as a work space. There were plywood shelves everywhere and an old couch and long desk-height working surfaces to lay out images, with hanging electric lights and strategically placed task lights, as well as a hole cut in the ceiling topped with a stovepipe with a vent sawed into the wood floor to create a simple ventilation system. The roof had been painted white to keep the interior cool.

The book he had in mind was an illustrated one, and he had created an office that let him place images on the walls held in place with magnets, making it possible to see each chapter as a visual whole, with cards of the associated text. It was a work space where he could lay things out, walk away from them for two months, and come back and they'd be exactly where they had been left—under a fine layer of dust. The whole thing was an extension of his *Whole Earth Catalog* production days. He wanted to produce a camera-ready product in which he completely controlled the appearance as well as the content, delivering a finished product to the publisher. When he was done, he could simply put a padlock on the container and have a complete archive of the project.

When Brand had described his new work space to his agent, Brockman had replied that he thought the idea was creepy and that he couldn't conceive of working in a library without windows, to which Brand replied, "The library is a window."

O

ON OCTOBER 17, 1989, BRAND WAS IN SAN FRANCISCO ON HIS WAY back to Sausalito, driving by the Marina Safeway, when he felt as if he simultaneously had four flat tires. He slowed down and heard car alarms going and saw mud squirting out of the ground and then a large cloud of dust hanging over the Marina neighborhood.

He was in the midst of the magnitude 6.9 Loma Prieta earthquake. The Marina, which was home to a young and affluent community, was the hardest hit neighborhood in San Francisco, with more than sixty buildings damaged or destroyed. Four people died, four buildings were destroyed by fire, and seven buildings collapsed.

Afterward, Brand would write about what had happened in a *Chronicle* series[5] as well as a longer piece in the *Whole Earth Review*.[6] Both articles focused on the mistakes that were made by rescuers and the lessons that needed to be learned. He wrote about having the right tools in the aftermath of an earthquake, pointedly noting that the professionals needed the army of volunteers who were prepared to help.

In the moments after the quake hit, he parked his car and walked toward the rising column of dust. He found he was of three minds: one was full of pure feline curiosity, the second was garden variety altruism, and playing in the very back of his mind was a scene from a Jimmy Stewart movie, with him as "the alert good guy walking up the street toward trouble."[7]

Soon he came upon two four-story apartment buildings that had collapsed to half their height. With an off-duty police sergeant, he explored one of the buildings, climbing over splintered walls, aware of the stench of leaking gas. As he poked into collapsed rooms it felt frighteningly silent. Later he would chastise himself for not calling out—a woman had been trapped ten feet below him and might have been located if she'd been aware there was someone above.

He would write about the horror of watching a building ablaze with people trapped inside. He interviewed the husband of Janet Ray, who'd been unable to escape. William Ray was convinced that the disorganization of the response and lack of water for firefighters had led to the death of his wife.

In his articles Brand offered a clinical analysis of what went right and wrong. But he also admitted that his experience as a volunteer rescuer left him with another feeling—he was an adrenaline junky, and in his quiet everyday life he missed the drama.

THAT FALL, BRAND WAS INVITED TO JOIN THE BOARD OF TRUSTees of the Santa Fe Institute. Rooted in physics and founded by a group of scientists at the Los Alamos National Laboratory, it had tentacles and interests that spread throughout both the physical and social sciences and had a growing reputation for being at the forefront of several emerging disciplines, most notably complexity theory.

The offer had come from George Cowan, a chemist who had been involved in the Manhattan Project and who had founded the institute. Cowan had picked up Brand on his radar and came to visit in Sausalito to personally make the pitch. Brand, already completely awestruck by the institute, responded immediately that if he had to choose between a British knighthood and a board position at the institute, he would be torn, and since there would be no knighthood, he would be honored.

His focus remained on *How Buildings Learn*. He read *From Bauhaus to Our House,* Tom Wolfe's critique of modern architecture, and decided he was a fan, noting in his journal: "Tom Wolfe took his customary knitting needle, rammed it up the nostril of high-style architecture, hoping to engage the brain—and found no brain at all. Just face."

He threaded his research through the constant interruptions of traveling to speak and serving as a Global Business Network consultant. His goal was to turn architecture if not on its head, at least on its side. The design of buildings, he believed, was something that should happen across time rather than as an artistic snapshot frozen in time. From the outset, architects and their profession were in his crosshairs. "Frank Lloyd Wright houses are aesthetically paralyzed forever," he wrote in his journal one day. "His commercial buildings get demolished." He spoke with Peter Calthorpe and Sim van der Ryn as well as Christopher Alexander and Jane Jacobs. He also went back to his *Whole Earth Catalog* roots and interviewed J. Baldwin and Lloyd Kahn and went back even further to visit with his early mentor and boss Gordon Ashby.

He also formed a small brain trust of people that he could use as a sounding board as he played with the ideas for the structure and the content of the book. (Later he would formalize this approach, maintaining a small circle of intellectual fellow travelers he would refer to as his "guild.") People like Hillis and William Rawn, a Boston architect who specialized in designing for reuse, became part of a wide-ranging conversation mostly carried out by email. Nobody, however, was as valuable as Brian Eno.

Brand had discovered Eno when the musician appeared at the San Francisco Exploratorium in 1988. During a lecture, Eno mentioned how important Alexander's *A Pattern Language* was to his thinking, and afterward Brand tried to reach him backstage to tell him he not only shared a passion for Alexander but could introduce the two. Guards kept him away, however, and it wasn't until a year later that they would meet at a Global Business Network meeting. They quickly formed an intense friendship. Beyond their personal bond, Eno became Brand's role model.

Eno's career was almost as shapeshifting as Brand's. He had begun his career playing synthesizers in the English glam rock band Roxy Music but had later become a pioneer of ambient music and the producer of several landmark records by other artists, perhaps the most legendary being *Remain in Light* by Talking Heads. Eno had also long been interested in multimedia, setting up installations that combined film, text, and music.

Their bonding led to a prolific email correspondence (easier because they were on separate continents), initially carried out on the WELL. Years later Eno would abstract more than a hundred thousand words of just his side of their conversation in a single year for his memoir, *A Year with Swollen Appendices: Brian Eno's Diary*.[8] That, however, would scarcely touch the surface: when Brand archived their email during the 1990s, the printout totaled several thousand pages.

After one shared Bay Area dinner, he decided that he wanted to work like Eno—"hungrily and artfully." In particular they shared a common aesthetic. It was Eno who pointed out that in British country homes, the servants' quarters were more adaptable and human scale than the impressive but overbuilt manors themselves.

In writing about buildings, Brand decided to adopt the voice of a biologist. He wanted to convey the idea that buildings are analogous to living things. In his mind, good architecture would convey the "vernacular" life of a structure after it was built. Building design should evoke an ant colony or a termite nest, built from the bottom up and continuously evolving. It was an idea that had its roots in the same do-it-yourself, grow-from-scratch philosophy with which he had started the *Whole Earth Catalog* several decades earlier.

His neighbor Peter Calthorpe had convinced him that great architecture was not art—which was not to say the result couldn't be beautiful. The Bradbury Building that had originally been built in 1893 in Los Angeles, designed in a manner that combined spectacular craftsmanship (particularly ironwork in this instance), with visible stairwells around a dramatically lit center space, was diametrically opposed to what Brand had seen at I. M. Pei's Media Lab building, where the open space was arranged to obstruct casual interaction, while prioritizing privacy made for isolation and sterility. The Bradbury's celebration of transparency was also stunningly unlike what Pei had aimed for; the physical workings of his building were secreted from view, and—concealed and ignored—tended to be much more prone to failure. The Bradbury Build-

ing was a living organism; the Media Lab, a lifeless one. The Bradbury
had been revived several times, including an eerie back-to-the-future
cameo in the movie *Blade Runner.*

The sort of building Brand favored was one, like MIT's Building 20,
that shaped and reshaped itself over time. Brand saw that sort of practice all over, but not among modernist architects and their clients. With
them, formidability took precedence over humanity. The failure of architects and planners to look beyond the moment, to consider what
would happen, say, once the snow piled up on the roof (the Frank Lloyd
Wright example)—this was not so different from what Brand had seen
in the army, what he had seen in the American government's policies toward the natural world, what he had seen in the Sausalito evictions, and
on and on: the triumph of inefficiency disguised as proficiency, the demand for automation at the price of autonomy. Modernism put the building first and residents last.

Although it would be mentioned only in passing in his book, his own
Gate 5 Road community was his touchstone. When he traveled to his
friends' "nicer" homes, he would find them walled in like sterile deserts
by zoning restrictions, "stuck in a place where nothing ever changes."[9]
The neighborhood immediately around his office included lots of recycled buildings, and because it sat between county and city jurisdictions,
planning codes were fuzzy. Certainly, it fit Calthorpe's New Urbanist
worldview: from Brand's boat, it was a short walk not only to his office but
to everything he needed to live, from a supermarket to a gym to every
kind of supply imaginable, all "scruffy" and neighborly and walkable.

W HEN HE FOUNDED THE WELL, BRAND HAD CREATED SEVERAL
rules to ensure a more convivial online community accountable
for its actions by placing responsibility on each individual. He decreed
that "you own your own words." This was meant to avoid the threat of
libel suits against the WELL, absolving it from liability for any troublesome content. Unfortunately, the impact of YOYOW (as it was called by

many WELL beings) veered badly from his original intent, quickly turning into what Brand would later describe as "copyright insanity, where people thought that their precious words should not be copied in other contexts."[10] He tried to alter the impact by editing his original aphorism to read "You are responsible for your own words," but to little effect.

In 1990 the WELL user Mike Godwin had proposed what became known as Godwin's law: "If an online discussion (regardless of topic or scope) goes on long enough, sooner or later someone will compare someone or something to Adolf Hitler or his deeds, the point at which effectively the discussion or thread often ends." By allowing people to post under "pseuds," Brand had opened the door for users, in some cases their true names only lightly veiled, to write things that they would never dream of saying to someone face-to-face. Personal spats were carried on online, and on frequent occasions the WELL's board had to step in much as a teacher would discipline schoolchildren.

Like many of the early participants in cyberspace, despite seeing the glimmers of the dystopian culture that was spelled out in the cyberpunk science fiction books he devoured, Brand would remain generally optimistic about the digital revolution. It was a sad admission that virtual communities were fundamentally different—often more sterile, anonymous, and less empathetic—than traditional groups in the physical world. The virtual world had some wonderful attributes, but there was a darker aspect as well that was just coming into view. Brand had an early inkling of some of the forces that would radically reshape the world three decades later when the cyberworld the WELL presaged became the world's predominant communications channel. What would drive Brand away from the WELL foreshadowed on a microscale the online culture of trolls, filter bubbles, disinformation, surveillance, and censorship that has come to deeply trouble the entire world in the past decade.

Brand's disaffection was complicated by the fact that the WELL's growth had been strong up until 1991 and then slowed visibly, and it was not clear whether the slowdown was due to the national recession or some other reason. What was clear was that despite grand visions of expansion, the WELL had national competition with much deeper pockets, such as America Online, which already offered email, conferences, and much more than Brand's squad could provide, as well as a much larger online community.

At the same time, it was indisputable that the WELL was feeling growing pains. The system had launched on a Digital Equipment Corporation VAX minicomputer, which it had outgrown. Relatively quickly, however, even its replacement had been overwhelmed, and system performance became a continual complaint.

John Coate, who was managing the WELL, came to the board with a well-thought-out plan for growth. His concept was to compete with AOL, but rather than create a centralized service, his strategy was to do it by federating a large number of community-oriented systems that would each reflect their local community. The board quickly rejected the idea, which would have required a large infusion of capital to grow it at scale.

At the end of 1991, the bickering and customer frustration boiled over in a WELL discussion (online conversations were organized as "topics") named Backstage, which was meant to be a place to discuss the WELL's inner workings. Invited to join the discussion, Brand and Kelly showed up and immediately were set upon by an angry e-mob of WELL users, many blaming the board for the system's shortcomings.

Several weeks later, in January, Brand was still upset about the tone and ferocity of the attack, which he would later characterize as a mugging. When he sat down at his computer to log into the WELL, he found his hands were trembling. He created a new topic titled "System Scapegoat—does The WELL always need one?" He began by describing in his view what had taken place:

> It was a lynch mob looking for an evildoer. I spoke, got misheard, knocked flat, and then piled on. The Board was obviously trying to kill the WELL, went the apparent fantasy, and I was a poor simpleton who needed straightening out fast. I learned not to in-

terrupt a mob delusion with contrary facts—it just makes them madder. . . . I stayed for a few days and then realized in e.e. cummings' poetic words, there is some shit I will not eat. . . .

For the next three days, the WELL community went back and forth about Brand, whether he was to blame and what should be done. The tone of the conversation ran toward what the virtual world called cyberspace had become: sharp-edged, often strident statements full of certitude. Even empathy when it was in evidence felt hollow.

A variety of people, including Howard Rheingold and Kelly, who posted a long description of how he would like to invent "WELL 2.0," came to Brand's defense, but the spirit of the conversation was exemplified by a user who went by the pseud Axon:

> Well, heaven forfend that i should bully saint stewart. i'm afraid that this latest exercise in obfuscation fails to sustain my interest. it's just a lot of handwaving. now that i've been branded a Torturer i suppose i can die happy. my only intent is to help.

After three days of reading and occasionally responding, Brand decided that it wasn't the personalities, it was the medium itself that was failing—online communication was flawed, and he had been unable to find a way to solve it.

Increasingly, he wanted to lose himself in finishing the writing of *How Buildings Learn*, and the WELL had become a huge distraction.

On January 16, 1992, he posted word of his departure:

> Anyway, ahem. <Chink, chink> Thank you. Um . . . this discussion persuades me that I can't be on the WELL Board, these days, and have a life. The choice between those is easy. I'd love to engage in the fascinating problems of civilizing this part of the electronic frontier, but I've made other commitments that have to come first. The hurly-burly of online combat is fun, but it takes too much time, to not much effect. . . .

Four hours later he added a brief footnote:

Just a quick really gone comment. Thanks for the good thoughts.
I'll be around in fun and work conferences, but I've become a liability in any discussion of WELL behavior or business, so I'd
best take my ears as well as my big mouth away. Beaming on out
of here . . . blessings wherever they'll do any good . . .

After he sent the last message, he felt wretched about his decision.
He spoke with Rheingold, who offered several ideas for how to continue
on the board in a more restricted fashion and suggested ways for him
to deal with the mob. For a while he felt better because Rheingold was
writing a book about online communities and it seemed to resonate in
some basic way.

Then he went back and reread some of what had been said and felt
depressed again, torn between giving up on something he cared about
and being rejected by something he had created. In the end, he realized
that there was no way to continue in the role he had been funneled into.

After he finalized his decision, he began to feel a sense of relief. At
least momentarily he felt lonely, and he contemplated inviting friends
to dinner more frequently. In his journal, he noted that it had been a
long time since he had checked in about his emotions. Later, he told *New
Yorker* writer John Seabrook, "I never loved the WELL again, nor fully
trusted its process."[11]

IN NOVEMBER OF 1992, BRAND WENT TO DC SEVERAL DAYS EARLY
to attend a GBN meeting without checking to see exactly when he
needed to arrive. He found himself trapped in a hotel room near the
Pentagon, reading a biography of Darwin, while Phelan was back on the
West Coast at her parents' home suffering from intense back spasms.
Realizing that if he had stayed home he could have tended to her, Brand
felt increasingly guilty and stupid.

It was a terrible year for Phelan, placing new strains on their marriage. Her award-winning *Equestrian Catalog* company was in the process of being undone by its board of directors and their unwillingness
to commit capital to help survive a recession. While the business had
grown from $50,000 in revenue the first year to $3.2 million in its last,
it was not enough. In the end, in early 1993, she would move out of the
warehouse space in Sausalito she had occupied for five years, closing it
down. (Brand referenced the space in *How Buildings Learn* as an example of a business that grew organically in a building originally built for
manufacturing Liberty ships during World War II.)

The year was further complicated by a failed pregnancy, which left
Phelan distraught. Discussions about adoption would prompt the most
stressful period in their relationship, especially because Brand had Noah
and seemed less devoted to trying again. Phelan considered leaving the
relationship but ultimately decided that she cared deeply enough for
Brand to forgo motherhood. (Eventually, they would learn that genetic
issues made a successful pregnancy impossible.)

O

BRAND HAD LONG STRUGGLED WITH BOREDOM—HE STARTED
things but had difficulty staying long. While he wrote *How Buildings
Learn* he was already regularly daydreaming about what he might do for
his next act.

One day, while he lay stretched out in a dentist's chair, listening in-
tently to the radio on headphones while the man above him drilled for
a crown, Stewart Brand began to contemplate the apocalypse. The Rod-
ney King riots had just erupted in Los Angeles, and he felt impatient
with both civilization and himself. What would last? What wouldn't?

Brand had long been fascinated by the role that libraries played as
the pillars of civilization, and they had been frequent subjects in both
the *Whole Earth Catalog* and the *CoEvolution Quarterly*. The favor had been
returned, and now Brand was a frequent speaker at libraries and to groups
of librarians.

Indeed, library fantasies were a frequent escape. During a Caribbean vacation on Nevis, he saw an old stone church with huge beams that had been converted into a public library, and he pictured his own private library in a similar stone building with several large open rooms with high ceilings and a roof with dormers, including a plain ceremonial writing room like something from Machiavelli's era. In the summer of 1990, his mother told him that she was planning to will all of her books to him. He immediately began to draw a detailed sketch of the space he would create to host her books, which he believed included many of the best contemporary works of the past half century. After Phelan shuttered her business, he considered that they might become innkeepers at an "intellectual motel" constructed for GBN as well as his friends, to give them access to his growing collection of books. How, he wondered, might he re-create the conversations that Baker Roshi had fostered at the Zen Center or capture the experience of staying at Chan Chich Lodge in Belize, an eco-tourist retreat that had enchanted him?

A great library would need an equally compelling bar, he decided. Momentarily, that left him nonplused while he tried to figure out what set his idea apart from a yachting club, or the Masonic Lodge for that matter.

His musings would likely have remained fantasies, however, if not for the unexpected email that arrived in Brand's inbox several months after his visit to the dentist. In it, Danny Hillis sketched out his vision of a mechanical clock designed to run for ten thousand years. Hillis, still making the world's fastest supercomputers but increasingly frustrated with society's seemingly ever-shortening time horizon, was searching for some way to encourage humanity to take a long view. He had settled on a clock. Excited, he emailed his friends:

I think it is time for us to start a long-term project that gets people thinking past the mental barrier of the Millennium. I would like to propose a large (think Stonehenge) mechanical clock, powered by seasonal temperature changes. It ticks once a year, bongs once a century, and the cuckoo comes out every millennium.

It would be, Hillis said, the world's slowest computer.

He also reached out beyond his immediate circle, taking bemused reporters on long walks to outline his plan. As it turned out, however, Brand was the only one who responded positively to his scheme.

Hillis had realized that his plan dealt with time only as it is conceptualized in Newtonian physics. But he understood that there was another kind of time, which he described as "evolutionary information" time, and his clock as a stand-alone device wouldn't capture that broader concern. Do you have an idea, he wrote, how I might be able to encompass both aspects of time?

A library, Brand responded.

It was the beginning of the Clock Library project, which, thanks to Brian Eno's naming wizardry, would become the Clock of the Long Now, the project that would become the focus of Brand's life up through the present day.

The roots of the phrase came from Eno's riff on Brand's original question about a photograph of the Earth from space, which gave us "the big here" and changed everything. How would it be possible to deliver a sense of a "long now" and have a similar impact on future generations? he wondered.

Significantly, while Hillis was exploring the idea of designing a clock, it was Brand who realized that creating an organization capable of serving for ten millennia to maintain the clock was just as thorny an engineering challenge.

During the time that he was beginning to think about the clock, Brand kept one foot in the world of the future as well. In November 1993, Brand attended a GBN-led meeting at Science Applications International Corporation, a Reston, Virginia, company with close links to the Pentagon and the intelligence agencies.

The following month the *New York Times* would publish the first news article describing the "World Wide Web,"[12] but there was already a growing buzz about the potential impact of the internet. The previous year Kevin Kelly had worked hard to help the WELL connect to the network, both making it easier to send email globally and making it possible for WELL users to reach the system without paying costly telephone charges. The GBN seminar had the bland title Agenda for Information Systems Seminar and had been pulled together by Andy Marshall, a Pentagon strategist who was the director of something called the Office of Net Assessment, a small think tank inside the giant bureaucracy that was tasked with developing future military strategies. Marshall, who had originally been recruited into the National Security Council by Henry Kissinger during the Nixon administration, was a legendary figure with a reputation for strategic thinking, and he was early to see the impact of computing and computer networks on the military. The seminar brought together a diverse group of military planners who were thinking about the concept of "information warfare" with a group of GBN consultants that included Brand; Hillis; Schwartz; Paul Saffo, a lawyer and geologist who had left practicing law to become a "forecaster" at the Institute for the Future; Esther Dyson, a well-known computer industry pundit; as well as computer scientists Joseph Traub and Tom Malone.

The first Iraq war was still a fresh memory, and there was a consensus that information technology, an area where the United States had an unassailable lead, would be a strategic advantage in the next war. The Pentagon officials talked about the coming era of "global situational awareness" and the development of a worldwide targeting grid. The assumption was that "information dominance" would assure American superiority.

Saffo, who was an inveterate and careful notetaker at meetings, listened as Brand spoke on the second day. He cited cyberpunk writers like William Gibson, who pointed out that technologies often played out in unforeseen ways, asserting that "the street finds its own uses for things," and Vernor Vinge, who had forecast the advent—only decades in the future—of a "singularity" in which machines would outthink humans. Brand warned the group that we were moving from the world of global cities, from a world where citizens had lots of privacy, to the online global village, where we would have no privacy.

After the meeting, Brand thought about why he had felt so energized during the seminar. He realized that he had not gotten out very
much in the past year as he pursued his buildings book. Now he felt a
renewed pull toward software, computing, and the brand-new internet.

Afterward, with Noah, then thirteen, Brand visited Hillis in Brook-
line, where the supercomputer designer had a rambling old home that
he had restored to capture a sense of life during the nineteenth century.
Noah was going through a precocious stand-up comic phase, and so
while Brand and Hillis tried to have a serious conversation about building an organization to care for a clock destined to last for ten millennia,
he frequently interrupted with childish banter.

Hillis had been thinking no further than his passion project, but
both men were fascinated by the story of the rotting beams in College
Hall of New College at Oxford. When the beams needed to be replaced
in the eighteenth century, it was discovered that a fourteenth-century
builder had planted trees in anticipation that they would be needed centuries in the future.

The clock, they decided, would need to be located in a remote location, and some kind of facility would need to be built to house the clock's
caretakers—perhaps, Hillis thought, something akin to a monastery
with buildings designed to last for more than a thousand years. The two
men even fantasized about creating a religion based on the clock and
the idea of long-term thinking to ensure that the device would be maintained across centuries. In Brand's mind, it would begin as a practice,
perhaps like Zen, intended to undertake a variety of long-term ecology
research projects meant to extend beyond a single generation. (The religion notion was quickly shelved after Brand reconsidered the value of
transforming the clock into an object of worship.)

Brand and Hillis continued to discuss possibilities, and the clock
project slowly evolved into a massive mechanical mechanism buried inside a mountain for tracking "deep time." With the desire that it would
become a positive icon, a symbol to inspire societal reflection, they
decided that visiting the clock should require the equivalent of a pilgrimage.

It would be almost four years before Hillis would publicly describe the clock project at a symposium at the Getty Museum in Los Angeles, and another two years before an initial prototype was completed. Drilling began on a mountaintop in Texas on the first ten-thousand-year clock in 2009.

O

PHELAN HAD SUGGESTED BRAND SPEND AN UNINTERRUPTED month finishing *How Buildings Learn*, so he took refuge in a rented cabin in the hills outside of Willits, a Northern California town a two-hour drive north of Sausalito. But even on his own, he found seemingly endless ways to distract himself from the painful task of writing. One day he spent the afternoon working his way along the banks of a creek swollen with spring runoff. When he returned to his cabin, he brewed mango tea and enjoyed a cigar and then had a long phone conversation with Peter Schwartz. That evening he burned a TV dinner and then decided to fry a steak instead. Afterward, he settled down and spent time exploring System 7, the latest version of Apple's operating system for the Macintosh. It was only after he returned to Sausalito several weeks later that he was finally able to finish.

Rather than the light tables, glue, and scissors he'd employed to assemble the *Catalog*, this time he used software that afforded him the ability to control every aspect of the design and content of his book. The finished book would contain 350 photographs of the evolving buildings that he used to make his case, all of his experience in photojournalism paying off in the composition of his images and the elegance of his layouts. (*How Buildings Learn* would also include several uniquely Brandian touches, including the author's address, on the premise that "publishers are ill-equipped to forward mail and authors need contact from readers to make corrections for later printings and to make things happen in the real world.")

After seven years of research and writing, *How Buildings Learn: What Happens After They're Built* was published in 1994. As Christopher Alexander's *A Pattern Language* and Jane Jacobs's *The Death and Life of Great American Cities* had for their respective authors, *How Buildings Learn* established Brand as an independent and original thinker on design, urbanism, and ingenuity. Of all five of his books, this would be the one in which he would take the most pride.

That said, when it was published, the book made much less of a splash than *The Media Lab*. The *San Francisco Chronicle* gave it a prominent review, but overall it received a far more modest reception than his previous book. Those who appreciated it, however, adored it. Soon after *How Buildings Learn* was published, Jacobs wrote to journalist Katherine Fulton, then completing a profile of Brand for the *Los Angeles Times Magazine*:

> The reason I used the word "awed" back in the beginning of this letter is that I so much marveled at Stewart Brand's originality of thought, his common sense raised to the point of wisdom, and his enormous and always specific knowledge that when I closed the book I knew I was reading a classic and probably a work of genius.[13]

Genius or not, Brand soon learned that, unlike in the technology world, there were few hefty speakers' fees available from architects, who proved to be a generally miserly group. Perhaps because unlike in his previous book, he was not extolling the profession but indicting it. Indeed, *How Buildings Learn* found a much more receptive audience among urban planners than architects. It appeared just a year after the formation of the Congress for the New Urbanism, founded by a group of progressive urban planners, including Brand's friend Peter Calthorpe. The book sparked a discussion that led to the idea of what are described as "form-based codes" in the planning world—the idea that rather than designing planning codes for land use, regulations should focus on the relationship between facades and scale and public space to create more livable communities.

At the root of that idea was what Brand had originally discovered when interviewing architect Frank Duffy, who described buildings in terms of "shearing layers" that changed at different rates, which became a concept that Brand would later generalize as pace layering.

In 1990, Brand visited London and interviewed Duffy, an architect who had thought systematically about the rate at which buildings change. Brand knew that over a half century, typically only a third of the cost of a building was spent on its original construction; instead, the vast majority of expenses went to remodeling, maintenance, and operation. Duffy had identified four aspects of building design—services, skins, structure, and site—that changed at different rates as buildings aged.

After *How Buildings Learn* was published, many of the responses Brand received suggested that the layering concept could be generalized—for example, it applied neatly to the world of software and computer systems as well. (Hardware is popularly understood to be "frozen" software, but the resulting distinction is that over time components of a computer system age at a different rate or pace.)

Two years later, visiting London with Brian Eno, Brand would repurpose his layering framework to understand how civilization evolves—or learns—with a hierarchy that changes at varying rates, beginning with the slow (for a natural biological system) and moving up through culture, governance, infrastructure, and commerce to that which changed most quickly (fashion). He had argued that features that emerged in the quickly changing layers were instrumental in how buildings became "wise" over time and eventually loved. With Eno he would argue that each layer constrains the faster changing adjacent one, moderating the rate of change and creating a healthy civilization. In grasping how the concept related more generally to systems, Brand had created an analytical tool that has since been widely adopted in different fields.

Unsurprisingly, *How Buildings Learn* also found enemies in the architectural community, initially preventing its publication in England. In addition to criticizing I. M. Pei, he had written critically about the buildings of other world-famous architects, such as Frank Lloyd Wright and

Richard Rogers, the then recently knighted codesigner of the Pompidou Centre in Paris. Although Brand had included some positive comments about Rogers, he had also noted that putting the plumbing on the outside of a building, no matter how iconic it was, had created a costly maintenance nightmare.

Rogers threatened to sue Brand for libel both in the United States and in England. In the United States, Brand's publisher told Rogers's lawyer to get lost, but British libel laws created a more significant challenge. Brand first tried to reach out to Rogers to compromise, offering to change the language to assuage the architect's ego, but Rogers began shouting at him and the phone call proved futile. Brian Eno, who was a friend of Rogers's, then tried to negotiate on Brand's behalf, to no avail.

Rogers was able to prevail, and Penguin UK initially chose to cancel publication of the book, but not before the kerfuffle showed up in the British press, where Brand stood his ground: "What I was saying is that I liked the look of his buildings—I am kinky for high tech—but putting the insides on the outside is still wrong," Brand was quoted by the *Evening Standard*.[14]

In due course, Brand would have the last word when in 1997 he hosted, wrote, and narrated a six-part BBC documentary series titled *How Buildings Learn*, with a musical score provided by Eno, that ignored Rogers.

# Chapter 10

# Float Upstream

I N THE SPRING OF 1994, STEWART BRAND FOUND HIMSELF BACK on the edge of Stanford University, camping out in an office belonging to Microsoft cofounder Paul Allen. He was there thanks to David Liddle, chairman of the board of the Santa Fe Institute. The two had bonded when they assisted with a group effort to come up with a new mission statement for the institute. When Brand, who was then new to the Santa Fe Institute board, opined, "My experience is that mission statements are treated as a substitute for leadership, and there is no substitute for leadership," Liddle had looked at him with an *aha!* glint and soon began to give him more responsibilities.

Two years earlier, with Allen's backing, Liddle, who was a Xerox PARC veteran, had established Interval Research Corporation, with the goal of reinventing the future of computing with an impact similar to PARC's influence in the 1970s. Interval was near a cluster of former Hewlett-Packard offices in the Stanford Industrial Park in the foothills looking out over Silicon Valley, just a quarter mile from the Xerox laboratory. That PARC was the model was no surprise. By the midnineties it had already become a defining legend in Silicon Valley, due in no small part to Brand's 1972 *Rolling Stone* "SPACEWAR" article, which had featured PARC researchers like Robert Taylor and Alan Kay.

A tall former college basketball player, Liddle had studied electrical engineering at the University of Michigan. Before creating Interval, he had commercialized an early and pricey personal workstation for corporate workers known as the Star and then left Xerox to cofound Metaphor, which developed a similar computer targeted at professionals in corporate marketing departments, which he sold to IBM in 1991.

He had gone to great lengths to shield his researchers from the prying eyes of Silicon Valley technology writers, but he believed that Brand, thanks to his book on the Media Lab, would be a friendly observer, and he offered him an inside look in the form of a once-a-week consulting role, where he would have unfettered access to the researchers.

Brand had recently published an interview with the feminist author Camille Paglia in *Wired*, and Kevin Kelly, who had become the first executive editor of the magazine that would soon become known as the voice of the dot-com era, assigned him a piece on Interval. Later Brand would acknowledge that an article written by a paid consultant was "awkward," but he spent months making the commute to Silicon Valley, interviewing and photographing the scientists and engineers who were engaged in a collective effort to leap past the world of desktop personal computing.

Liddle's strategy was to hire the Valley's best and brightest technologists and give them the freedom to come up with innovative ideas. He referred to his midcareer stars as being "post-managerial" talents and set up a laboratory with the intent of cross-fertilizing the research of almost fifty engineers and scientists. For several months, Brand would show up at the laboratory, sit in on meetings, and then meet with Liddle. He soon discovered, however, that his potential conflict of interest situation was further complicated by the fact that Interval had created a multimedia contest named New Voices and New Visions with Voyager, an innovative CD-ROM publishing company, and *Wired*.

In August he was still finding his way, showing up regularly and taking photographs as well as interviewing the researchers. He quickly found out that the goal of generating another computer revolution might not be as simple as putting a group of smart people together in a lab, which in terms of figuring out what to write left him feeling at sea. He had also come to dread the long commute from Sausalito in the swollen dot-com commuter traffic.

Interval's lofty goal of becoming the next PARC was even more complicated because the ground was shifting underneath its researchers. The web browser had burst on the scene at the end of 1993, and during 1994 and 1995 it was becoming clear that the commercial internet was going to reset the technology world and eventually disrupt entire industries.

In August, Brand's dilemma was abruptly solved when *Wired* ran a caustic profile of Allen, poking fun at the billionaire, describing his Microsoft-generated wealth as accidental and referring to Interval as a "rich-guy toy."[1] Allen didn't say anything, but a furious Liddle killed the *Wired* partnership, including Brand's profile.

That was fine with Brand. He felt as if he had been let off the hook for an article he wasn't sure how to write. Ending his role at Interval allowed him to simultaneously step back from the internet boom that was consuming Silicon Valley.

As a GBN consultant, Brand had a front-row seat to witness the commercial internet boom. Yet while he played the role of futurist for corporate clients, his heart was increasingly elsewhere. "Even churning fields like computing and communications seem grimly predictable," he wrote in his journal.

While all around him his friends and colleagues were becoming obsessed with chips of silicon with clocks that ticked billions of times each second, Brand's attention turned increasingly to the Long Now Clock. "My interest in history the last few years makes current events seem ever less eventful," he wrote. He was searching for something worth doing that would take him in a different direction from the internet gold rush. It mirrored earlier periods in Brand's life when he had pursued new ideas early on and then gotten bored and moved on just as the mainstream was arriving to take part in a fad or trend.

He found particular pleasure in exploring the design work that had once captivated him as a young student at the San Francisco Art Institute. While he had been researching and writing *How Buildings Learn*, he had taken responsibility for guiding the interior design of GBN's new office in Emeryville, across the Bay from San Francisco. In attempting to create a convivial and productive office setting, he borrowed heavily on ideas that he had become familiar with at MIT, where Marvin Minsky had introduced him to the notion of "cave and commons"—small private offices that surrounded a shared common area for informally exchanging ideas.

A second opportunity to work as an architect came from the Santa Fe Institute. Liddle's friendship with Brand had survived the consulting job, and he was an admirer of *How Buildings Learn*. A wealthy donor had given the institute an opportunity to bring in a well-known Chicago architect to remodel the Hurley estate, once the grand twelve-thousand-square-foot family home of General Patrick Hurley, a World War I hero, which would serve as the institute's new quarters, but Liddle instead contracted with Brand to work with a different architect and bring some of the ideas he had developed in his book to life.

Once again borrowing on the cave and commons approach, Brand focused on transforming the Hurley estate to house roughly ninety people around four design points: conversation, community, concentration, and adaptivity. The new campus was considered a resounding success, but it would be his last foray as an architect.

UNTIL HE TURNED FIFTY-SEVEN, BRAND HAD INCREASINGLY BEcome more sedentary. In his thirties and forties he had tried hang gliding and windsurfing, but as he aged, he'd given up running and other fitness activities. That changed abruptly one day when he was out for a hike with Phelan and found he couldn't keep up. Mortified, he decided to fix the problem and joined a band of fitness fanatics who would torture themselves at the mercy of Jim DiRuscio, a forty-one-year-old former Army Airborne School veteran. DeRuscio greeted his clients in the dark at five a.m. and took them through an intense workout three times each week. The workout, known as Hit the Hill, consisted of intense three-hour training sessions based around running up Mount Tamalpais with a fifty-pound pack filled with water bottles strapped to their backs on some days and grueling outdoor weight-lifting sessions on others. Afterward, Brand would frequently fall asleep during GBN meetings, but his new fitness regimen would become a lifelong commitment. At eighty he was still attending CrossFit classes several times each week.

Although he would occasionally think about a new book project growing out of some of the GBN work that he had dubbed "Futurity," the clock and creating an organization to build it remained his central focus. For this purpose, Brand gradually pulled together a band of fellow travelers including Doug Carlson, Kevin Kelly, Peter Schwartz, Paul Saffo, Brian Eno, Esther Dyson, and Mitchell Kapor, who had founded Lotus Development Corporation, a maker of spreadsheet software.

Several years earlier, Kapor and John Perry Barlow (who, in addition to being a Grateful Dead lyricist and a WELL member, was a cattle rancher living in Pinedale, Wyoming) had bonded over the plight of a group of teenagers who had run afoul of the Secret Service for breaking into computers. Several months later, the men met for dinner with Jaron Lanier; Saffo; John Gilmore, a technologist and privacy activist; and Brand. From the dinner came the Electronic Frontier Foundation, dedicated to civil liberties in cyberspace, and Brand was asked to join the board, along with Gilmore, Dyson, and Steve Wozniak. He served on the board for several years, but in the fall of 1994, Brand sent Kapor an email equating the Clock Library to a fragile startup venture that required all of his energies and politely stepped off it.

In July 1995, Brand met with Hillis, who was vacationing at Jacqui Safra's vineyard in Napa Valley. (Safra was an heir to the Syrian Swiss banking family.) The two men sat by the pool brainstorming. Hillis worried that the clock might prove a futile gesture if humanity somehow destroyed itself in the next century. At the time, many computer technologists were concerned that the accelerating pace of computing might lead to a time, described as "the singularity," when a machine superintelligence would surpass human intelligence.

Brand responded that he believed the clock could serve as a counterbalance to rapidly advancing technologies such as semiconductor

chips and nanotechnology. While generally optimistic, later that summer he returned to his family's Lakeside camp at Higgins Lake, where
he came face-to-face with the shortcomings of institutions that became
stagnant over time. Immersion in the Brand clan, with their grudges
and bad history, made him reckon intimately with how institutions that
were more than a generation old could become "dysfunctionally convo-
luted, aggressively introspective and both oblivious and irrelevant to the
rest of the world."

Some of that was particular to his own relationship with his family,
but there was another concern as well. In November when he discussed
the Clock Library during an impromptu session with Kevin Kelly at the
Hackers Conference (still an annual event a decade after they'd founded
it), the Long Clock idea proved to be a hard sell. Many of the hackers
believed that an obsession with the past was lethally stultifying. "Everything that needs to be remembered is being collected and stashed easily accessible" was the consensus. "We are freed to innovate."

The digerati's disdain didn't dissuade Brand—once again, he decided to go where the others weren't.

While he was enjoying his freedom to focus on something other than
the dot-com mania that had seized the nation in 1995, Phelan had a different reaction. After the demise of her *Equestrian Catalog*, her first im-
pulse had been to shun the idea of ever doing another start-up. Then she
saw a web browser and quickly realized that the internet would transform the market for consumer medical information. Planetree, where she
had worked after leaving her job at the *Quarterly*, had created the first
free public medical library in the United States. Now Phelan was certain
the web would transform health care. She went to the Planetree board of
directors and alerted them that medical information would be online in
the future. She then immediately launched an online consumer health
service named Direct Medical Knowledge, and Brand became a non-
voting board member. In just four years she would sell the company to
WebMD, just months before the peak of the dot-com bubble. It would
leave the couple worth millions of dollars and contemplating their future as philanthropists—only briefly, as it would soon turn out.

O

HIS FATHER'S RUSSIAN FUR HAT ALMOST KILLED THE TOUR OF Big Ben.

On a windy January morning in 1995 Brand and Brian Eno had shown up at the clock tower at Westminster for a backstage visit arranged by the new British prime minister, Tony Blair.

Both men were wearing fur hats. Eno was a Russophile, and Brand was garbed in a sinister-looking black cape to match his dad's hat. He had inherited his father's affinity for hats—Bob Brand had kept a large, colorful collection at Higgins Lake—and after his death, Brand had nabbed the Russian hat plus a hooded monk's robe that he had worn at the Demise Party.

As a result, the suspicious Westminster guards grilled both men before letting them join a group of Argentinian students climbing the 334 steps to the top of the clock tower. Eno noticed that Brand knew a lot more about the clock than their guide, who specialized in witty anecdotes rather than a detailed understanding of what was the world's most accurate four-faced clock when it was completed in 1859.

Right before the top of the hour, the guide told them to keep their heads away from the walls—the force of the bell had a way of snapping people's heads back when they were unprepared—and promptly at eleven a.m. Big Ben struck. It was possible to feel the full force of the bell and even revel in it. Brand likened the power of the bell to feeling the impact of a rocket.

It had been a technological marvel in the nineteenth century, "exact to the second" thought to have been unachievable in a tower clock at the time. As a result, Big Ben had become a reference point for Hillis and Brand as they plunged into the intricacies of clock design.

By the autumn of 1995, the Clock Library had an office in the Presidio, the historic former army base with picturesque rows of neat white barracks and officers' quarters, which had recently been transferred to the National Park Service, as well as a bank account and nonprofit status.

(It was also where Brand had been assigned to do his reserve training all those years ago when he'd been allowed early military leave to take classes at the San Francisco Art Institute and San Francisco State.)

Soon they would begin to acquire their first round of financial backers, including Microsoft's Nathan Myhrvold and Amazon founder Jeff Bezos; Shel Kaphan, Amazon's first employee and a former Whole Earth Truck Store worker; Jacqui Safra; and Priceline founder Jay Walker.

Shortly after the Clock Library project was launched, Brand returned to the Black Rock Desert to attend the annual Burning Man festival, which since 1990 had been held annually on the Black Rock playa, which he had first visited in the early sixties. He came away momentarily certain that he had identified a mountain nearby that would one day house both the library and the clock. At roughly the same time, Hillis convinced him that while they might use twenty-first-century tools like computers and sophisticated machine tools to build the clock, maintaining it should not require any advanced technology.

By the following summer there was enough momentum that the group decided that in addition to a clock hidden somewhere in the desert, there should be an urban wing in San Francisco that would offer higher visibility for the project. Eventually, they moved their headquarters from the Presidio to nearby Fort Mason, another converted military outpost. They settled on the Long Now Foundation, given concerns that if they called it simply Long Now Foundation, it might be assumed that it was the family foundation of a Chinese gentleman.

Early on, the Long Now board met several times at Doug Carlston's Aspen home. Alexander Rose, who had grown up on the Sausalito waterfront and known Brand since childhood, had graduated from Carnegie Mellon with a degree in industrial design and returned to the Bay Area. Brand had described the Clock Library idea to Rose, but he also got him a number of job interviews at dot-com start-up firms. Rose went to the interviews, but he couldn't stop thinking about the idea of building the clock. He was determined to find something else to do besides designing more "plastic toasters" for a living, and he came back to Brand and told him he wanted to work on the clock.

The first board retreat at Carlston's house was scheduled in just a couple of weeks, so Brand invited Rose to join the group, where he met Hillis and spent the next four days designing pieces of the clock. Rose bonded with Hillis at the retreat and would soon take on the role of the project manager.

The official coming-out event for the Clock Library took place at the Getty Museum in Los Angeles in early 1998. Brand moderated a two-day event—Time and Bits, Managing Digital Continuity—exploring what the digital transition would mean for storing and using human knowledge.

Hunting for a remote location where the clock could be permanently installed, he had set up a war room in the back room of his Sausalito office on Gate 5 Road. There was a large table in the center of the room, and he had pinned maps to the metal-clad walls with magnets. For a while Hillis was passionate about locating the clock in Cheyenne Mountain, where the North American Air Defense Command, or NORAD, is headquartered. For his part, Brand was continuing to explore the idea of finding a place for the clock in the mountains just south of the Black Rock Desert.

One day he was busy briefing Roger Kennedy, soon to become a Long Now board member, on the Nevada desert. Kennedy, whom Brand had met through a mutual acquaintance, was a polymath who had been a popular director of the National Park Service during the Clinton administration. He had also been a Justice Department lawyer, an NBC news correspondent, a banker, an unsuccessful candidate for Congress, and an author.

While Kennedy listened to Brand describe what attracted him to the Black Rock Desert site, he studied the eastern side of the large Nevada map that Brand had laid on the table.

Finally, he stopped Brand and pointed to a remote mountain range just south of where Highway 50—the "loneliest road in America"—crossed the Utah border.

"Have you ever been here?" he asked. "This is a new national park—Great Basin National Park. They're so isolated, I think they'd be rapturous to have something like the Clock."

Kennedy was pointing to the Snake Range, one of the most remote places in the United States. Midway between the Rockies and the Sierra Nevada, it had everything Brand was looking for—a 4,900-year-old tree, deep caves, and rock that would protect a mechanical clock for millennia. The closest town was Ely, Nevada, which was forty miles from Great Basin National Park.

"I'm dying to just jump in a car and go investigate," Brand wrote the board.

At the end of September 1998, he drove his Range Rover across Nevada to Ely, where he met Danny Hillis and his then wife, Patti; Rose; Kennedy; Saffo; and Brewster Kahle. Kahle, a New Yorker who had studied computer science at MIT and then worked for Hillis at Thinking Machines, had come to the West Coast to create Alexa, an early search engine he had sold to Jeff Bezos for $250 million in Amazon stock. Concerned that the ephemeral digital information that made up the World Wide Web would be easily lost, in 1996 Kahle had created the Internet Archive as a nonprofit repository for the world's digital information. Given his preservationist inclinations, he seemed well suited for this new mission.

A local real estate agent drove them to Spring Valley, a remote farming region in the shadow of Mount Washington, an 11,767-foot-high peak in the Snake Range, just one of the park's fourteen peaks over 11,000 feet. To Brand, Mount Washington looked like it should have been dubbed "Flag Mountain"—from the valley floor it looked like a beautiful white and green pennant streaming to the south. Together they explored deep into a mine shaft that had been cut into the base of the mountain.

The next day Becky Mills, the head park ranger, took them on a walking tour of the trail that led up Wheeler Peak, the highest mountain in the Snake Range. Brand was swept away as they walked through groves of bristlecones, hardy survivors that matched the timescale that Long Now was attempting to capture.

"Thousands of years old, each big one is," he wrote later in his trip report. "Gazing at a bristlecone is a glimpse into another time zone—the one we want to occupy."

On Monday everyone except for Brand and Rose flew home, and the two of them went on to hunt for a property before driving back to California. On their return, they stopped at Spencer Hot Springs, one of Brand's old haunts, where they coincidentally shared a plunge with several dozen Burning Man veterans.

Afterward, Brand wrote that they had "Found The Place."

The next year they bought the abandoned property they had discovered on their expedition. Jay Walker; Bill Joy, a cofounder of Sun Microsystems; and Mitch Kapor donated the money to buy 180 acres of virtually inaccessible mining land suitably populated with bristlecone pines. Safra's contribution would be used to fund the first clock prototype and Myhrvold would underwrite a second model, a development project that would take until 2005.

Ultimately, it would be Jeff Bezos who would offer the Long Now project the financial wherewithal to create a full-scale model of the clock. Bezos had been enthused by reading *How Buildings Learn*, and Brand had gotten to know him through Shel Kaphan, who was just leaving Amazon in 1999. Bezos and Brand had met in Hawaii at a party Bezos organized to celebrate Kaphan's birthday and his impending departure. Bezos was immediately intrigued after Brand described the Long Now project, in part because he felt his business philosophy was predicated on a set of ideas that resonated with the effort: "It's All About the Long Term," he had written in his first letter to Amazon shareholders in 1997. "We believe that a fundamental measure of our success will be the shareholder value we create over the long term."[2] In 2003, Bezos jumped in as a funder.

Several years later, while he was looking for land for his space company, Blue Origin, Bezos traveled several times with Brand to the Long Now Nevada site. The Amazon founder argued that the site, surrounded by National Forest and Bureau of Land Management property, would prove to be a bureaucratic nightmare for building an offbeat undertaking like the clock and that a better idea would be to install the first clock on a 165,000-acre tract he was buying in Texas, bounded by the Sierra Diablo Mountains, the least occupied region of the United States. Bezos

offered the Long Now Foundation land and underwrote the construction of the clock, beginning in 2005. In 2009, construction began on a
massive version of the instrument inside a 6,000-foot-high mountain on
the edge of his vast ranch and spaceport in West Texas.

Brand was about as enthusiastic and as happy as he had ever been.
It was exciting to imagine building a sturdy mechanism and an organization robust enough to last for more than three hundred human generations. However, occasionally doubt continued to intrude on his
fantasies. In December of 1999, while on vacation in Morocco, staying
in the luxurious La Maison Bleue hotel in Fez, the reality of his undertaking took on a darker veneer. While the city radiated history, a dingy
arts and crafts museum in a rundown palace by the Blue Gate had a
more decrepit feel. A light rain was coming down when he visited the
museum, which held exhibits of tools, musical instruments, embroidery,
and wood ornaments. Everything felt shabby and neglected. The roof
of the museum leaked. It felt as if the artifacts had been kept in a basement for decades. They were "faded, rough, dingy, worth throwing out,"
he wrote in his journal. The "eternal burden of maintaining something
for 10,000 years" took on a new, more oppressive reality for him. "I see
why people would rather just start over," Phelan said. "I would."

Brand pushed the doubts aside. He was committed.

SHORTLY BEFORE THE STROKE OF MIDNIGHT AT THE TURN OF THE
new millennium in 2000, a crowd of thirty people gathered in the combined offices of the Long Now Foundation and Brewster Kahle's Internet
Archive. By now the Presidio's residential housing had been converted to
upscale rentals for the dot-com workforce. For several years it would also
become home to both Brand and Phelan and to Paul Hawken when they
rented homes that had been former officers' quarters. Kahle lived in the
Presidio as well.

Hillis and Rose had assembled the first prototype of the clock, and
at a party held to celebrate the end of the century, it loomed above the

assembled guests as an elegant, predigital-era artifact. Hillis had committed to the idea that the clock's bell—designed to ring once a century—would chime for the first time in 2000. Rushing to finish the prototype, they had cut it so close to the wire that they had not had time to test the clock before the new millennium arrived, and so there was a good deal of trepidation as the moment neared.

There was no countdown, only the quiet announcement by one of the young clockmakers who worked for Hillis, "One more minute . . ."

"The pendulum rotated. Rotated back. The escapement ticked. Then the governor balls atop the right-hand power helix spun up," Brand wrote in an email message to Hawken. "The adder rings were rotating! Clickety-click, clickety-click, and then TWO of the Geneva hexagons rotated, and up on the Clock face in the brass index rectangle '99' became '00' and '019' became '020'—and . . ."

Somebody cheered and was hushed and then through a painful silence a faint motor could be heard, and then a prolonged *bonngggg*. And then silence followed by a longer *bonnnnggggg*.

A decade after Hillis had sent out his email to friends, the clock was alive.

O

A FEW MONTHS AFTER THE CLOCK CHIMED, A DIFFERENT BELL rang. After rapidly filtering into every aspect of life in the Bay Area, the go-go internet boom ended abruptly in March of 2000. While Brand had tried to stay away from the greed and acquisitiveness that had accompanied the boom, he was inescapably tied to it both professionally and personally.

After selling Direct Medical Knowledge to WebMD, Phelan was briefly a dot-com multimillionaire. Flush, she and Brand began planning their next phase of life as philanthropists. They set up a foundation and began thinking about an organization that might parallel the Point Foundation that had grown from Brand's first brush with wealth generated by the *Whole Earth Catalog*. Those dreams would end abruptly.

Through Paul Hawken, they had been introduced to Reed Slatkin, the initial investor in and cofounder of EarthLink, a high-flying internet service provider. Slatkin was a former Scientology minister who between 1986 and 2001 bilked roughly eight hundred investors in a Ponzi scheme that was outdone only by Bernie Madoff.

Brand and Phelan had the misfortune of being among the very last people to invest with Slatkin. Late in 2000, he had dinner with the couple on the deck of the *Mirene*, appearing to carefully weigh whether to let them in on his remarkable portfolio. It was not something he would usually do, he told them. But since they were close friends of several of his other investors, he would make an exception.

Phelan was skeptical, but Brand felt that it would be foolish to miss the opportunity. Before the end of the year, the couple would transfer more than $1.6 million to Slatkin's account, almost half of their stock portfolio. Their money was then immediately transferred into another Slatkin account that was used to pay off previous investors. In May of 2001, the scheme was shut down by the Security and Exchange Commission, but ultimately Brand and Phelan would see less than 10 percent of their money returned. It was a crushing blow that wounded both of them deeply. Brand was mortified that he had been so gullible, believing against logic that someone had beaten the system and was going to invite them along for the ride. They told few of their friends about the financial setback, and it would contribute to a chill in their friendship with Hawken. There was little joy in the fact that a year later Slatkin was arrested and would ultimately spend more than a decade in prison.

AT THE PEAK OF THE DOT-COM BUBBLE, SILICON VALLEY—WHICH had expanded to San Francisco in a burst of multimedia and then e-commerce activity—was awash in enthusiasm for technology-centric ideas for changing the world. That zeal was reflected in the Long Now Foundation's expanding focus. The challenge was to figure out ways to get beyond dependence upon philanthropy, to come up with unique ideas that would produce revenue.

Initially, Brand thought that he had come up with a brilliant moneymaking scheme to support the Long Now Foundation. "Long Bets" grew out of an impromptu workshop debate between Peter Schwartz and Amory Lovins's wife, Hunter, over how quickly electric cars would be adopted. Listening to their argument, Brand decided there needed to be a place for that kind of intellectual disagreement with real money on the line and with the follow-through it takes to make it happen.

Kevin Kelly, still executive editor at *Wired*, got the magazine involved in the idea. That in turn brought in the lawyers, who pointed out that if the nonprofit Long Now Foundation became a bookie by taking a cut on the bets, the Internal Revenue Service would certainly frown on the idea. Eventually, Long Bets found its way around the legality issue by establishing a philanthropic fund to disburse the winnings.

Despite its shortcomings, the project was launched after Jeff Bezos became its first backer. Mitch Kapor and Ray Kurzweil (a high-profile inventor who gained increasing recognition for his belief in the inevitability of the singularity) placed the first bet over the question of when a computer program would successfully pass the Turing test, an idea first proposed by the English mathematician Alan Turing to determine whether a computer could be programmed to exhibit such humanlike intelligence that an observer would be unable to distinguish its answers from those of an actual person. Several other projects, including efforts to catalog all living species and all languages, were launched at around the same time, with varying degrees of success.

However, it was Long Now's focus on finding a location for the clock that would have a most significant impact on Brand. In 2002, he had taken a field trip to Yucca Mountain, a multibillion-dollar US government effort to construct a storage facility for the waste generated by the nation's nuclear power plants. The visit, and Brand's participation the next year in a GBN study performed for the Pentagon on the possibility of abrupt climate change, reshaped his view on technology and the environment.

Brand thought of himself as being "mildly" antinuclear before touring Yucca Mountain with Hillis, Schwartz, Kelly, Rose, and Pierre Omidyar, the founder of eBay. Eno had come to Nevada as well but was unable to get a security clearance in time to participate in the tour and so he joined the group for dinner afterward in Las Vegas.

There had been some thought that the tunnel-boring machine used at Yucca Mountain might be repurposed for the clock, but instead, the group came away from the tour with a deep sense of the folly of bureaucratic-style planning for a time period that would stretch over ten thousand years. That was the time period the government had set to protect nuclear waste, and it was clear even from a short tour that the Yucca project was plagued by technical, geologic, and political problems. It was also a clear warning to the clockmakers, who had multimillennial-scale ambitions.

Afterward, the group repaired to Las Vegas, universally appalled that the government had spent as much as $16 billion on a hole in the ground that was obviously destined for failure. "It was a grotesque expenditure," Brand wrote, "based on 1950s ideas, a deeply political set of gestures meant to reassure critics who are largely uninterested in science and distrustful no matter what." The group rallied that evening, however, deciding they were smart enough to circumvent the pitfalls of what they had just seen.

The question that Brand never pursued was the fact that Yucca Mountain was on sacred ground for the tribes that had occupied land around the site for more than a millennium and would be a major factor in the opposition to the facility. He would later acknowledge their wisdom, however, when he publicly broke with the environmental movement and quoted his friend Gary Snyder: "There is something to be learned from the Native American people about where we are. It can't be learned from anybody else."

Schwartz, who had the best sense of big energy economics among the group from his time working at Shell, was certain that what was seen today as waste would, in as soon as half a century, be seen as a valuable energy source to be reclaimed. For his part, Hillis decided that he could have built the same hole for no more than a couple of hundred million dollars, based on the idea of keeping the waste safe for just a hundred years while time was taken to think about the long-run disposal of substances that might remain radioactive for much longer than ten thousand years.

Brand decided the problem was focusing exclusively on long-term planning. "What Long Now pushes is almost the opposite, long-term THINKING—where you set in motion a framing of events so that a process is made intensely adaptive, preserving and indeed increasing options as time goes by," he wrote.[3]

Both Brand and Schwartz came away from their visit convinced that nuclear waste was not the boogeyman it was painted by the environmental movement. Waste had long been a significant issue for Brand. He had run antinuclear articles when he edited the *CoEvolution Quarterly*. Now he decided he may have gotten it wrong.

O

LYING ON HIS BACK ONE NIGHT, ASLEEP, BRAND TURNED ONTO his right side and with a lurch felt the world go into a terrible spin. It wouldn't stop.

It was terrifying, and he cried aloud, "Help." Phelan woke up and before long she had Brand, who was in a frightened state, in the emergency room of Marin General.

After checking all of his symptoms, the medical staff told him he was experiencing vertigo, and that it was common and not surprising for a person in his early sixties. They taught him the Epley maneuver, a treatment he could do by himself to control vertigo, and sent him home.

But he wasn't cured. Once again, he found himself beset by fearful mood swings that would occasionally turn to panic, for no reason that he could easily identify. Driving across the Golden Gate Bridge, he would again become acrophobic. He had trouble appearing in public. Lines at the airport suddenly petrified him. He had a nagging fear that vertigo would suddenly appear, causing him to keel over.

Increasingly afraid of his own brain, Brand soon found he was again struggling with bouts of depression. Not since the late sixties and early seventies had he faced a similar dark time. Then he had gone into therapy, left his first wife, and found his way into a new relationship to extract himself from the gray world he was inhabiting. Now, at the end of 2002, a similar dark cloud enveloped him. He had just turned sixty-four; he felt as if he were drifting a bit, and professionally he was at a loss. "I feel the lack of a new trail I'm hot on that illuminates everything else," he wrote in his journal.

Work didn't help. On occasion he would take an Ativan, a common antianxiety drug, to help with the vertigo, but he began hunting for different cures. Acupuncture did nothing and left him feeling as if he were dealing with quacks and medical fraudsters. He applied himself to perfecting what he cared about—focusing on Long Now, taking care of the *Mirene*, his marriage, friendships, and spending his free time on hiking and fitness—but that still left him feeling as if there were a hole in his life.

Phelan told him, "You have no reason to be depressed," and he agreed with her. Finally, a neurologist prescribed Zoloft, a commonly used antidepressant, which pulled him back out of the depths. Once he found he was no longer in a place where he was afraid of completely losing it, everything else was manageable.

While he continued to try to sort out his depression, a second piece to the nuclear puzzle came together that year when Schwartz asked Brand to participate with a GBN team that had been awarded a $10,000 contract by the Pentagon to prepare a report outlining a scenario portraying the impact of abrupt climate change.

The assignment had again come from Andy Marshall at the Pentagon's Office of Net Assessment. There was already a growing body of evidence that abrupt shifts in the Earth's climate had happened relatively frequently, and the new scientific data had caught Marshall's eye. The previous year the National Academy of Sciences had published a report that speculated that rapid climate change could have a dramatic impact on human society and ecosystems.

Schwartz had been aware of the potential impact of climate change on humanity going back to 1977, when SRI had done an early report on the subject. He plunged into the scientific climate literature and soon concluded that climate change could be both abrupt and extreme.

Pennsylvania State geologist Richard Alley, who had coauthored a 1998 article in *Nature* on abrupt climate change, was involved in the preparation for the Pentagon meeting, and Brand used his own involvement as an excuse to meet and interview a number of climatologists, including Stephen Schneider, a well-known Stanford biology professor known as a hardcore environmentalist and "climate warrior."

A draft of the document was prepared by June of 2003, coauthored by Schwartz and Doug Randall, an independent scenario-planning consultant who worked with GBN. The final report was delivered in October,[4] but it had a more dramatic impact after Marshall gave GBN the go-ahead to have *Fortune* publish an article about their findings.

Run as a cover story in January 2004 with the headline THE PENTAGON'S WEATHER NIGHTMARE, the report described a scenario in which climate effects would be felt as early as 2010, claiming:

> Global warming, rather than causing gradual, centuries-spanning change, may be pushing the climate to a tipping point. Growing evidence suggests the ocean-atmosphere system that controls the world's climate can lurch from one state to another in less than a decade—like a canoe that's gradually tilted until suddenly it flips over.[5]

The report generated a dramatic international reaction. At the height of the Bush era, everyone was shocked to see the Pentagon stepping forward and highlighting climate change as a national security issue. For Brand and Schwartz, it sealed the deal on nuclear power. Brand's faith and optimism made him confident that a new generation of nuclear power could lower the risk of catastrophic meltdowns such as those seen at Three-Mile Island and Chernobyl.

Schwartz went public first beginning in 2003 with an article in *Wired* arguing that nuclear could be a "stopgap" while other sustainable energy sources matured.[6] Two years later he gave a more full-throated endorsement of nuclear power with another article, coauthored with Spencer Reiss, that proclaimed "Nuclear Now!"[7]

In 2003, two renegade environmental activists, Michael Shellenberger and Ted Nordhaus, had created the Breakthrough Institute to promote technological solutions to environmental problems, departing from the environmental movement's opposition to nuclear power. The next year they published a manifesto titled "Is Environmentalism Dead?" that touched off a heated debate about nuclear power within the American environmental movement, with Shellenberger and Nordhaus being attacked by a range of mainstream environmentalists led by Carl Pope, the president of the Sierra Club.

There had long been warring factions within the American environmental movement. As far back as the nineteenth century, John Muir and Gifford Pinchot had very different views about caring for the environment. Muir believed that wilderness should be *preserved*–untouched. Pinchot, who founded the United States Forest Service, believed that the environment should be *conserved*–that humans were very much a part of and not separate from the environment.

Influenced by his encounters with American Indians in Oregon in the early 1960s, Brand had always aligned himself with the conservationist wing of the environmental movement as distinct from the preservationists as characterized by Muir. Now, living and working within the technology-centric and corporation-aligned Global Business Network, he grew increasingly sympathetic to the Pinchot side of the debate—those who argued that in addition to exploiting the environment, humans were responsible for caring for it as well.

Historian Andrew Kirk has described what he refers to as "conservative preservationists." They were more likely to have a politically conservative outlook or to be apolitical, they were not anticapitalist, and they were comfortable with technology. It is a worldview that he traces back to Teddy Roosevelt's original "wise use" conservationism, and he places Brand and Whole Earth countercultural environmentalism in the context of a reinvigoration of a deep strain of pragmatism in American environmentalism. So while in the late 1960s and early 1970s Brand had been instrumental in the creation of a new environmental movement, when he broke ranks, he would insist that he had not left behind that commitment to environmental stewardship and responsibility. His dispute, he argued, was over means and not ends.

Untangling Brand's rejection of the mainstream environmental movement after 2004 reveals that portions of his new worldview were clearly rooted in his long-held antipathy of the Left—going all the way back to his 1950s anticommunist ideals centered around individual freedom—as well as his rejection of neo-Luddite environmentalists who raised the specter of the unintended consequences of technology. He would remain consistently hostile to those who warned prematurely about the dark side of new technologies and ultimately became a fan of what would be called "intended consequences." It would become part of his philosophy; he would eschew the power of Bateson's exhortation to be aware of both sides of a paradox.

In May 2005, Brand added his own voice to the growing debate with an article titled "Environmental Heresies," which had been solicited by Jason Pontin, a Hit the Hill comrade who had edited *Red Herring* during the dot-com boom and then moved to Cambridge to edit MIT's *Technology Review*.

Influenced by Peter Calthorpe and then by Robert Neuwirth's book *Shadow Cities*, which offered a generally positive interpretation of slums and shantytowns, Brand had been giving a Green Cities talk for a number of years, pointing out that, for environmentalists, urbanization could represent an opportunity. Now he asserted that within a decade, genetically modified organisms would be embraced by the environmental movement. He took aim at his former mentor Paul Ehrlich and his argument in *The Population Bomb* declaring that urbanization was lowering population growth trends dramatically. Finally, he tweaked his friend Amory Lovins, calling him out for convincing environmentalists that a battery-powered Green car was possible but refusing to do the same with nuclear power.

Within weeks, Brand's arguments had attracted a national audience.

*New York Times* environmental reporter Felicity Barringer put him on the paper's front page under the headline OLD FOES SOFTEN TO NEW REACTORS.[8] "It's not that something new and important and good had happened with nuclear," Brand was quoted as saying. "It's that something new and important and bad has happened with climate change." The article was bookended with a reference to Schwartz's *Wired* article, noting that his critics had described it as "right-leaning Cheney-worshipping drivel."

The debate launched Brand onto a speakers' circuit that was far more lucrative than city planning organizations. In May 2006 he gave a talk to a group of "nuclear power heavyweights" at the Nuclear Energy Institute, making the case that they were all environmentalists now, thanks to climate change. "I use audio with my slides these days," he wrote Eno. "I hit them with an atom bomb explosion (they jumped!) and showed how all of us see a mushroom cloud when we look at those toroidal cooling towers. Then later I converted the same cooling tower photo into a beautiful view up a tree trunk, with a robin caroling away. The engineers were moved." Brand had indeed moved a long way from the revulsion to colossal machinery that he'd felt when visiting the space shuttle while working for Jerry Brown during the 1970s.

O

AFTER SHE LAUNCHED DIRECT MEDICAL KNOWLEDGE, PATTY Phelan had quickly discovered that it was difficult for a female chief executive of a start-up to get her calls returned.

She thought for a while about new names to get around the problem. On vacation in Maine, Brand helpfully suggested that she could name herself "Elecktra," the name of the person whose room they were staying in. Instead, at the age of forty-four, she chose to begin using her middle name, Ryan, which in phone messages gave no hint of gender. It also gave her more gravitas when she was addressing rooms filled with male venture capitalists.[9]

Quickly realizing that she did not feel comfortable working with the WebMD executives who had acquired her company, she departed and began working with the Long Now Foundation. There she cofounded and eventually became chief executive of the All Species project, Kevin Kelly's idea to create a compendium of all the species on the planet and make them available in a public archive. They had estimated that the undertaking might cost as much as $1 billion to $3 billion, but by the end of 2002, unable to get adequate foundation funding in the wake of the dot-com bust, it had largely shut down.

Two years later, Phelan, taking advantage of the falling cost of genetic sequencing, founded DNA Direct, one of the first companies to offer genetic tests directly to consumers. She soon met Richard Rockefeller in a registration line at one of the many medical conferences she was frequenting. Afterward, she and Brand became close friends with Rockefeller and his wife.

In Rockefeller, whose *New York Times* obituary headline read A ROCKEFELLER KNOWN NOT FOR WEALTH BUT FOR HIS EFFORTS TO HELP,[10] Brand saw the clearest expression of what it takes to be a responsible leader. In his mind, it wasn't about social class, it was about how a responsible human being should behave in the world.

Brand's philosophy of wealth remained complicated. He had long viewed himself as having an upper-class background, having been educated at Exeter and Stanford and with a small amount of family money available when he needed it.

Cofounding the Global Business Network paid him a good salary for twenty years and left him with a reasonable 401(k) retirement fund, but all of his other ventures had been established as nonprofits, and the wealth generated by the *Whole Earth Catalog* had gone to create the Point Foundation.

Phelan had also not inherited much, and after they purchased the *Mirene* and in 2005 a small dairy farm ten miles north of Sausalito next to the Petaluma River as a weekend retreat, virtually all of their wealth went into maintaining their homes. After meeting Phelan, Brand had sold the property on Cape Breton Island.

Brand's experience at the Demise Party had left him believing that money did not inspire creativity. However, his growing fame in the 1970s and cofounding GBN increasingly put him in contact with great wealth, giving him a window into a rarefied world of power and privilege. Being flown to Aspen on Doug Carlston's plane for a Long Now board retreat or spending time with Bezos in Hawaii was intoxicating. It confirmed his increasingly elitist point of view, but at the same time, he had continued to believe that privilege came with a responsibility to society.

He saw people he knew who were "creative and in business becoming extremely wealthy quickly while they were still young. Their money suddenly became a huge issue in their life—ruining their marriage (or creating one), torquing their values and character, and distracting them mightily." He decided they had no one to look to for guidance besides one another.

Moreover, while others saw great poverty in the world, Brand, preternaturally optimistic, decided that affluence was not inherently bad. Life-changing wealth comes at many levels, he concluded. The rich could be far more of a boon to society if nudged in the right direction, and the world could use that.

He had been looking for a new book project for several years. At one point after visiting with Jane Jacobs in 2003 he had considered writing *Great Grand: The Value, Values, and Duties of Elders.* Then he reconceived his idea, focusing on the wealthy rather than the aging.

"Title is HOW TO BE RICH WELL," Brand wrote Brockman in an email message.

"Tell me more," Brockman responded.

A week later, the perfect Brockman moment, in the form of a laudatory profile, appeared in the *New York Times*, written by John Tierney, a *Times* columnist with a libertarian bent.

"Stewart Brand has become a heretic to environmentalism, a movement he helped found, but he doesn't plan to be isolated for long," Tierney wrote. "He expects that environmentalists will soon share his affection for nuclear power. They'll lose their fear of population growth and start appreciating sprawling megacities. They'll stop worrying about 'Frankenfoods' and embrace genetic engineering."[11]

Despite the fact that Tierney had neatly summarized an obvious book idea, Brand was still focused on writing about how to be wealthy and socially responsible. The next day Brockman circulated a two-page proposal written by Brand for *How to Be Rich Well.*

"Billions of people in the world are now climbing out of poverty. Toward what?" he wrote. "They look to the rich to find out. The rich are a great public asset—as goals, as models, and as creators—when they do their job right, when they figure out how to be rich well." Brand's proposal described a book that would advise the rich on lifestyle and ethics, confident that they would feel free to talk about their lives and their wealth if he agreed not to name them. "I know the rich will speak to me because I've tried a few," he wrote. "They trust me it turns out and they are highly interested in teaching—and in learning from others—how to be rich well."

He would begin with the old money families he knew well: "Richard Rockefeller and his sister Peggy, William Randolph Hearst III, Juan Enriquez, Jacob Safra, Richard Thierot, Nion McEvoy, Nicholas Negroponte, possibly the Duchess of Devonshire . . ." To that list he added his "new-money friends": "Sir Richard Branson, Jeff & Mackenzie Bezos, Pierre and Pam Omidyar, Mitch Kapor, Nathan Myhrvold, Susie Tomkins Buell, Garrett Gruener, Sergey Brin, Larry Page, John Doerr, Steve Jobs."

In justifying his proposal, he noted that in 1982 he had decided that the world headed where the rich led it, and if they failed to lead, it went nowhere. In response he had run several seminars on Creative Philanthropy, trying to help a group—described as "Doughnuts"—of largely unhappy heirs to do something useful with their fortunes.

It is true that many people in America do admire the wealthy, but Brand's proposal was tone-deaf to what Thorstein Veblen had described as "conspicuous consumption"—witness pronouncements such as, "Managing luxury is a huge and tricky issue, worth a couple of chapters, ranging from the burden of multiple houses to the lethality of helicopters." In particular, the generation who venerated Brand generally didn't share his access to or adulation of great wealth, and it is likely that the book, had it been published, would have tarnished his reputation.

Fortunately, the proposal landed with a resounding thud. Many publishers wrote back to express an interest in purchasing a book written by Brand—just not that one.

It would take Brand just a week to put a new proposal together and not much longer than that for Brockman to sell the proposal to Viking Penguin, the publisher of the *Whole Earth Epilog*.

The auction brought a hefty advance of $475,000.

It would also be his easiest book to write. Beginning in the spring of 2007, he buried himself in his pro-technology vision of environmentalism— committed to countering the deleterious impact that humans were having on the Earth by embracing technology rather than rejecting it. He centered his argument around four themes meant to redefine Green: massive urbanization, globalization, biotechnology, and decarbonized energy, based on his conviction that conservation and renewables would be grossly insufficient.

Initially, he titled the book *Think Globally, ACT Globally*, but after discussion, he settled on *Whole Earth Discipline*. Brockman came up with the subtitle *An Ecopragmatist Manifesto*, which Brand would come to regret. He wanted his book to be seen as journalism and not as a polemic.

In July 2009, several months before publishing *Whole Earth Discipline*, Brand stood before a US State Department audience to preview his book, which he described as a supplement to the *Whole Earth Catalog*. He didn't look like a counterculture icon, but he also wasn't wearing an East Coast business suit. He wore a dark brown dress shirt and a bright turquoise tie tucked into his shirt. He began by establishing his environmentalist credentials by reciting the *Outdoor Life*'s Conservation Pledge that he had taken as a youngster.

"I still believe that," he told the audience.

Then, for the benefit of his State Department audience, he added, "The rise of the West is over." Rapidly growing cities in the developing world were the world's future, he explained, and that will be environmentally positive. As the rural exodus occurs, the natural world could reclaim the territory that has become depopulated. Furthermore, as soon as people arrive in cities, they have fewer children, defusing the population bomb.

Brand then shifted to climate. "My fellow environmentalists on this subject have been irrational, anti-scientific and very harmful," he asserted, also bashing Greenpeace and Friends of the Earth for their opposition to GMO foods in Africa. "We are as gods and HAVE to get good at it," he concluded.

Almost four decades after Earth Day, this was a new Stewart Brand. John Brockman had chosen to label *Whole Earth Discipline* as a "manifesto," and now, once again perhaps somewhat reluctantly, Brand had become an activist.

In the sixties, he had shared Ken Kesey's contempt for the antiwar movement, and afterward, he retained an antipathy for the New Left. His efforts at environmental activism in the sixties, *Life-Raft Earth*, and in organizing the Life Forum in Stockholm at the United Nations Conference on the Human Environment had led him to believe that activism was ineffective. Now, however, he set out to convert the environmentalists, whom he condemned as backward-looking romantics, not just to a scientific view of the world but to an engineering sensibility. His new mantra had become "What we call natural and what we call human are inseparable." In one sense, it was a worldview he had first stumbled upon when he visited the Indian reservation in Oregon, and he had now returned to it, but with a twist even he would not have predicted in his *America Needs Indians!* days. It would now put Brand on a collision course with people who had once been his close friends.

WHEN IT WAS PUBLISHED IN OCTOBER 2009, *WHOLE EARTH DISCI-pline* was viewed as a rallying cry by some and as treason by others. By endorsing vast cities, GMO foods, nuclear power, and geoengineering to forestall global warming, he threw down a gauntlet that left many Greens feeling betrayed.

As he wrote, he savored the notion of going against the grain but in one sense felt the book was a reaffirmation of where he had begun with the *Whole Earth Catalog* four decades earlier. In the fall of 1968, he had set out by arguing, "We are as Gods and might as well get good at it." *Whole Earth Discipline* opened with his new epigraph: "We are as Gods and HAVE to get good at it," underscoring his belief in the existential threat of climate change.

Nevertheless, *Whole Earth Discipline* also marked a fundamental break with his youthful libertarian, do-it-yourself philosophy. It was his earlier perspective that had been called out as the "California Ideology" in 1995 by two British academics in a critique of what they called "dot-com neoliberalism"—a blending of hippie libertarianism and conservative economics.

That view of Brand was further documented by sociologist Thomas Streeter and communications theorist Fred Turner, who both trace an intellectual through line from the *Whole Earth Catalog* to the emergence of the libertarian online culture that emerged beginning in the mid-1980s.[12] None of the critics appeared to realize, however, the degree to which Brand had changed his mind about the value of "good" government from his year spent consulting for Jerry Brown in Sacramento long before the emergence of a digital culture. Brand felt that Huey Johnson had been "shockingly effective" as natural resources secretary for California in the early '80s, proving "how much lasting good can be accomplished by intelligent and integrated policy at the state level." Streeter and Turner also ignored the degree to which he felt that his effort in building a workable, not-for-profit digital culture upon the WELL had been a failure. They also failed to make the distinction—made later by Jonathan Taplin in his 2017 book, *Move Fast and Break Things: How Facebook, Google, and Amazon Cornered Culture and Undermined Democracy*—between Brand's original technological utopianism and Silicon Valley–centered digital libertarianism that emerged with a group of Stanford-educated young Turks known as the PayPal Mafia during the dot-com era.

In *Whole Earth Discipline*, Brand made a decidedly non-neoliberal argument: "The scale of the climate challenge is so vast that it cannot be met solely by grassroots groups and corporations, no matter how Green. The situation requires Government fiat to set rules and enforce them."[13]

Brand retained a passionate commitment to the ideas of Aldo Leopold, an author, a philosopher, and an ecologist who is generally seen as the father of the science of wildlife management. He treated Leopold's *Sand County Almanac*, a collection of essays enumerating a "land ethic" regarding the responsibility that people hold toward the land they inhabit, as a "holy writ." Brand did not see himself as abandoning that obligation. Rather, he was calling for a new "turquoise" movement that would blend Green activism with what he described as a blue planet–friendly attitude toward science and technology.

In framing the urgency of the coming climate crisis, Brand turned first to his friend James Lovelock, the iconoclastic environmental theorist who had first proposed the Gaia theory in the *CoEvolution Quarterly*. He had known Lovelock as a gentle optimist and was struck when he read the text of a talk he had given to the Royal Society in October of 2007. The melting of the arctic ice, Lovelock had argued, was introducing more global heating, adding that large regions of the Earth would become uninhabitable decades sooner than the then current climate models estimated. The only answer was to go to a wartime emergency footing. The world needed to prepare for climate refugees and consider geoengineering solutions to buy time.

Brand called Lovelock for more detail.

"People don't realize how little time we have. The planet is on the move," Lovelock said, explaining that global warming was on track to raise the average temperature of the planet by 5 degrees Celsius.

How many humans could such a world support? Brand asked.

"Oh, I think it's less than a billion," Lovelock replied. "It will be too hot for things to grow." To Brand, this reinforced his call for radical solutions that went well beyond what the bulk of the environmental movement seemed willing to consider.

To further make his case, Brand relied on Saul Griffith, an MIT-trained materials scientist and a MacArthur Genius Grant winner who had invented an ingenious wind power system based on tethered high-flying kites. Brand had met Griffith at a TED Conference dinner in 2008 and Griffith had introduced him to the idea of Renewistan, the land area needed for renewable power sources like wind and solar to generate enough power for the world's population.

Griffith told Brand that simple calculations had proved to him that it would take the landmass of a country the size of Australia and require a committed, massively expensive development project involving every country on Earth. It was only a slightly less pessimistic assessment than Lovelock's. "I'm not trying to be pro-nuclear," Brand quoted David MacKay, a British physicist and Griffith ally, as saying, "I'm just pro-arithmetic."[14] (Significantly perhaps, a number of Green critics of Brand and Griffith would quarrel with Griffith's arithmetic;[15] meanwhile, Schwartz claimed that Lovins, who sat on the other side of the fence, had been loose with the facts in a debate they'd had on nuclear power.)

As Brand became visibly pronuclear, his disagreement with Lovins over the issue had grown increasingly chilly. Brand attempted to explain that he was trying to keep a relatively open mind on the issue of nuclear power. Lovins responded: "The more you tell me about your nuclear beliefs, the more I think you've been hoodwinked and are putting your reputation ever more seriously at risk."

At that point, Brand, who had attended Lovins's wedding and published several of his articles in the *CoEvolution Quarterly*, read into the letter that Lovins was actually threatening his reputation and decided they were no longer friends.

In his research Brand also relied heavily on the perspective of Gwyneth Cravens, a novelist and *New Yorker* writer who had been an antinuclear activist before changing her mind and writing *Power to Change the World: The Truth About Nuclear Energy* in 2007. When he began his research, he called Cravens to ask her why she had changed her mind.

"Two things," she replied. "Baseload and Footprint."[16] Part of her argument was the same one made by Griffith; the other part focused on the most challenging aspect of wind and solar—they did not produce power continuously.

On the day *Whole Earth Discipline* was published, *Grist*, an online environmental site, posted "Stewart Brand's Nuclear Enthusiasm Falls Short on Facts and Logic," a full-throated attack by Lovins on a draft of the chapter of the book focused on nuclear power.[17] The article was a distillation of a much longer twenty-thousand-word paper, "Four Nuclear Myths," which Lovins helpfully appended.

He began by stating, "I have known Stewart Brand as a friend for many years," but there was nothing affable about his article, which concluded that despite Brand's reputation as an environmental icon, his ideas would worsen the climate and increase security risks.

Brand felt he had been ambushed. He complained that Lovins had not read his entire book, only an early draft of the chapter on nuclear power—which someone had passed on to Lovins without telling Brand. He attributed the widespread belief about the extraordinarily high capital costs of nuclear power to Lovins. He felt Lovins had completely missed the heart of his argument, which was about the potential of technology innovation. He even pointed out that the potential of smaller "microreactors" could actually be the most effective way to implement one of Lovins's cherished goals of distributing the generation of electric power. Brand argued that by his "obsessive ongoing vendetta against nuclear," Lovins was actually contributing to increased reliance on coal and gas.[18] A once warm friendship ended.

A number of Green activists treated Brand's book with respect, if not agreement. Denis Hayes, the founder of Earth Day, noted Brand's "trademark" technological optimism in placing his faith in next-generation nuclear power plants: "With a serious commitment, Generation IV reactors might be commercially available by the 2030s, by which time global warming will have cooked our goose if we haven't already built an economy relying heavily on solar energy, affordable storage, and smart power grids."[19]

WHOLE EARTH DISCIPLINE CREATED A REASONABLE AMOUNT OF buzz in the new world of social media when it appeared in print. Brand also heard from many of his friends and GBN associates who were uniformly enthusiastic. Traditional environmentalists were not the only ones upset, however. John Gilmore, the cofounder of the Electronic Frontier Foundation, a libertarian and a committed privacy activist, sent a note saying he believed that Brand had fallen into a "doomsday" trap, and he accused him of forsaking his libertarian principles, writing that "it [the threat of global warming] has inspired in you exactly the same response—to throw away your reason and your principles. . . ."

In the mainstream media world that Brand still valued, however, there was much less attention—indeed deafening silence—a fact he blamed on Lovins's attack. He felt that because Lovins was revered as the voice of Green science, his criticism had led to a de facto boycott by reviewers. Even the *Chronicle*, his hometown daily print newspaper, chose not to review it, which he took as a personal affront. The other daily paper he read, the *New York Times*, would not mention *Whole Earth Discipline* in its book review section for five months, and then only parenthetically in a skeptical essay on a spate of books subtitled "manifestos." There were several offers to debate Lovins on radio and television, which Brand tersely declined, replying that his chapter would serve as his response now that both of their positions were on the table.

The lack of media attention translated into slow sales, disappointing his publisher—a striking contrast to the reception the book received in the United Kingdom when it was published there several months later. The complete gamut of newspapers reviewed the book, treating it in a serious fashion.

Less than a year later, however, shortly after the paperback had been published with a new subtitle—*Why Dense Cities, Nuclear Power, Transgenic Crops, Restored Wildlands, and Geoengineering Are Necessary*—Brand ran into a more challenging confrontation in London.

The British television network Channel 4 had produced *Where the Greens Went Wrong*, a documentary based largely on the themes of *Whole Earth Discipline*, and had invited Brand to appear on a panel in London the day after the film was scheduled to air. At the same time, the Italian translation of *Whole Earth Discipline* was published, and he was invited to tour in Italy, where he was a popular figure. The plan was to begin his book tour in Italy, fly to England for the debate, and then return to Italy.

The trip was fraught from the beginning. Phelan's sister Kathleen was dying of cancer, and in the middle of the European trip, Phelan learned that she had had a seizure and was near death. Distraught, she immediately began planning her flight back to the United States. (Kathleen Phelan died a month later.) Meanwhile, Brand's friend *Whole Earth* editor and environmentalist Peter Warshall was also suffering from cancer and close to death. "I am 72 next week," Brand wrote Eno. "I feel like our cohort lives in the graveyard these days."

Even before it aired, the documentary itself had created an uproar among environmentalists when it became apparent that it was effectively a hit piece against the Green movement. Earlier in October, Brand had received a panicked email from Adam Werbach, the former president of the Sierra Club, who felt that he had been persuaded to do an interview under false pretenses and was "doing everything in my power" to get himself removed from the film.

"I have always had the utmost respect for you, Stewart; and even when I've disagreed with you I've been inspired by your earnest efforts to help those struggling for a better planet to improve their work," he wrote. "But I fear that this documentary has become something altogether different."

The British press had a field day: "There's almost nothing so delicious on the small screen as a hippy eating humble pie; except maybe a whole parade of them, exhibited over an hour, and packed into a documentary prosecuting the intellectual crimes of a pressure group who claim to speak for the zeitgeist," chortled a reviewer in the *Independent*.[20]

To make matters worse, in Italy Brand was horrified to discover, only after a reporter confronted him with evidence he was unaware of, that the publication of his book had been secretly sponsored by the Italian nuclear power industry. Brand had tried his best to stay away from this sort of thing. Back in February 2008, the main pronuclear industry group in the United States had approached him with its plan to award

him its annual award for "outstanding contribution to technical development, an improved regulatory climate or public acceptance of nuclear energy." Brand politely declined, explaining that such an award
would undermine his independence: "To nuclear audiences I talk Green,
to Green audiences and the public I talk nuclear (and other things). My
identity is with Greens, not with nuclear. That's how I can be effective
and it's the reality. You can see how an award would confuse that."

In Italy, it was too late.

The evening debate in London was a one-sided slaughter, with Brand
and Mark Lynas, a British journalist, environmental activist, and like-
minded Green critic who had also appeared in the documentary, having the worst night of their lives. Brand's nemesis was *Guardian* columnist
George Monbiot, an argumentative Oxford-educated environmentalist,
who had decided that Brand's role in GBN proved that he was a corporate
shill and basically a "spokesman" for a small group of oil and chemical
companies. Monbiot spent the evening shouting at Brand and Lynas, asserting that Brand had made false claims in his book. Brand responded
in a brief correspondence after the debate that GBN was a scenario-
planning consulting firm and that it was worthwhile having these companies hear from a "hard over" member of the Green movement. When
Monbiot continued to castigate him in his *Guardian* column, Brand chose
to ignore him. (Several years later Monbiot would reverse himself and
take a pronuclear stance.)

The day after the debate Brand flew back to Italy to complete his
book tour. In Rome, he wandered by himself to the Colosseum in the
rain, marveling at what a cruel place it was. He made a series of presentations to various audiences, capped by a three-hour drive to Naples to
speak to fourteen people in a bookstore. His talk was interrupted by a
fierce young woman who revealed a scar on her neck, asserting that cancer, caused by nuclear power and by implication Brand himself, had
taken her thyroid.

"I offered sympathy, she had nothing to receive it with," he wrote
Eno. "What have we done, frightening so many people so badly about
so little?"

Back in the United States, Brand continued touring to promote the paperback edition of *Whole Earth Discipline*. When Steve Jobs had given his celebrated 2005 commencement address at Stanford, encouraging the graduates to "stay hungry, stay foolish," he had introduced Brand to a new generation of college students. As a result, he was still a sought-after campus speaker.

In March he traveled to Oregon to speak at Oregon State University and at Reed College, where he invited his son, Noah, who had moved to Portland, to attend his lecture. He had now been giving his talk for almost a year and a half, and in Corvallis, he again began by noting his Green credentials and concluded that we are now terraforming the Earth and we can't stop terraforming the Earth and "so our only choice is to terraform it well."

The following evening, he gave the same talk at Reed to a more skeptical group of students. A small group of polite protesters showed up, their enthusiasm dampened by an evening downpour. In an article headlined ROGUE ENVIRONMENTALIST DRAWS PROTEST, Lucy Bellwood, a student journalist who covered the talk for *Reed Magazine*, wrote:

> Fielding questions about the dangers of radiation from the audience, Brand exclaimed, "Well, Reed has its own reactor. These people have been around it. And they're not weird [Cue: whoops and hollers from reactor operators in the audience] . . . yet." But despite the ripple of laughter that ran through the crowd, many attendees clearly remained skeptical about the virtues of nuclear power.[21]

As fate would have it, his talk took place just two days before the tsunami and resulting catastrophe at the Fukushima Daiichi Nuclear Power Plant, and Bellwood concluded her piece by noting, "If Brand had delivered the same lecture after that, it seems likely that the debate would have gotten even more heated."[22]

Even seven years later, the Fukushima incident remained so painful for Brand that he recalled incorrectly that his lecture and the accident had happened on the same day. The grim situation began to unfold slowly and as the magnitude of the calamity at the plant became clear, reporters in the United States scrambled to find expert commentary. Brand, back in Sausalito, largely avoided fielding the dozens of calls and emails that flooded in, while continuing to follow the disaster and discussing it online with his Long Now coworkers.

Alexander Rose pointed out that while the tsunami and dam failure had killed thousands, at that time none had died from the nuclear power plant accident. Brand initially turned down all media requests—a good move, Paul Saffo agreed, given that "Stewart has been very wise to stay silent on the topic (the worst time to argue for nuclear power is while the crisis is still unfolding)."

Early on, Brand declined with humor a local San Francisco radio interview a reporter on deadline might have missed, responding by email: "Thanks. Can't. At dermatologist getting treated for radiation damage from sun."

Two weeks after the meltdown, however, he chose to sit down with *Foreign Policy* to stand fast in his defense of nuclear power. He reiterated his belief that nuclear energy was still essential for the challenge of baseload power when the sun wasn't shining and the wind wasn't blowing. He also argued that solar panels had not yet shown a Moore's law rapid improvement associated with silicon chips. (The cost would fall by almost 90 percent during the ensuing decade.)

When asked about Fukushima, he credited the media with a more rational response to the accident than during Three Mile Island and Chernobyl. He expressed optimism that the accident wouldn't completely sideline nuclear power as an option in the United States, making the point that half of US nuclear electricity was now coming from recycled warheads, an environmental win-win. "It's kind of cool," he enthused.

But as world public opinion shifted dramatically away from nuclear power in the wake of the accident, Brand was increasingly isolated. Beginning in 2001, there had been mounting talk of a "nuclear renaissance." It had been tempered by the 2008 recession, but there had still been forecasts of a doubling of nuclear power by 2035 worldwide.[23] After Fukushima, however, countries in both Asia and Europe backed away, while in the United States the low cost of natural gas undercut any meaningful new investment in nuclear energy.

Brand has remained unmoved. He repeated over and over again that the climate situation hadn't changed and that made nuclear energy necessary. During the summer of 2011, he agreed to debate Winona LaDuke, a well-known environmentalist and Ralph Nader's Green Party vice presidential candidate in both 1996 and 2000. The debate was to be held in July at the David Brower Center in Berkeley, literally the home court of the "romantic" environmental movement that Brand had condemned in *Whole Earth Discipline*.

*Chronicle* columnist Jon Carroll teased him, writing in an email: "Oh man, are you in for it now. An Indian environmentalist and photographer. In this movie, you play the Nazi."

Before the evening began, Brand had sent LaDuke a description of his *America Needs Indians!* show and the pages from *Whole Earth Discipline* that recounted his experience on the Oregon reservation in the 1960s. That evening he wore a blue work shirt with his sleeves rolled up to his elbows and a necklace with his eagle bone whistle from his days as a peyote roadman.

The debate, moderated by *Earth Island Journal* editor Mark Hertsgaard, almost didn't happen—a "security threat" after a "powerful but unnamed member of the environmental community had objected to offering a platform to Brand because of his pro-nuclear stance" came close to jeopardizing the event—but eventually things went forward.[24]

He was clearly entering hostile territory. LaDuke and Brand covered a gamut of topics that had been discussed in *Whole Earth Discipline*, from nuclear power to GMO foods to geoengineering. "Brand appears to inhabit a parallel Earth that is free of the specter of corporate control or haunted by the profit motive; where new technology arises spontaneously, pure and without intent," wrote Gar Smith, the correspondent for the *Berkeley Daily Planet*.[25]

When he was asked about Fukushima, Brand responded with a

forced analogy, recalling the *E. coli* outbreak that had killed scores and sickened thousands in Europe, then exclaiming, "But you didn't hear anyone demanding that organic farms should be shut down!"

Afterward, in the question period as the final audience member of the evening reached the microphone Brand leaned forward with his hands on his knees and smiled with pleasure.

"Hi Peter!"

Peter Coyote, one of those who had partied with Brand at the Trips Festival more than half a century earlier, stood at the microphone and challenged his "We are as gods" hypothesis.

"Idiot savants," Coyote proposed instead. "We're highly developed, we have these great skills, this technological cleverness, which is completely untethered to wisdom. My question is that what I find disturbing and a little sociopathic about your perspective is the absence of doubt."

Coyote continued for more than three minutes, questioning Brand's faith in technology and condemning him for endorsing a technology that would leave a poisonous residue for millennia.

Brand squinted into the stage lights and responded with a sad smile.

He leaned back in his seat and countered: "Try making that speech in Dharavi, that slum in Mumbai where a million people are getting out of poverty. Remember voluntary simplicity? The *Whole Earth Catalog* espoused it. . . . Guess what, that wasn't even the standard for us. You and I are living better now than we did in that decade. And everybody else wants to do that too. Voluntary simplicity is a great idea, but it does not get taken up by people. Involuntary simplicity is where they're stuck and what they want to get the hell out of. And I'm with them." He stopped, raised his eyebrows at the audience, and the evening was over.

# Epilogue

THEIR COLORS ARE BRILLIANT.

Set against a black background that conveys the frozen motion of an M. C. Escher print, sixty-three plants and creatures populate an oil canvas that depicts species that have become extinct since the colonization of the New World began in the 1700s.

*Gone* was painted by Stewart Brand's former employee and Sausalito neighbor Isabella Kirkland in 2004. It depicts life-size images of birds such as the laughing owl, the Oʻahu ʻakialoa, and the Carolina parakeet, as well as insects and animals like Darwin's rice rat, tucked into an impossible jungle that includes extinct plants such as Hawaiian ferns and mariposa lilies.

Hidden in the painting is a barely noticeable iridescent butterfly, the Xerces blue. Once found exclusively in the sand dunes of the Sunset District in San Francisco, it is thought to have become extinct sometime before 1943. It has the dubious distinction of being the first butterfly to vanish due to destruction of its habitat as a consequence of urban development.

Yet while it hasn't received the attention and controversy of proposals to bring back extinct species such as the woolly mammoth or the passenger pigeon, Xerces may become the first species to be brought back from extinction.

The idea of bringing Xerces back to life goes back at least to 2000, but in 2021 it came significantly closer to reality. A restored Xerces

would become a powerful symbol of what French philosopher Bruno Latour describes as "Gaia 2.0."[1] We have entered the Anthropocene era,
and having completely reshaped our planet, we are faced with the reality that consciousness of humanity's impact on the world entails responsibility as well. It is that realization that is the heart of Stewart Brand's
contribution and his continuing relevance.

Revive & Restore was launched as a Long Now Foundation project
in early 2012 after Brand and Phelan sponsored a small symposium at
the Harvard Medical School, hosted by geneticist George Church,
called Bringing Back the Passenger Pigeon. While watching Church
demonstrate new gene-editing techniques, it confirmed for them that if
it was possible to revive the passenger pigeon, it would be possible to
bring back other species as well, offering a path to restoring biodiver-
sity, essential for combating the effects of global warming.

Not only would it be possible to bring back the woolly mammoth
(which the project's backers assert will help insulate the Siberian tundra
from climate change); it would now be possible to rescue species endangered by the impact of global warming by modifying their genomes to
add resilience. Efforts to change the genetic structure of coral to protect
against the bleaching effect of the sun are an example of efforts to save
entire species.

The project would complete the arc defined by the opening sentence
in the *Whole Earth Catalog*: "We are as Gods and might as well get good
at it." At the age of seventy-three Brand set out to make good on his original promise, orchestrating an active conversation between biologists,
ecologists, urban planners, and government officials about using modern biotechnology tools to reintroduce the species into the remaining
dunes in San Francisco's Golden Gate Recreation Area.

As he had with *Whole Earth Discipline*, he immediately stepped into
an ethical thicket. The idea of de-extinction had already been defined
in the public mind by the novel (and then the movie) *Jurassic Park*, and
similar controversies soon swirled around the potential unintended consequences of passenger pigeons and woolly mammoths. Another coun-
terargument came from Brand's early mentor Paul Ehrlich, who argued

that de-extinction is a waste of resources that could be invested in saving endangered species—a crusade that Brand and Phelan are also committed to. (Ehrlich later hedged, acknowledging that there might be some educational value in restoring the butterfly.)[2]

Yet the proposed revival of Xerces is perhaps the clearest way to illustrate Brand's pragmatic approach and his optimistic philosophy, a literal evocation of the "butterfly effect" that suggests the possibility that the smallest change in the environment can have an immense and nondeterministic effect. It evokes Brand's access-to-tools philosophy as well as Engelbart's augmentation philosophy.

It also stakes out the boundaries of Brand's techno-optimist philosophy. The idea of the butterfly effect was at the heart of Ray Bradbury's 1952 science fiction short story "The Sound of Thunder," in which a hunter uses a time travel machine to journey back in history to shoot a *Tyrannosaurus rex*, only to accidentally crush a butterfly by stepping off the path. He returns to a completely different and much nastier world from the one he left.

A decade later, a meteorologist running a computer model of a weather simulation made a tiny alteration in an initial value and found that it generated a completely different output. That led to the notion that a tornado might trace its origin to the flapping of a distant butterfly's wings weeks earlier—perhaps a Xerces.

Throughout his life Brand has chosen to view the glass as potentially more than half full—he has believed that a guiding human hand will make it possible to systematically enhance outcomes, whether they are terraforming the Earth or reducing the impact of climate change. His faith in humanity's identity as tool makers and users was at the heart of his confrontation with his old friend Peter Coyote, and his campaign to revive and restore species is a gamble that technology can be used to gain more than linear leverage in a positive direction.

However, there is an important difference between the "We are as gods and better get good at it" in the 1968 *Catalog* and the "We are as gods and *have* to get good at it" in *Whole Earth Discipline* four decades later. In the preface to the *Catalog*, Brand was addressing individuals, young individuals of his own generation, many of whom had set out to reinvent society by moving back to the land. In the second case, "we" refers to civilization. To the entire human species.

In part the shift in framing is about growing up. Brand as a thirty-year-old had a different outlook on the world than he did three decades later when he and Hillis formed the Long Now Foundation. The idea of a library capable of persisting across many generations, which Brand brought to the project, is at its heart the concept of creating and maintaining a planetary culture for human civilization.

It is a complement to his original assertion about the value of seeing the whole Earth. In a world increasingly defined by the twin political forces of nationalism and identity politics, Brand's singular insight into the value of seeing the whole Earth is a significant counterpoint. A planetwide approach had been given form by the League of Nations after World War I and the United Nations at the end of World War II. More recently the global election of an International Corporation of Assigned Names and Numbers (ICANN) board of directors in 2000 was a small but meaningful example of global democracy.

Briefly in vogue in the 1970s, the notion of a "planetary consciousness" is an idea that Brand's work first provoked in the 1960s. A half century later it remains his signature contribution.

O

IN DECEMBER OF 2018, BRAND PAUSED TO REST AT A SWITCHBACK near the end of a steep trail just a quarter of a mile short of a vertical cliff face on the side of a six-thousand-foot-high mountain in southern Texas.

He showed few of the physical limitations of someone his age. Intellectually and physically active, he remained a public intellectual engaged in crusades that in some cases stretched back throughout much of his life.

As he paused on the path, leaning forward hand on knee, he was exhausted from a three-hour hike that would end just ahead at a tunnel that looked like the entrance to some historical mining operation. The opening in the rock face instead leads to the clock of the Long Now, housed in a five-hundred-foot-tall cylindrical space hollowed out inside the mountain thanks to the largesse of the Long Now Foundation's largest benefactor, Jeff Bezos, one of the world's richest people.

The first version of the clock was nearing completion. With Hillis, Brand entered the chamber housing the clock through a roughly hewn door set at the base of the cliff. Like all visitors, he then made a rock scramble in a darkened cavern, passing through a *Star Trek*–style airlock to reach the base of the clock itself. A circular staircase chiseled from the rock wall that forms a giant cylinder in which the clock hangs ascends to a domed observation room hundreds of feet above near the top of the mountain. Halfway up is a platform where those who come to see the clock can physically wind it by laboring on a platform against rotating large wooden arms. Although the clock is designed to keep time without human intervention, winding is necessary to display the time in an observation room. The act of winding implies the idea of long-term responsibility that both Hillis and Brand are trying to convey with their machine.

Responsibility can take many forms. Social responsibility, for example, is an ethical ideal that mandates that an individual should act to benefit society as a whole. Brand and Hillis are attempting to turn social responsibility on its side, encouraging society itself to consider its obligation to uncountable future generations.

Brand has said that when he first encountered Hillis's clock idea, he was simply thinking about the long term, "wondering what things might make it worth wondering about."

It had begun as the idea that a library is a tool with which to communicate knowledge and wisdom to the future. Today, Brand's message is not merely about blind devotion to tools. The evolution in his thinking has been in his departure from his devotion to Buckminster Fuller's engineering-centric approach and his embrace of Gregory Bateson's grasp of the significance of paradox and the importance of understanding "whole systems." Brand's sensibility thus emerged from his encounters with Fuller, Bateson, and Doug Engelbart, as well as from studying with Paul Ehrlich and learning about the dynamism at the edge of adjacent populations, confronting the evils of bureaucracy while in the army, and realizing that buildings are more than a snapshot in time from visiting Building 20 at MIT.

Despite his reticence, it is an activist's sensibility.

It was what Brand told a group of students at the Parsons School of Design on the first Earth Day in 1970: "When we realize that we are as gods, then we will know that we have to assume a god's responsibilities, and that we are able to. Many good things are very simple."[3]

# Acknowledgments

This book was made possible by the generous support of three separate institutions, each of which explores in different ways the societal impact of technology. After leaving the *New York Times* in 2017, I joined the Computer History Museum. The museum's director at the time was John Hollar, and he graciously hired me as a staff historian. In the fall of 2017, I became a fellow at Stanford University's Center for Advanced Study in the Behavioral Sciences, affording me the opportunity to interact with a remarkable group of social scientists, historians, philosophers, and computer scientists. Director Margaret Levi and associate director Sally Schroeder run a remarkable program, set in the hills west of Stanford, that is both a retreat for scholars and increasingly a forum for exploring the way advanced technologies are intertwined with our economy, political system, and culture. Beginning in the fall of 2018, I was a fellow and writer-in-residence at the then brand-new Stanford Institute for Human-Centered Artificial Intelligence. Codirectors Fei-Fei Li and John Etchemendy are at the forefront of a growing movement to think more deeply about the consequences of artificial intelligence.

I embarked on this project at the suggestion of Kevin Kelly, who is both a longtime friend of Stewart Brand's and a classic example of someone whose life was set in a completely new direction by his encounter with the *Whole Earth Catalog*.

Brand has donated his papers, correspondence, and journals to the Stanford University Library Special Collections, and over the course of four years I spent countless hours in the Field Reading Room in Green Library. Michael Keller, Stanford vice provost and head of Stanford Libraries, gave me access and support during an earlier project, and I want to give special thanks to Stanford librarians Henry Lowood, Tim Nowack, and Peter Chan, who assisted me during the current project. Leslie Berlin, project historian for Stanford's Silicon Valley Archives, offered help on methodology and was also a sounding board. Mark Seiden was a careful reader who offered suggestions on grammar, style, and fact. Scott Moyers, my editor at Penguin Press, was a patient and thoughtful guide. Geoff Shandler helped wrestle a sprawling manuscript into something more compact and more readable.

Joseph Monzel and Paula Terzian aided with research at various points in the project. Fred Turner and Katherine Fulton, both of whom have previously written extensively about Stewart Brand, were kind to help with questions and by making their research materials available. Jason Sussberg and David Alvarado of Structure Films shared research materials with me while they prepared a documentary, *We Are as Gods*, about Brand's life.

While doing interviews, research, and writing, I had the benefit of a number of thoughtful friends and colleagues including Adam Fisher, Lynnea Johnson, John Kelley, Art Kleiner, Randy Komisar, Glenn Kramon, Steven Levy, Steve Most, Michael Schrage, Jason Vest, and Gregg Zachary.

John Brockman and Katinka Matson have been my literary agents since 1989. In this case they performed double duty, because they were part of the story.

And to Leslie, always.

## Author's Note

This biography relies on a series of weekly interviews I recorded with Stewart Brand at his office in Sausalito beginning in the fall of 2016. Eventually, over the course of four years, I spoke with him more than seventy-six times, sometimes for three hours or longer. I also conducted dozens of interviews with his friends, colleagues, employees, and acquaintances.

At the same time, my biggest lesson as a biographer came from my hours spent in the Stanford University Special Collections reading room. One thing I discovered after beginning my research was that for much of his life Stewart Brand has been a pack rat and that he has fortuitously kept his journals, correspondence, and papers, simply tossing things into a conveniently located shipping container until the librarians arrived. The lesson is that there is simply no substitute for contemporaneous documents. Although Brand showed no signs of slowing down during our interviews, his memory for events that in some cases had happened more than six decades earlier was understandably sketchy at times.

Because this is not intended as a scholarly volume, I have chosen not to footnote quotes taken from my interviews and from Brand's correspondence. My apologies to future historians and researchers. I will do my best to curate both the transcripts of my interviews as well as the material that I have digitized from Stanford to make it available to others where appropriate.

# Brandisms

"'These are the poetical efforts of a prose writer trying to plant a burr in the mind as it goes by in great haste,' he says. 'The mind collects stories, collects experiences, and it's pretty good at managing aphorisms and slogans and little rhymes.'"

—Stewart Brand as quoted by Katherine Fulton,
*Los Angeles Times Magazine*, October 30, 1994

( )

"Why haven't we seen a photograph of the whole Earth yet?"

—San Francisco, March 1966

( )

"We are as Gods and we might as well get *used* to it."
"We are as Gods and we might as well get *good* at it."

—First and second printings of the first
*Whole Earth Catalog*, October 1968

○

"Who are they: (Who are we?) Persons in their late twenties or early thirties mostly. Havers of families, many of them. Outlaws, dope fiends, and fanatics naturally. Doers, primarily, with a functional grimy grasp of the world. World thinkers, dropouts from specialization. Hope freaks."

—Alloy Conference, La Luz, New Mexico, March 1969

○

"I do believe that information is replacing laws."

—Journal, June 2, 1969

○

"Obviously truth is not to be found in the past, nor likely in the future, next summer, when the buses race from Stonehenge to the Pyramids, maybe. Truth was good while it lasted."

—Comments on errors in his account of the
Great Bus Race, Summer Solstice 1969

○

"When You Don't Know Where to Go Next, Go Back and Start Over. Drop out of specialization. Develop rudimentary skills good for any situation."

—"Stay Loose," *Earth*, December 1971

○

"When we realize that we are as gods, then we will know that we have to assume a god's responsibilities, and that we are able to. Many good things are very simple."

—Parsons School of Design, Earth Day 1970

○

"If we tried to teach infants to talk, they would never learn. I suspect it is the same with ecology. It must be learned. It is being learned. If you try to teach it to people, you will only teach them to hate it."

—Letter to the House Committee on
Education and Labor, April 14, 1970

○

"We won't be successful until we become a religion."

—Comment made frequently to friend Michael Phillips in the early 1970s

○

"When you design a tool, the best you can do is fashion a prototype and hand it over to the local evolutionary system: 'Here, try this.'"

—Journal, June 23, 1971

○

"The Hippie/long-hair thing has brash panache, but it lacks world-based substance. Dope ain't enough, it raises hopes and dashes them in a kind of fond isolation."

—Journal, July 16, 1971

○

"It says here that people working at TV stations during interviews and such watch the program on studio monitors rather than right over there real. Signifying: Reality is only Reality if it is shared. The more the realer."

—Journal, December 14, 1971

○

"We can't put it together. It is together."

—Back cover, *The Last Whole Earth Catalog*, 1971

○

"Don't talk population control to a man making love."

—Journal, January 15, 1972

○

"We are a generation longing for a Pearl Harbor."

—Journal, March 1972

O

"Ready or not, computers are coming to the people. That's good news, maybe the best since psychedelics."

—Opening sentences from "SPACEWAR: Fanatic Life and Symbolic Death Among the Computer Bums," *Rolling Stone*, December 7, 1972

O

"After burning our bridges we reported before the Throne to announce: 'We're here for our next terrific idea.' The Throne said, 'That was it.'"

—Introduction to the *CoEvolution Quarterly*, April 1974, on why he is restarting the *Catalog* and the Supplement

O

"Ecology maintains. Coevolution learns."

—*Whole Earth Epilog*, wraparound, *Harper's*, April 1974

O

"Stay hungry. Stay foolish."

—Back cover, *Whole Earth Epilog*, 1975 (first found in personal journal, March 1966)

○

"An old joke says that the lake with the longest name in the world is called—in a single native American word—'You-fish-on-your-side-we'll-fish-on-our-side-nobody-fish-in-the-middle.' 'Nobody fish in the middle' is a formula for a perpetually livable planet."

—From the *CoEvolution Quarterly*, as quoted in
the *Los Angeles Times*, September 22, 1975

○

"New York City is without a middle class, which accounts for the excellence of its graffiti and the absurdity of its financing."

—Journal, April 1977

○

"On the one hand information wants to be expensive, because it's so valuable. The right information in the right place just changes your life. On the other hand, information wants to be free, because the cost of getting it out is getting lower and lower all the time. So you have these two fighting against each other."

—First Hacker's Conference, November 1984

○

"'It's a small world after all,' sang Disneyland. We didn't know it was a threat. Now the world is becoming Disney World."

—Journal, while preparing to leave for Africa, June 1985

( )

"Charisma is theft. Commitment is a trap. If the group says, and means your life, 'you're either on the bus or off the bus,' get off the bus."

—Journal, June 1985

( )

"You Own Your Own Words" (YOYOW)

—Whole Earth 'Lectronic Link Guidelines, Spring 1985

( )

"America Needs Indians, I echoed on the uptick. America Needs Computers, I echoed myself, on the down-tick, but still worth doing. Late, I insist, because I didn't fall back and scan like I am now."

—Journal, from Africa, July 1985

( )

"When it comes to the point you're doing it mainly for the money, take steps toward moving on. (Like Whole Earth for me currently.) At a minimum pick up something else you aren't doing for the money."

—Journal, November 1985

O

"Keep it simple stupid, is a good way to keep it stupid."

—Journal, March 1986

O

"A major source of learning, maybe the major source, is other people's mistakes."

—Journal, August 1986

O

"Cows are slave deer."

—Journal, June 1987

O

"Once a new technology rolls over you, if you're not part of the steamroller, you're part of the road."

—*The Media Lab: Inventing the Future at MIT*, 1987

O

"We shape our buildings, then they shape us." (Winston Churchill) "Then we shape them."

—Journal, September 1987

○

"Most of adulthood and its skills consists of adventure prevention."

—Journal, July 27, 1988

○

"The Library is a window."

—Response to John Brockman, who thought the idea of creating an office without windows in a shipping container was "creepy," 1988

○

"If I get to heaven, and it's not a library, I'll be very disappointed. On the other hand, if I get to hell, and it's a library, I won't be surprised."

—A speech to the Illinois Library Association, April, 4, 1989

○

"I find things and I found things."

—Response to Brian Arthur, who had asked who Brant was when the two met at the Santa Fe Institute, 1989

○

"Hm. Things have shifted from most of the world looking older to most of the world looking younger."

—Journal, St. Louis airport, May 1990

○

"Some buildings are born great. Some become great. And some have greatness forced upon them."

—Journal, the *Mirene*, May 1990

○

"Rushing is at the root of all lack of quality."

—Journal, June 1990

○

"Buildings don't just have a history, they are a history."

—Journal, June 1990

○

"Judge a new building not just by what it is, but what it is capable of becoming. Judge an old building by how it has played its options."

—Journal, Victoria and Albert Museum, London, September 5, 1990

○

"Judge architects by what their buildings may become. Judge organizations in part by what their building is becoming."

—Journal, October 1990

○

"Immediacy is addictive."

—Journal, November 1990

○

"Trying to go high road with low money is a waste of time. Trying to go low road with high money is a waste of money."

—Journal, November 1990

○

"Why are Levi's Blue Jeans a timeless garment. Because they show time, honestly and elegantly."

—Journal, *High Tide*, March 1991

○

"Statement buildings soon are, at worst, making statements no one wants to hear, or, at best, soon tire of hearing."

—Journal, *High Tide*, May 1991

○

"Buildings are like people—Not yet smart when they're beautiful, no longer beautiful when they're smart."

—Journal, June 1991

◯

"Who would you rather be, the guide or the rich client? They would rather be each other."

—Journal, *High Tide*, July 1992

◯

"Will the info environment ever stabilize? No sign of that yet."

—Journal, June 1993

◯

"If I were to write an autobiography, it would have to be Float Upstream. Based on the supposed Morley Clan observation, 'Toss any Brand into the river and they'll float upstream.'"

—Journal, *High Tide*, July 1993

◯

"Why is it so hard to get biotech people to talk about the future, when you can't get computer people to shut up about it."

—Journal, December 1993

◯

"Evolutionary design is healthier than visionary design."

—*How Buildings Learn*, 1994

○

"Summing up the '60s in a sentence: We rolled our own cigarettes."

—Journal, May 21, 2004

○

"The most proficient knot is one that can be easily untied."

—Journal, September 6, 2005

○

"We are as gods and we HAVE to get good at it."

—Epigraph to *Whole Earth Discipline*, where Brand
endorses nuclear power and GMO foods, 2009

○

"Living on a houseboat forces you to live small, which forces you
to live large."

—Interview, Sausalito, California, June 21, 2016

○

"I adore the un-urgency, the realization that it takes time to get
things right, and that there is plenty of time to do that. Just keep
bearing down."

—Email to the Long Now board of directors, September 20, 2019

# Notes

## Prologue

1. Stewart Brand has described his mystic experience on the roof in various publications. The most complete account is in Michael Katz et al., eds., *Earth's Answer: Explorations of Planetary Culture at the Lindisfarne Conferences* (New York: Harper & Row, 1977), 184–88.
2. https://archive.org/details/WholeEarthCulture

## Chapter 1: Shoppenagon

1. Laura Ream, *History of a Trip to the Great Saginaw Valley, June, 1871; by invitation of the Fort Wayne, Muncie, and Cincinnati railroad, and with the co-operation of the Bee line, Fort Wayne, Jackson and Saginaw, and Jackson, Lansing and Saginaw railroad companies* (Indianapolis: R. J. Bright, 1871), 57–59.
2. Robert M. Hendershot, "The Legacy of an Ojibwe 'Lumber Chief': David Shoppenagon," *Michigan Historical Review* 29, no. 2 (Fall 2003): 40–68, https://www.jstor.org/stable/i20174030
3. Arthur W. Rosenau, *Lakeside: A History of Lakeside Association on Higgins Lake in Roscommon County, Michigan* (Roscommon, MI: Lakeside Association, 1979), 38.
4. Edward Hoagland, *Heart's Desire: The Best of Edward Hoagland* (New York: Simon & Schuster, 1988), 45.
5. Rosenau, *Lakeside*, 20.
6. W. Lloyd Warner, "A Sociologist Looks at an American Community," *Life*, September 12, 1949, 108–19.

## Chapter 2: On the Golden Shore

1. https://www.nytimes.com/1985/10/28/books/books-of-the-times-110794.html
2. Ahmed M. Kabil, "The New Myth: Frederic Spiegelberg and the Rise of a Whole Earth, 1914–1968," *Integral Review* 8, no. 1 (July 2012): 44.
3. Steven M. Gelber, "Sequoia Seminar: The Sources of Religious Sectarianism," *California History* 69, no. 1 (April 1990): 36–51. (Published by University of California Press in association with the California Historical Society.)
4. Myron J. Stolaroff, *Thanatos to Eros: 35 Years of Psychedelic Exploration* (Berlin: Verlag fur Wissenschaft und Bildung, 1994), 20.
5. Michel Oren, "USCO: 'Getting Out of Your Mind to Use Your Head,'" *Art Journal* 69, no. 4 (Winter 2010): 85.
6. "Woman Found Suicide at the Y," *San Francisco Examiner*, May 13, 1961, 5.

## Chapter 3: Acid

1. Jeffrey J. Kripal, *Esalen: America and the Religion of No Religion* (Chicago: University of Chicago Press, 2007), 85.

## Chapter 4: American Indian

1. Sherry L. Smith, *Hippies, Indians and the Fight for Red Power* (New York: Oxford University Press, 2012), 5.
2. Alfred Frankenstein, "A Landmark of a Flop," *San Francisco Chronicle*, November 13, 1963, 6.
3. Paula L. Wagoner, "The Search for an Honest Man: Iktomi Ȟicala as an Ethnohistorial and Humanistic Conundrum," in *Transforming Ethnohistories: Narrative, Meaning, and Community*, ed. Sebastian Felix Braun (Norman: University of Oklahoma Press, 2013).
4. Calvin Fentress, "The Next to Last Book on Earth," *Washington Post*, November 16, 1980, https://www.washingtonpost.com/archive/lifestyle/1980/11/16/the-next-to-last-book-on-earth
5. Ansel Adams, personal letter, February 11, 1964, Stewart Brand Papers, Stanford University Special Collection.
6. Lyndon Baines Johnson's inaugural address, January 20, 1965, https://avalon.law.yale.edu/20th_century/johnson.asp
7. "Wyoming Indian Powwow Opens," *Lawton Constitution*, July 28, 1964, 10.

## Chapter 5: Multimedia

1. Sherry L. Smith, *Hippies, Indians, and the Fight for Red Power* (New York: Oxford University Press, 2012), 5.
2. Smith, *Hippies, Indians, and the Fight,* 5.
3. Alfred Frankenstein, "Surrealist Film Show on Indians," *San Francisco Chronicle*, November 23, 1965, 44.
4. Tom Wolfe, *The Electric Kool-Aid Acid Test* (New York: Farrar, Straus and Giroux, 1968), 222.
5. http://rockprosopography101.blogspot.com/2009/08/december-18-1965-big-beat-palo-alto.html
6. Phil Lesh, *Searching for the Sound: My Life with the Grateful Dead* (New York and Boston: Little, Brown, 2005), 67.
7. Donovan Bess, "Opening the Golden Gate," *Realist*, December 1966, 8.
8. Bess, "Opening the Golden Gate," 8.
9. Laurel Graeber, "So Where Is the Lava Lamp Now?," *New York Times,* April 6, 1995, sec. C, 6.
10. Mike Fleiss, dir., *The Other One: The Long Strange Trip of Bob Weir* (Seoul: Next Entertainment, 2014). Distributed by Netflix, 2015.
11. Charles Perry, *The Haight-Ashbury: A History* (New York: Vintage Books, 1984), 41.
12. Peter Richardson, *No Simple Highway: A Cultural History of the Grateful Dead* (New York: St. Martin's Press, 2015), 60.
13. Andrew G. Kirk, *Counterculture Green: The Whole Earth Catalog and American Environmentalism* (Lawrence: University of Kansas Press, 2007), 42.
14. Michael Malone, "Stewart Brand: From Hippie Prince to Software Savant," *San Jose Mercury News West Magazine*, September 16, 1984, 16.
15. Yale Joel, "Psychedelic Art," *Life*, September 9, 1966, 19.
16. Kate Daloz, *We Are as Gods: Back to the Land in the 1970s on the Quest for a New America* (New York: PublicAffairs Books, 2016), 92.
17. James Hurd Nixon, "The Experimental College: A Bit of a Memoir," unpublished memoir, June 2017.

## Chapter 6: Access to Tools

1. Ed McClanahan and Gurney Norman, "The Whole Earth Catalog," *Esquire*, July 1970, 120–21.
2. https://www.digibarn.com/collections/newsletters/peoples-computer/peoples-1972-oct /1972-10-PCC-cover-medium.jpg
3. Phil Garlington, "Hippies 'War' Declared," *Daily Gater*, February 10, 1967, 1.
4. John Markoff, "It's Moore's Law, but Another Had the Idea First," *New York Times*, April 18, 2005.
5. Stewart Brand, "History," *The Last Whole Earth Catalog* (Menlo Park, CA: Portola Institute, distributed by Random House, 1971), 439.
6. Brand, "History," 439.
7. Brand, "History," 439.
8. Jackson Lears, "Aquarius Rising," *New York Review of Books*, September 27, 2018, 8.
9. *Whole Earth Catalog*, Fall 1970, 1.
10. McClanahan and Norman, "The Whole Earth Catalog," 95.
11. Edmund Ronald Leach, *A Runaway World?* (Oxford, UK: Oxford University Press, 1968), 1.
12. John Markoff, *Machines of Loving Grace: The Quest for Common Ground Between Humans and Robots* (New York: Ecco Press, 2015), 107–8.
13. Thomas Rid, *Rise of the Machines: A Cybernetic History* (New York: W. W. Norton, 2016), 171.
14. *Whole Earth Catalog* Supplement, January 1969, 19.
15. Joe Kane, "This Year's Brand," *Esquire*, July 1983, 69.
16. Tom Wolfe, *The Electric Kool-Aid Acid Test* (New York: Farrar, Straus and Giroux, 1968), 2.
17. Steven Levy, *Insanely Great: The Life and Times of Macintosh, the Computer That Changed Everything* (New York: Viking, 1994), 42.
18. Philip Morrison, *Scientific American* 220, no. 6 (1969): 142, http://www.jstor.org/stable/24926396
19. Andrew Kirk, *Counterculture Green: The Whole Earth Catalog and American Environmentalism* (Lawrence: University Press of Kansas, 2007), 68.
20. Kirk, *Counterculture Green*, 73.
21. McClanahan and Norman, "The Whole Earth Catalog," 95.
22. Alan Cline, "Whole Earth Man Shocks Probers," *San Francisco Examiner*, May 2, 1970, 3.
23. Cline, "Whole Earth Man," 3.
24. John Markoff, *What the Dormouse Said: How the Sixties Counterculture Shaped the Personal Computer Industry* (New York: Viking, 2005), 282.
25. Thomas Albright and Charles Perry, "The Last Twelve Hours of the Whole Earth," *Rolling Stone*, July 8, 1971.

## Chapter 7: CoEvolution

1. http://hrlibrary.umn.edu/instree/humanenvironment.html
2. Ross Gelbspan and David Gurin, "Woodstockholm '72: The Subject Is Survival," *Village Voice*, May 11, 1972, 29.
3. The first use of the term was in an advertisement for the Hewlett-Packard 9100A desktop calculator in *Science* magazine in October of 1968. However, Brand's use of it in describing Alan Kay's Dynabook in his 1974 work *II Cybernetic Frontiers* was the first modern use. The 9100A was programmable but only for numeric applications.
4. Stewart Brand, "SPACEWAR: Fanatic Life and Symbolic Death Among the Computer Bums," *Rolling Stone*, December 7, 1972, 33.
5. Paul Krassner, *Sex, Drugs & the Twinkie Murders* (Port Townsend, WA: Loompanics Unlimted, 2000), 55.
6. Paul Krassner, "A Question of Lifestyle: Fond Memories of My Old Roommate Stewart Brand," *Los Angeles Times Book Review*, 15, https://www.latimes.com/archives/la-xpm-1994 -12-18-bk-10177-story.html

7. Krassner, "A Question of Lifestyle," 15.

8. Susan Elizabeth Lewak, "Sustainable Gardens of the Mind: Beat Ecopoetry and Prose in Stewart Brand's Whole Earth Publications," PhD diss., University of California at Los Angeles, 2014, iii.

9. Quoted in Kevin Kelly, *Out of Control* (New York: Perseus Books, 1994), 85.

10. Russell Schweickart, "No Frames, No Boundaries," in *Earth's Answer: Explorations of Planetary Culture at the Lindisfarne Conferences* (New York: Lindisfarne Books/Harper & Row, 1977), 5.

11. Miriam Pawel, *The Browns of California: The Family Dynasty That Transformed a State and Shaped a Nation* (London: Bloomsbury Publishing, 2018), 229.

12. Orville Schell, *Brown* (New York: Random House, 1978), 140.

13. Bill Press, *From the Left: A Life in the Crossfire* (New York: St. Martin's Publishing Group, 2018), 101nia.

14. Stewart Brand, "Local Dependency," *CoEvolution Quarterly*, Winter 1975, 5.

15. Thomas Albright, "Stewart Brand Really Knows How to Throw a Party," *San Francisco Examiner*, August 20, 1978, 15.

## Chapter 8: Anonymity

1. Art Kleiner, "A History of Magazines on a Timeline," *CoEvolution Quarterly*, Spring 1979, 59.

2. Calvin Fentress, "The Next to Last Book on Earth," *Washington Post*, November 16, 1980. https://www.washingtonpost.com/archive/lifestyle/1980/11/16/the-next-to-last-book-on-earth

3. Stewart McBride, "Stewart Brand and His Five Pounds of Ideas for the '80s," *Christian Science Monitor*, January 15, 1981.

4. Sue M. Halpern, "Private Jets and Road Kills," *Nation*, December 27, 1980.

5. Katherine Fulton interview with the author, September 1994.

6. Stewart McBride, "Uncommon Courtesy: Stewart Brand Launches a School of 'Compassionate Skills,'" *Christian Science Monitor*, August 19, 1982.

7. Stewart Brand, "Uncommon Courtesy: Triple Training," *CoEvolution Quarterly*, July 1983, 38.

8. Stewart Brand, "Gossip," *CoEvolution Quarterly*, April 1984, 134.

9. Stewart Brand, "SPACEWAR: Fanatic Life and Symbolic Death Among the Computer Bums," *Rolling Stone*, December 7, 1972, 33.

10. Brand, "Gossip," Spring 1983, 152.

11. Brand, "Gossip," Fall 1984, 143.

12. Barbara Kelley, "Whole Earth Founder Strolls into 'Empowering Field of Mind Tools,'" *Christian Science Monitor*, December 17, 1984.

13. Stewart Brand, "Why and How This Magazine Is Non-profit," *Whole Earth Software Review*, Spring 1984, 125.

14. Stewart Brand, "A Book-in-Progress," *Whole Earth Software Review*, Spring 1984, 1.

15. Michael Malone, "Stewart Brand: From Hippy Prince to Software Savant," *San Jose Mercury News*, September 16, 1984, 6.

16. Stewart Brand, "Apocalypse Juggernaut, Hello," *Whole Earth Catalog*, January 1970, 21.

17. William Gibson, *Neuromancer* (New York: Ace Books, 1984), 52.

18. David Crook, "Spreading the Silicon Gospel," *Los Angeles Times*, February 28, 1984, 32.

19. Crook, "Spreading the Silicon Gospel."

20. Katie Hafner, *The Well: A Story of Love, Death & Real Life in the Seminal Online Community* (New York: Carroll & Graf, 2001), 9.

21. https://en.wikipedia.org/wiki/Usenet

22. John Markoff, "Whole Earth State-of-the-Art Rapping," *New York Times*, August 15, 1989, A14.

23. Ken Kelley, "The Interview: Whole Earthling and Software Savant Stewart Brand," *SF Focus*, February 1985, 76.

24. Steven Levy, *Hackers: Heroes of the Computer Revolution* (New York: Anchor Press/Doubleday, 1984), 27–32.
25. Fentress,"The Next to Last Book on Earth."
26. John Markoff, "Up to Date," *BYTE*, March 1985, 355.
27. Steven Levy, "Hackers at 30: 'Hackers' and 'Information Wants to Be Free,'" *WIRED* BackChannel, https://www.wired.com/story/hackers-at-30-hackers-and-information-wants-to-be-free/
28. Levy, "Hackers at 30."
29. Stewart Brand, *The Media Lab: Inventing the Future at MIT* (New York: Viking Penguin, 1987), 190.
30. Lawrence Hunter, "Gadgets for Utopia," *New York Times*, September 27, 1987.

## Chapter 9: Learning

1. "Oil: The 30-Year Anniversary of the 1986 Collapse," *Oil & Gas 360*, January 29, 2016, https://www.oilandgas360.com/oil-the-30-year-anniversary-of-the-1986-collapse/
2. Peter Schwartz and Peter Leyden, "The Long Boom: A History of the Future, 1980–2020," *Wired*, July 1997.
3. Howard Rheingold, "An Exercise in Perspective," *Deeper News*, September 22, 1989.
4. Rheingold, "An Exercise in Perspective."
5. Stewart Brand with John Aiello, "The Rescue That Failed—Disorganization, Empty Hydrants Killed Janet Ray," pt. 1 of series of 3, *San Francisco Chronicle*, April 16, 1990, 32.
6. Stewart Brand, "Earthquake Lessons: The Crucial Difference of Volunteers," *Whole Earth Review*, Spring 1990.
7. Brand, "Earthquake Lessons."
8. Brian Eno, *A Year with Swollen Appendices* (Boston: Faber and Faber, 1996).
9. Rheingold, "An Exercise in Perspective."
10. Katie Hafner, *The Well: A Story of Love, Death & Real Life in the Seminal Online Community* (New York: Carroll & Graf, 2001), 11.
11. John Seabrook, *Deeper: My Two-Year Odyssey in Cyberspace* (New York: Simon & Schuster, 1997), 161.
12. John Markoff, "A Free and Simple Computer Link," *New York Times*, December 8, 1993, D1.
13. Jane Jacobs personal letter to Katherine Fulton, August 25, 1994.
14. Londoner's Diary, "Jolly Rogers Sails into Architect Row," *Evening Standard*, September 9, 1996, 8.

## Chapter 10: Float Upstream

1. Paulina Borsook, "The Accidental Billionaire," *Wired*, August 1994, http://www.paulina borsook.com/PDF-disk-1/The%20Accidental%20Zillionaire_WIRED.pdf
2. Jeff Bezos letter to shareholders, 1997, https://www.sec.gov/Archives/edgar/data/1018724 /000119312513151836/d511111dex991.htm
3. Stewart Brand, *Whole Earth Discipline: Why Dense Cities, Nuclear Power, Transgenic Crops, Restored Wildlands, and Geoengineering Are Necessary* (New York: Penguin Books, 2010), 78.
4. Peter Schwartz and Doug Randall, "An Abrupt Climate Change Scenario and Its Implications for United States National Security," Global Business Network, October 2003.
5. David Stipp, "The Pentagon's Weather Nightmare," *Fortune*, January 12, 2004.
6. Peter Schwartz and Doug Randall, "How Hydrogen Can Save America," *Wired*, April 2003.
7. Peter Schwartz and Spencer Reiss, "Nuclear Now!" *Wired*, February 2005.
8. Felicity Barringer, "Old Foes Soften to New Reactors," *New York Times*, May 15, 2005, A1.
9. Katherine Seligman, "The Social Entrepreneur," *San Francisco Chronicle*, January 8, 2006.
10. Joseph Berger, "A Rockefeller Known Not for Wealth but for His Efforts to Help," *New York Times*, June 23, 2014, A22.

11. John Tierney, "An Early Environmentalist, Embracing New 'Heresies,'" *New York Times*, February, 27, 2007, https://www.nytimes.com/2007/02/27/science/earth/27tier.html

12. Thomas Streeter, "'That Deep Romantic Chasm': Libertarianism, Neoliberalism, and the Computer Culture," in *Communication, Citizenship, and Social Policy: Rethinking the Limits of the Welfare State*, eds. Andrew Calabrese and Jean-Claude Burgelman (Lanham, MD: Rowman & Littlefield, 1999), 49–64; and Fred Turner, *From Counterculture to Cyberculture: Stewart Brand, the Whole Earth Network, and the Rise of Digital Utopianism* (Chicago: University of Chicago Press, 2006).

13. Stewart Brand, *Whole Earth Discipline: An Ecopragmatist Manifesto* (New York: Viking Penguin, 2009), 20.

14. Brand, *Whole Earth Discipline: Why Dense Cities*, 307.

15. John A. Matthews, "The Myths of Renewistan," chapter 13 in *Global Green Shift* (New York: Anthem Press, 2017), 145.

16. Brand, *Whole Earth Discipline: Why Dense Cities*, 80.

17. Amory Lovins, "Stewart Brand's Nuclear Enthusiasm Falls Short on Facts and Logic," *Grist*, October 14, 2009, https://grist.org/article/2009-10-13-stewart-brands-nuclear-enthusiasm-falls-short-on-facts-and-logic/

18. Brand, *Whole Earth Discipline: Why Dense Cities*, 97.

19. Denis Hayes, "An Environmental Provocateur," *Stanford Social Innovation Review*, Winter 2010, 18.

20. Amol Rajan, "Last Night's TV: What the Green Movement Got Wrong," *Independent*, November 5, 2010, https://www.independent.co.uk/arts-entertainment/tv/reviews/last-night-s-tv-what-green-movement-got-wrong-channel-4ego-strange-and-wonderful-world-self-portraits-bbc4-2125523.html

21. Lucy Bellwood, "Rogue Environmentalist Draws Protest," *Reed Magazine*, March 17, 2011, https://www.reed.edu/reed_magazine/sallyportal/posts/2011/rogue-environmentalist-draws-protests.html

22. Bellwood, "Rogue Environmentalist Draws Protest."

23. *Impacts of the Fukushima Daiichi Accident on Nuclear Development Policies* (Paris: OECD Publishing, 2017), 3.

24. Gar Smith, "Where Is Nuclear Energy Going? A Debate," *Berkeley Daily Planet*, July 27, 2011, https://www.berkeleydailyplanet.com/issue/2011-07-27/article/38192?headline=Where-is-Nuclear-Energy-Going-A-Debate

25. Smith, "Where Is Nuclear Energy Going?"

## Epilogue

1. Tim M. Lenton and Bruno Latour, "Gaia 2.0," *Science* 361, no. 6407 (September 14, 2018): 1066–68.

2. John Markoff, "The Dust on a Butterfly's Wings," *Alta Journal*, Spring 2021, Issue 15, 30.

3. "Talk of the Town," *New Yorker*, May 2, 1970.

# Illustration Credits

# Index

# DR. SCHWEITZER

# of LAMBARÉNÉ

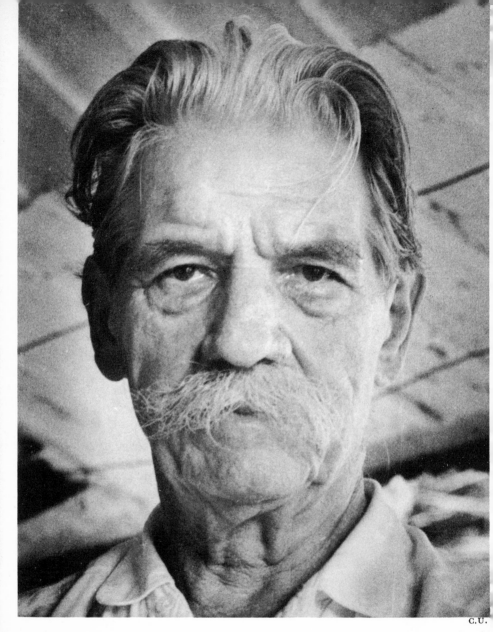

C.U.

"It is good to be reminded now and then that even in a world struggling with the momentous issue of war and peace the individual has problems."

# DR. SCHWEITZER

## of LAMBARÉNÉ

by Norman Cousins

With Photographs by Clara Urquhart

Harper & Brothers, Publishers, New York

To My Father

# Author's Note

This book is in the nature of a personal appreciation. It does not seek to be either an historical analysis of an eminent contemporary or a detailed biographical treatment. It is concerned with the carrying power of a symbol and with some of the people who are part of it. It was constructed from notes taken on a trip to Africa. Though most of these notes are about a man at Lambaréné, some of them are in the nature of digression. Lambaréné is a good place for digressions, especially those of a retrospective turn.

A word about the photographs. The initials C.U. belong to Clara Urquhart. Mrs. Urquhart is not to be taxed with the responsibility for the photographs that carry no initials; these were taken by the author.

Mrs. Urquhart, who was with me in Lambaréné, has given me the benefit of her own recollections and has made important suggestions about the manuscript. Erica Anderson, who made the major film about Dr. Schweitzer, checked the facts in this manuscript and spared me the agony of a number of errors. Nicholas Balint helped check the proofs. Sallie Lou Parker picked up after me graciously and generously, and put up with an author whose changes on manuscript necessitated at least a dozen retypings. To all these, and to a forbearing wife and daughters, I give acknowledgments and thanks.

N.C.

Outdoor cathedral; Sunday morning service on Hospital Row.

# I

AT THE END of dinner each evening at his jungle Hospital in Lambaréné, French Equatorial Africa, Dr. Albert Schweitzer would fold his napkin, announce the number of the hymn to be sung, get up and walk over to the upright piano on the other side of the room. He would arrange the hymn carefully on the music stand, study it for a moment, then start to play.

I doubt whether I shall ever forget my shock and disbelief when, the first evening of my visit, I saw him approach the upright. Earlier in the day, while exploring the Hospital on my own, I had wandered into the dining room where Dr. Schweitzer and his staff of fifteen eat each day. The first thing that caught my eye was the piano. It must have been at least fifty years old. The keyboard was badly stained; large double screws fastened the ivory to each note. I tried to play but drew back almost instantly. The volume pedal was stuck and the reverberations of the harsh sounds hung in the air. One or more strings were missing on at least a dozen keys. The felt covering the hammers was worn thin and produced pinging effects.

Before coming to Lambaréné, I had heard that under equatorial conditions of extreme heat and moisture one doesn't even try to keep a piano in tune; you make your peace with the inevitable and do the best you can.

Even so, when I saw Dr. Schweitzer sit down at the piano and prop up the hymnbook, I winced. Here was one of history's greatest interpreters of Bach, a man who could fill any concert hall in the world. The best grand piano ever made would be none too good for him. But he was now about to play a dilapidated upright virtually beyond repair. And he went at it easily and with the dignity that never leaves him.

I knew then that I would never be able to put out of my mind the image—painful in one sense, exalting in another—of Schweitzer at the old upright in Lambaréné. For here was the symbol, visible and complete, of everything he had given up in order to found a hospital in Africa. Renunciation by itself may mean little. What is renounced and the purpose of the renunciation—that is what is important. In the case of Albert Schweitzer, renunciation involved a distinguished career as organist and pianist; it extended to the intimate study and analysis of the nature of music in general and the organ in particular; it embraced a detailed understanding of the life and meaning of Johann Sebastian Bach. In all of this work there was the meticulous pursuit of perfection. Yet this did not exhaust the renunciation. There was a record in theology, philosophy, and history, in each of which Schweitzer had made major contributions as teacher and author. Solid foundations had been built for a lifelong career in any of these fields.

I wrote a moment ago that I felt not only pain but a certain inspiration in the image of Schweitzer at the old piano. For the amazing and wondrous thing was that the piano seemed to lose its poverty in his hands. Whatever its capacity was to yield music was now being fully realized. The tinniness and chattering echoes seemed subdued. It may be that this was the result of Schweitzer's intimate acquaintance with the piano, enabling him to avoid the rebellious keys and favoring only the co-operative ones. Whatever the reason, his being at the piano strangely seemed to make it right.

And, in a curious way, I discovered that this was to be true of almost everything else at Lambaréné. Schweitzer's being there made it right. Much of what you saw for the first time at the Hospital seemed so primitive and inadequate as to startle. But when Dr. Schweitzer walked through the grounds, everything seemed as it should be. More than that: the profound meaning of Lambaréné suddenly came to life. And I was to learn that there was a reason behind everything at Lambaréné.

And I was to get close to one of the things that drew me to Lambaréné—not Schweitzer's purpose, which was clear enough, but the sources of that purpose, about which I had long wondered.

"YOU MUST COME to Lambaréné," Emory Ross had said early in 1955. "There is something there that I can't capture for you in words but it will mean much to you when you come near it."

Emory Ross was the head of the Schweitzer Fellowship in the United States. He was a former missionary who was now devoting his life to advancing the cause of the African people. He had helped to build schools in various parts of Africa; he had brought promising African students to American colleges and universities; he had raised thousands of dollars for Dr. Schweitzer's Hospital in Lambaréné.

After being with Emory Ross for only a minute or two, you felt you were with a country doctor whose knowledge and skill could put him at the head of almost any hospital in the country but who preferred to sit at the non-specialized bedsides of people who needed him. Emory Ross's manner was extraordinarily kind and reassuring; you never left him with the feeling that half of what you wanted to talk to him about had been left unsaid.

One of the things Emory Ross and I had discusssed at those early meetings was Dr. Schweitzer's unfinished literary works. There were at least two books on which he had been working off and on for more than a quarter of a century. One was concerned with philosophy and history; the other with theology.

Dr. Ross said he believed the manuscripts were fairly close to completion, but that the Hospital's demands on Schweitzer's time were increasing with each passing year. So far as Dr. Ross knew, months would go by without the Doctor's being able to touch his manuscripts. Worse still was the fact that there were no carbon copies. The Doctor wrote in longhand on faded sheets of paper. He was in the habit of hanging them on nails in his room; Dr. Ross spoke of the hazards of moisture, wandering goats, possible fire, and just plain loss.

"I tremble when I think of what would happen if some of that manuscript should come undone," he said.

We discussed various means of persuading the Doctor, then in his eighty-first year, both to take the time necessary to finish his books and to provide for the safety of the manuscripts. We also considered various methods for making duplicate copies, and we agreed to investigate the comparative merits of microfilming, duplicating machines, and plain photography. None of these devices, of course, could be used without the permission and co-operation of the Doctor himself. And here we anticipated trouble. Emory Ross emphasized that Dr. Schweitzer was a perfectionist who was severely reluctant to part with anything he wrote that was not absolutely final and complete. Dr. Ross was contemplating a trip to Lambaréné, but he said he didn't feel in a position to press the issue.

"It's too easy for the Doctor to say 'no' to me. You've got to come and put it to him. It won't be easy. If his mind is made up about something, it takes some real powers of persuasion to move him. Even if we don't succeed, it is important that you come. It will be an experience you'll never forget."

I could feel the beginning of an irresistible tug; but I had only recently returned from Japan in connection with the project to provide reconstructive surgery for some young women from Hiroshima who had been disfigured by the atomic bombing, and my time away from the magazine was limited. And so I told Dr. Ross that much as I wanted to go, I couldn't quite be sure I could do so.

In the weeks that followed, Dr. Ross and I met with other members of the Schweitzer Fellowship. I recall with particular pleasure a luncheon with Erica Anderson, Jerome Hill, and Eugene Exman. Miss Anderson had just returned from Lambaréné where she worked on a film biography of Dr. Schweitzer. Miss Anderson had also just completed, in collaboration with Eugene Exman, vice-president of Harper & Brothers, a picture-and-text book about Dr. Schweitzer and the Hospital.

Emory Ross explained the purpose of the luncheon to the small group. He spoke of the invaluable literary treasures now at Lambaréné and the loss to the world if they should be damaged or lost.

Then he asked: "What do you think the chances are that we could persuade the Doctor to have the manuscripts duplicated?"

"I can tell you," Miss Anderson said, "that the Doctor is suspicious of anyone who arrives at Lambaréné with a lot of mechanical equipment."

"If you approach the Doctor about it directly, it's my guess he'd give you a flat 'no,' " Eugene Exman said. "He just doesn't like to be prodded about his manuscripts—either with respect to finishing them or taking precautions for their physical safety. You've got to give this real thought."

Jerome Hill nodded assent.

As I listened, a picture formed in my mind of Schweitzer as an austere and remote figure who could be approached only with the greatest care. I found it interesting that even those who knew him best had to conjecture about his response to situations. They had to plan any action involving him as carefully as they would the strategy for a military campaign.

"Perhaps we ought also to find out what Dr. Schweitzer's friends in Europe think," Dr. Ross said. "I am sure Emil Mettler in London would know something about the manuscripts. And Dr. Schweitzer's daughter in Zurich ought to know something about it. J. D. Newth in London, who is Dr. Schweitzer's English publisher, may be able to give us a lead."

Erica Anderson's eyes were sparkling with the challenge.

"I think maybe the best thing to do would be just to go there," she said. "After you are there a while, tell the Doctor what you want to do. But just don't turn up with a lot of equipment showing. I made the mistake of doing that once and I ought to know."

I recalled that several years earlier Erica Anderson had received a flat refusal from Dr. Schweitzer when she wrote to him about her hope that she might do a film story of his life. But he invited her to come to Lambaréné just the same—without the film equipment. She went and got him to change his mind.

Incidents were then related at the table that gave me a contrasting picture of the Doctor to the one I had had earlier. I had associated him with warmth and responsiveness. The new image seemed somewhat aloof and austere. And a paradox began to emerge. Later, I was to learn at Lambaréné, that this was only one of several paradoxes about the man whose life embraced at least four full careers.

THE RECEPTIONIST at *The Saturday Review* announced Mr. Newth from London, representing A. & C. Black, publishers. He said he was on a short visit to the United States and had learned from Emory Ross of the prospect of our trip to Lambaréné.

I told Mr. Newth, as I had earlier told Emory Ross, that eager as I was to go, the matter was far from settled. I explained the circumstances that were holding me back.

"We are still hopeful that we can publish the Schweitzer manuscripts during his lifetime," he said. "I don't want to prod, but it might be helpful if you could manage the trip. A new voice is needed to talk to the Doctor. He has heard the same arguments from his old friends for so many years that it is too easy for him to wave them aside. We need fresh reinforcements. I hope you will do it."

What Mr. Newth said was most persuasive; I thanked him and told him I would keep the matter open. As the weeks passed, however, my reluctance to leave the family and the magazine so soon after the Far East trip became strengthened. Besides, the project for the Hiroshima Maidens was now in full swing, with dozens of operations yet to be performed by Dr. Arthur Barsky, Dr. Bernard Simon, and Dr. Sidney Kahn. The medical program, too, under Dr. William Hitzig, was far from complete.

When the time neared for Dr. Ross's own trip to Lambaréné, he spoke to me again about the possibility of my accompanying him. The chances were now nil. But he exacted a promise that I would go when I could. Dr. Ross would take no photographing equipment with him but would attempt to clear the way for my visit when I could make it. Meanwhile, I was to write to Dr. Schweitzer, urging him to give favorable consideration to our project.

Several weeks later, a reply arrived from Lambaréné. The Doctor was most cordial and thanked me for my interest, but said that he just did not have enough time to do anything about the manuscripts. He invited me to visit him at the Hospital when I felt free to do so. A postscript referred to an editorial I had written some time earlier, called "The Point about Schweitzer." In that editorial I had differed with recent visitors to Lambaréné who had complained about the primitive aspects of the place. The point I tried to make was that the Schweitzer symbol was more important than modern facilities. Dr. Schweitzer's postscript said that he hoped to justify the kind things I had written about him.

When Emory Ross returned from Africa several months later, he brought back a report of cautious optimism. The Doctor didn't seem to want to talk about his manuscripts at first. But Dr. Ross managed to find him alone one afternoon and spoke to him fully about the concerns of his friends. Dr. Schweitzer's response enabled Emory Ross to come away with the feeling that the project now had an even chance at least.

By this time, the surgical program of the Hiroshima Maidens was well advanced. Nine of the girls had completed all their operations and would soon be ready to return to their families in Japan. The others would probably be ready to return as a group in the fall. My obligations were thinning out.

ONE DAY IN September, 1956, Mrs. Clara Urquhart, who had been associated with Dr. Schweitzer for many years and who had just come from Lambaréné, visited the offices of the magazine. She had much to relate about Schweitzer—about his work and manuscripts and, in general, about his state of mind. She had been with the Doctor at the time he read my letter concerning his books, and she had something to say in that connection.

"You must not be discouraged," she began.

Later, I was to discover that no five words were more characteristic of Clara Urquhart than "you must not be discouraged." She never underestimated the difficult but never made the mistake, either, of confusing the difficult with the impossible.

"You must not be discouraged," she said again, "just because the Doctor said 'no' to you in his letter. He is so overburdened with work at the Hospital that he almost automatically says 'no' to anything that would make additional demands on his time. A certain innate modesty often makes him seem negative. But I think he really wants to complete his books."

"Has he done any work on them recently?"

"Very little, if at all. I've been after him for years about it."

"Can anything be done? Should anything be done?"

"Recently," she said, "he promised me he would do some work on the manuscripts. He came to his room early in the afternoon and began to write. I returned an hour later and peeked into the room. The Doctor was no longer there. A breeze had blown some of the sheets of the manuscripts off the desk. An antelope had wandered into the room. Some of the sheets had been trampled upon. I had no way of knowing whether any had been eaten.

"I gathered up the papers and smoothed them out. Right then, I became determined to see this through.

"When the Doctor returned to the room I told him what had happened. He shrugged. I said that even though he was reluctant to finish the manuscripts, the least he could do was to attend to the physical safety of his papers.

"Unlike previous occasions, when he brushed that kind of talk aside, this time he said nothing. My guess is that he is about ready to change his mind. I hope you will accept his invitation to come to Lambaréné. And so, if you want an accomplice in the project to copy the manuscript, I'd like to volunteer."

You didn't have to be with Clara Urquhart very long to know that this slight, dark woman possessed a rare combination of intensity of feeling with calmness of manner. She knew how to establish rapport in ten minutes that would take some people ten weeks. Also, she had the art of absolute relevance. When she listened, you had the feeling that all her energy was being mobilized in absorbing every sound and capturing your total intent. And when she spoke, she would address herself with precision to your questions or the things that interested you—and frequently to the thought behind your questions.

She had brought with her some photographs she had taken at Lambaréné over the years. Many of them were being incorporated in a book about Schweitzer shortly to be issued in London. Several of the photographs were intimate portraits of Dr. Schweitzer. They showed a face of vast power and purpose. It was lean and strong, with lines associated with expressiveness and sensitivity rather than with age. The eyes were set wide apart; they were like steel lanterns in the rugged landscape of his face.

Other photographs were of the people associated with Schweitzer at the Hospital. These were people I wanted to meet, and I said so.

"When are you leaving for Lambaréné?" Clara Urquhart asked.

I said I thought very soon.

THERE WAS YET another reason why I felt compelled to go to Lambaréné.

Ever since the end of the war, there had been one voice which might have had a powerful effect on the biggest issue of the age, but that voice was silent. I knew that Albert Schweitzer had deliberately avoided political issues in order to confine himself to the fundamental and overriding moral questions of concern to the entire human community. But such an overriding question now existed. It was whether the conditions which made human life possible on this planet could be maintained. The means now existed that could alter or destroy those conditions.

Could anyone who was concerned about the problem of ethics in modern man exempt himself from such a question? Albert Schweitzer believed in the sacredness of man. Was there no moral issue if man's genes were to be twisted, if the air he breathed was to be fouled, if unborn generations were to be punished for his present failures?

This crisis could not be easily met by the political leaders of the nations. For the requirements of sovereignty too often came first. The political leader was the spearhead of that total sovereignty, its chief presiding officer. In order that the entire human grouping be served, it would become necessary to create something higher than the nation itself; indeed, it would be necessary to create the means whereby the nation would find security through means other than massive armaments or coalitions. And if something beyond the nation had to be advocated, the national leaders might not be the most logical spokesmen. A man like Albert Schweitzer might enable people to see the need for fashioning allegiances to each other as members of the human commonwealth. What man most lacked was a consciousness of his relationship to other human beings. He lacked adequate awareness of the gift of human life and what was now required to preserve it. Alongside the real threat to life on earth the razor-sharp distinctions he insisted on making between himself and others now served as a dead weight for his hopes.

If I was wrong about the nature of the problem; if I was wrong about the feeling I had that time did not work for us but against us, then the man who could convince me of this was Albert Schweitzer. But if Albert Schweitzer agreed that the problem was real and universal, then it was important and proper for him to speak. And if his reluctance to speak was the result of humility or doubts as to whether his words would be heard, then I could at least attempt to remove these doubts. I did not take lightly the privilege of being in a position to try.

NOW THE REAL planning for the trip began. I had two more meetings with Clara Urquhart before she left for London. She also came out to the house at a picnic for the Hiroshima Maidens and met my wife Ellen and the four girls.

At one of these meetings she had a long checklist of questions to be asked and things to be done.

"How's your French?" she asked. "The Doctor does not speak English though he reads it with little difficulty and understands far more than he admits."

"My French is about the way you describe the Doctor's English."

"What about German?"

"Nil."

"You must not be discouraged," she said. "The Doctor is accustomed to communicating with English-speaking visitors through an interpreter, in German."

When I asked who would do the interpreting, Clara Urquhart said this was all part of the work of the "accomplice's" job that she had volunteered to perform.

Her next question concerned our approximate date of departure.

I made some hasty calculations. The Hiroshima Maidens still in the United States would be returning in October or November. The busy season at the magazine tapered off just before Christmas. That would mean that I might be able to get to Lambaréné late in December or early in January.

Mrs. Urquhart's plan was to leave for London and Rome to see some friends, then to return to her home in Johannesburg for a month. We could meet in Johannesburg, perhaps, she said, then go to Lambaréné. Meanwhile, she would write to Dr. Schweitzer to say that we were definitely coming.

"You'd better make up your mind to stay for a full week or more," she said. "The Doctor says that short visits make him nervous."

I agreed to stay for as long as was required to do the job.

"Now," she said, "comes the most important matter of all. You've got to promise that you won't be disillusioned."

I smiled.

"You mean a hospital ward without bedsheets, lack of sanitation, and all that sort of thing?" I said. "Please don't let it worry you; I know all about it. It was this kind of argument that seemed to me all along to miss the main point about Schweitzer."

"There's something more important than that," she replied. "I'm talking about Schweitzer himself."

This startled me. "Why is there any danger that I or anyone else would be disillusioned about Schweitzer?"

"Some people are. They come to Lambaréné with an image of a sort of sweet saintly St. Francis feeding the birds and they see instead a driving man fighting the jungle and African lethargy and they do not remain for a sufficiently long period to see or sense the goodness and saintliness underneath. They go away feeling hurt and unhappy."

I was touched by her concern but still puzzled. What was there about Schweitzer that created "hurt and unhappy" feelings in people? Whatever the answer, I couldn't guarantee Clara Urquhart what my feelings would be after I met him.

"Of course not," she said. "I just want to be sure you'll stay long enough to get over some first impressions that may not be so favorable."

"Such as?"

"Such as the fact that those who do not know the Doctor will think that his manner toward the indigene or black is unfeeling and authoritarian.

"Such as the fact that his views seem to reflect little confidence in the Africans to whom he has given his life. Schweitzer has deeper and wider dimensions than anyone else I have met. If evaluated from a superficial viewpoint the image is distorted. For better or for worse Schweitzer is a patriarch. I remember saying to him that he was an enlightened despot—to which he replied: 'An enlightened despot is able to give the greatest amount of freedom.'

"If one fails to remember that his basic motivation is reverence for life—he might seem arbitrary in his dealings with those around him. Just wait and observe for the first few days."

Three months later, I was on my way to Lambaréné.

# II

THE AIR CONNECTION from Brazzaville to Lambaréné in French Equatorial Africa was probably the most hazardous regularly scheduled flight in the world. It was operated by Air France over jungle mountain country. But the safety record of this particular run was close to the top among the world's airlines.

The men who flew the DC-3 on these jungle hops had earned a high reputation in the society of world airline pilots. The African "airports" at which they stopped had no radio beams for guiding planes through fog and rain, no light towers or signal beacons flashing across the sky, no neon ground markers or flares, no paved landing strips to pick out from the air—just a clearing with a dirt or grass strip. There were no sprawling cities to help a pilot get his bearings, no air terminal buildings or anything approaching them; generally, just a single small structure that served as an open shelter. Jungle country tends to look pretty much alike when you fly north from the Congo. Yet these airmen had a homing-pigeon touch at the controls that was the talk of their trade.

While we were waiting at the Brazzaville airport for the announcement that our plane was ready, Clara Urquhart called out to a tall, slender young man of about thirty who had just walked into the terminal building. She identified him as Dr. Frank Catchpool, from the staff of the Schweitzer Hospital.

Dr. Catchpool was obviously deeply pleased to see her. It developed that she had made the original arrangements with Dr. Schweitzer for Dr. Catchpool to go to Lambaréné. Frank Catchpool was an English citizen and a Quaker who, like many others, had been inspired by Dr. Schweitzer's example. When he first met Clara in London, he inquired about the Hospital and his chances for joining the staff. Clara wrote to Dr. Schweitzer and the matter was arranged.

Now, at the Brazzaville airport, Clara was able to chat with him for the first time since he had been at Lambaréné. We learned that he had been at the Hospital a little more than a month. He had come to Brazzaville five days earlier, he said somewhat ruefully, because of a little dog.

It all began about a week ago, he explained. One of the managers of a nearby French lumber camp had brought his pet dog, breed anonymous, to the hospital. The dog had been suffering for some weeks with a persistent cough. Dr. Schweitzer's Hospital turns away no patients, regardless of color, species, or previous condition of servitude. And so the manager put his dog under Dr. Schweitzer's care. The staff held a consultation; the consensus was that a bone was stuck in his throat. There being no X-ray machines in operation at Lambaréné, the diagnosis could not be confirmed. In any event, it was decided to operate.

Dr. Catchpool, who had had some previous experience, volunteered to apply the anesthetic. He cautioned the other surgeons and nurses about the astounding power in the sudden moves of even a small animal at the time the anesthetic is applied.

His apprehensions were all too accurate. Just as the anesthetic was applied to his mouth, the dog jerked free and bit Dr. Catchpool on the arm. A second attempt was successful, and the small obstruction was removed.

That evening, Dr. Catchpool's arm began to swell and showed discoloration. There being no antirabies serum at the Hospital, Dr. Schweitzer ordered both Dr. Catchpool and the dog to depart immediately for the hospital at Brazzaville, the former to receive antirabies injections, the latter to be put under observation. And now, four days later, both patients were returning to Lambaréné, the doctor having had his shots, and the dog having developed no symptoms of rabies. The dog was now in a crate which was already checked in and waiting to be loaded.

"I wish I could say I didn't feel pretty silly about this whole business," Dr. Catchpool said. "Dr. Schweitzer must have a rather poor opinion of me for having allowed myself to get into this mess. Here I am at the Hospital only a few weeks and I get immobilized by a little dog."

Clara said she was certain that Dr. Schweitzer had only the most sympathetic understanding of the situation.

"That brings up something else," he said. "I'm afraid I don't know where I stand with the Doctor. We've hardly spoken, except for the most routine things. I haven't wanted to go directly to him and tell him about the kind of work I'd like to do at the Hospital or to discuss other things on my mind. I just haven't got the heart to take up a moment of his time.

"You sound discouraged; you mustn't be," Clara said. "You recall that I asked you not to form any judgments until you had been there at least a month. Give yourself a little more time. Many of the things that are troubling you now will fall into place. What kind of work have you been doing at the Hospital?"

Dr. Catchpool grinned.

"I'm the chief electrician and engineer," he said. "I diagnose faulty wiring and operate on sluggish generators."

"I'm sure what you are doing is most essential," Clara said. "Besides, if I may say so, some of this at least is your own fault. You insisted that I say nothing to the Doctor about your excellence as a physician or your very high recommendations. The Doctor is disposed to take people at their own evaluation of themselves. And I'm sure that when you got to Lambaréné, you persisted in underrating yourself. But you must not be discouraged. Your chance will come."

An attendant came over to announce that our plane was ready.

Dr. Catchpool went off to look after the dog. Clara and I collected our hand luggage and boarded the plane. A moment later Dr. Catchpool joined us.

The air distance between Brazzaville and Lambaréné is perhaps no more than four hundred miles on a straight line. But the planes fly a zigzag route in order to cover the various jungle air stations en route to Lambaréné. The flight therefore generally requires from five to six hours. I memorized the colorful names of the jungle towns at which we stopped so I could tell Ellen and the girls about them when I returned—names like Djambala, Mayumba, Tchibanga, Mouila.

At each stop, curious children from surrounding villages would gather near the open shelter close to the landing strip. They would cluster together and stare with open-eyed wonder at the gleaming steel bird. At Tchibanga two children at the edge of the crowd were having a short game of catch with a rubber ball. I walked toward them and made the kind of throwing motion that indicated I wanted to be invited to join in. One of them tossed me the ball. I threw it back, then he smiled broadly and put some muscle into his next pitch.

I looked at the other children. They seemed astonished to observe a white man obviously enjoying himself in a game with black children. Then, suddenly, a clamor went up as at least fifty youngsters called out to join the fun. I would toss the ball into a sea of waving arms and out it would come again. One boy, about twelve or thirteen, sprang high in the air and caught the ball with one hand. Then he ran off at a distance of about forty yards so that he could show me how far he could throw. This in turn set up a cry from at least a dozen others who wanted to do the same. Almost before I knew it, children were streaking all over the airport clearing. It made me think of one of those bull-fighting festivals when the spectators come swarming down into the arena to try their luck with the brave bulls.

The two African policemen stationed at the field seemed agreeable enough when the game of catch had started innocently a few minutes earlier. But now the point of diminishing returns in their good will was just being reached. I beckoned to the youngsters that the game was over, handed the ball to its young owner, and thanked him.

Then I opened my camera and asked my ball-playing companion for the privilege of his photograph. Instead of ending the clamor, this merely set a new stage for it. In a matter of seconds, the crowd closed in on me and pinned me to the spot as each child called out for his right to be photographed.

All this while, Clara and Dr. Catchpool stood on the far side of the crowd, close to the plane. They were enjoying themselves hugely. Clara's expression as much as said: "You got yourself into it; now get yourself out of it."

Finally, the two African policemen made it clear to the youngsters that the big steel bird was ready to go into the sky again, and they helped to detach me from their midst. I doubt whether I shall ever forget the high-pitched deafening yells of *"au revoir"* and the wildly waving arms of the children as I boarded the plane.

"I hope you noticed how the children drew back at first when you approached but how quickly they responded when they saw that you wanted to make friends," Clara said after the plane had leveled off in flight.

It was, I agreed, very striking.

"You will find that this same thing is true of Africans of all ages almost everywhere on the continent," she said. "There is that initial hesitation. They are not sure what you want or intend to do. They are conditioned to react almost chemically against a white skin. But once you make it clear that you approach them as friends, the response is warm and hearty, almost overwhelming."

She paused, and her words seemed to hang in the air for a moment.

"Let me amend that a bit," she said. "I suddenly realize that this is the kind of observation I would have made almost automatically only a few years ago. But it is no longer true; at least, not to the same extent."

"No longer true in what way?" I asked.

"Things are changing in Africa, very fast," she said. "So fast that it becomes necessary to check one's ideas and reactions now and then just to make sure they are up to date. I am not sure that it is as easy now for a friendly white person to gain the good will and confidence of the Africans as it was only a short time ago.

"The atmosphere has changed. It's more tense. The color lines are hardening. More and more Africans are getting caught up in the nationalist movements. Just as many white people have a tendency to make generalizations about the characters and abilities of the blacks, so there is now a tendency by the blacks to make blanket generalizations about all whites. The feeling is growing that all whites are to be feared and opposed. And a white is identified as anyone who stands or appears to stand in the way of their eventual control of their nation.

"But even here, I've got to be careful not to overstate," she continued. "I don't want you to think that Africans don't smile any more or won't be friendly. Of course they will. What I'm trying to do is give you some idea that those of us who have lived in Africa a long time are aware of a tightening in the air. It's like a far-off storm. You continue to do everything you usually do in good weather but there's that uneasy feeling in the back of your mind that you'd better get ready for a sharp change."

AS SHE SPOKE I thought back to my experience involving those "uneasy" feelings a few days earlier in Johannesburg.

In planning my trip to Lambaréné, it had seemed foolish to travel that far without seeing even a little of the Union of South Africa. For distance, these days, is measured not by miles but by hours. And hours are translated into air units. A nonstop air hop is regarded as one unit. For example, in 1956—before the large-scale use of jets—New York to San Francisco took seven to eight hours but was nonstop and therefore one unit. New York to Paris was nonstop and one air unit (ten hours, flying East). Johannesburg to Brazzaville was one air unit (six hours).

". . . Suddenly, a clamor went up as at least fifty youngsters called out to join the fun. I would toss the ball into a sea of waving arms and out it would come again."

Arrival at Tchibanga: A white man out of the sky can be a frightening experience.

With only one unit separating me from South Africa, the decision to make the additional stop was almost automatic. In this I had the encouragement of Alan Paton who, on his visit to New York some weeks earlier, said that even a brief visit would be well worth the journey. The trials of more than a hundred South Africans under the new, extreme "treason" laws were coming up in Johannesburg and this was a good time to see history in the making.

Clara, who had lived most of her life in Johannesburg, arranged an intensive schedule. I was to meet people who were intimately involved in the problems of the Union. This would include persons of varying political opinions and backgrounds—all the way from owners of large gold mines to African writers and nationalist leaders.

One of the persons Clara was especially eager to have me meet was Henry Nxumalo, one of Africa's leading journalists. His writings appeared regularly in *Drum,* a monthly magazine, and in *The Golden City Post,* a newspaper, of which he was news editor. *Drum* was written by Africans and claimed the largest circulation of any non-European magazine on the continent.

Several people referred to the fact that Nxumalo was presently writing a book on South Africa for an American publisher. His growing importance, I was told, was largely the result of his crusading articles. Everyone I met who knew him said he was one of the soundest and most courageous among African observers. His achievements in journalism were prominently described in the book *Drum* by its former editor, Anthony Sampson.*

Consequently, I looked forward to meeting him at dinner the first night of my visit in Johannesburg.

Nxumalo didn't come to dinner. He had been murdered earlier in the day.

Right then, I learned one of the main facts about life in the Union of South Africa. I learned that there are two worlds. One is the world of graciousness, spaciousness, and infinite natural beauty

* Published in London by Collins (1956) and in the United States by Houghton Mifflin (1957).

and color, heightened in its loveliness by crisp air and sparkling sun. Then there is the other world, made entirely by people. It is taut, harsh, violent. The charming villas by day become places to be closely guarded at night. When two men approach each other after dark, each is apprehensive about the color of the other. Thus, passing a person on the street at night takes on the aspects of an encounter. For when the sun goes down the coolness seems to enter the human soul and the mood of the land hardens.

I knew I could never begin to understand South Africa unless I could understand the nature of this change, unless I could comprehend the proximity of the loveliness to the terror, and the interrelationship between the two. Perhaps if I could find out about Henry Nxumalo—why he wrote as he did, why he was feared and why he was murdered—I might learn a little about the two worlds.

The newspaper obituaries established that Henry Nxumalo, thirty-nine, lived with his wife and five children in the "location" called Orlando, some miles outside Johannesburg. (Africans are not permitted to live in the city itself; they live in "locations"— most of which are slum areas consisting of old huts and shacks and some of which are new housing developments with their well-built though fairly small homes. Orlando was one of the earliest of such developments.)

Nxumalo was born in Port Shepstone, on the East Coast near Durban. He came from a Zulu family. Both his parents died while he was a boy. Henry's first job was as a kitchen boy in Durban. Like thousands of Africans, he found himself lured to the big city, Johannesburg; and he became part of a giant paradox. Once having left the way of life of the village or the small town, the Africans who come to Johannesburg seldom want to return. They resent the white man's world not because they are forced to stay but because they are not fully accepted. What they seek is not freedom to return to the village but freedom to live decently and honorably in the city.

Henry Nxumalo came to Johannesburg as a youth and worked in a boilermaker's shop. After hours he wrote poetry, which he sent to the magazine, *Bantu World,* and much of which was published. This led to a job as messenger for the *Bantu World,* of which he later became sports editor. He enlisted during the war and became a sergeant. This brought him to North Africa. At the end of the war he went to London. When he returned to Johannesburg, he resumed his writing for the *Bantu World.* In 1948 he married a nurse. In 1951 he joined the staff of a new magazine called *Drum.* The same publishing house also put out *The Golden City Post,* of which he became news editor.

His exploits on *Drum* and, to a lesser extent, in *The Golden City Post,* made him perhaps the best-known non-European journalist in South Africa.

Why was he murdered?

The facts of the killing were elusive. The body had been found on the dirt sidewalk of a crowded location. Heavy footmarks indicated a struggle. Several persons apparently had been involved. A trail of blood showed that Nxumalo had crawled more than fifty feet from the scene of the attack before he died. Robbery did not appear to be a motive for the killing; his valuables were untouched. The police had no theories.

Most of Nxumalo's friends were reluctant to talk; they looked away and said nothing. I spoke to a leader in the YMCA who had known Henry Nxumalo as a boy. He said that from the beginning of their acquaintance Henry had talked about wanting to be a journalist.

"He came to see me after he got the job on *Drum* and told me the kind of thing he wanted to do. He wanted to expose the brutal conditions of the jails. He wanted to write about forced labor on the farms. He was not a revolutionary. He believed that many white people who were in a position to effect basic reforms really knew very little about what was happening. And he had confidence in his ability to reach their consciences with documented facts."

I learned that after Nxumalo started his series of articles under the byline "Mr. Drum" he began to get threats of various sorts. But not until now had any of them materialized. At that, no one

". . . Before I left, Mrs. Nxumalo handed me some photographs of her husband that had turned up while she was looking through his papers. They showed a young man with a strong, alert face. He was sitting at his typewriter."

Henry Nxumalo at home with his three children. The Nxumalo family had only recently moved into their house in the new Orlando "location" when this photograph was taken.

could say that any of these threatening letters had been received recently. Two close friends, however, mumbled their suspicion that it had been a political killing but would say nothing further.

I paid a visit to the publication offices of *Drum*. It was located just off one of the main streets in the heart of Johannesburg. *Drum's* reportorial staff was African, though the editor, Sylvester Stein, was white.

The arrangement of the office and the general atmosphere were more suggestive of a daily newspaper than that of a monthly magazine. I chatted with several men on the staff. They spoke freely about Nxumalo and his contempt for danger. They spoke, too, about his heartiness, his ready sense of humor, his joy of living.

"Despite everything that man went through, and it was quite a bit," said Zeke Mphahlele, himself one of the leading African writers, "Henry never became embittered. That's the most important thing for any writer, not to become embittered. Bitterness enters the eyes and keeps you from seeing the full story. It turns you away from the people you have to reach. All of us here envied Henry because he wasn't bitter. But we never knew how he managed it."

I met William Modisane, the music critic on the staff, who had been with Nxumalo only a short time before the murder. Mr. Modisane spoke about Nxumalo's part in building up the circulation of *Drum* until it became the biggest magazine by Africans for Africans on the continent. He told of the time Nxumalo committed a minor law violation in order to be put in jail, served his sentence, and then wrote about the actual conditions. There were official denials, but the reforms he called for came about.

Then there was the time that Nxumalo, disguised as a laborer, got a job on a farm in Bethal. He had heard that prisoners were being sent to Bethal in what amounted to forced labor. At Bethal Nxumalo was beaten up along with the prisoners. He worked alongside young men, not prisoners, who had been "recruited" on the understanding that they would be taken to an entirely different place but were shipped to Bethal and deprived of the right to leave.

The legal pretext for keeping them was that they had "held the pen." It is not necessary for Africans to sign their names to labor contracts. If they "hold the pen" over the contract, witnessed by a white man, the contract is considered to be binding.

The big need in South Africa is for human labor, and intensive efforts are made to recruit the Africans from the villages. Nxumalo was concerned both about the dishonesty of the recruiting and the dreadful conditions under which the men worked and from which there was no legal recourse. He escaped from Bethal, and presented the documentary evidence of what had happened. A large part of the continent was stunned by the disclosures and, once again, there were denials; but the basic reforms were made just the same.

Mr. Mphahlele and Mr. Modisane offered to take me to see various parts of Johannesburg that were indispensable for any understanding of the place. But first, I wanted to talk to *Drum's* editor, Sylvester Stein.

They brought me to Mr. Stein's office. Mr. Stein is a "European," as all white men, regardless of their geographical origin, are designated. I judged him to be about thirty-six. He was in the middle of a magazine deadline situation not unfamiliar to me. I watched sympathetically as he worked quickly to keep the production machinery moving. I could see that he knew his business.

Then, the copy cleared from his desk, Stein sat back and talked fully and openly. No, he didn't think the government had engineered the killing or was implicated in any way. Whatever *Drum's* political and social differences with the government—and they were substantial—he didn't feel that it would sanction murder as a technique for disposing of troublesome people.

"Things may be bad," he said, "but we're not dealing here exactly with a Hitler-type government. The Union is up to its neck in social injustices and problems, but it would be a serious mistake to confuse this government with the kind of dictator state that existed in Germany or Italy before the war or that exists under communism today. Exactly why Nxumalo was killed, I don't know. But you can be sure we're going to try to find out."

Stein went on to talk about Nxumalo's work for *Drum* and how, by exposing brutality and callousness, he had helped to effect important reforms. There was still some responsiveness to honest and dramatic fact-finding in the community at large. The executive and legislative branches of the government were responsible for the *apartheid* repressive legislation and restrictions. But a large part of the judiciary was still rigorously honest. And there was a considerable section of the white population which, even though it might go along with *apartheid* in general, balked about some of the specific aspects of *apartheid* in practice.

This was one of the many complexities that one came to recognize and respect in South Africa. Despite the iron will of the government about *apartheid,* the old forms of parliamentary government and judicial machinery were in large part preserved. This seemed at odds with the practices of the other branches of government, but you accepted it as part of the puzzle. Nor was the grand picture simplified by the sharp divisions among the whites—especially between the English and the Afrikanders (Dutch), with the latter now enjoying a preponderance in government. And even among the blacks, there were factions and frictions that not infrequently resulted in violence. This, too, was something that had to be taken into account before accepting or making any generalizations about the situation in South Africa.

Sylvester Stein's business was to deal with this complexity. It wasn't easy; the white world couldn't quite make up its mind about him because of his connection with the blacks. And the black world couldn't accept him all the way because he always had the privilege of retreating to the world which shut them out.

Stein's predecessor, Anthony Sampson, also white, had written about this predicament in his book.

"I could never explain [to whites]," he wrote, "that *Drum* was a job, and our readers were human like anyone else. I saw white people eyeing me as a crank; and I began to feel a crank. Probably I ostracized myself more than they ostracized me. In white society, I began to feel a slight chip on my shoulder.

"And as I penetrated farther into the world of the Africans, I found myself caught between the two camps of black and white. The contrast, from a cocktail party in the northern white suburbs to a drinking den in the southern black locations, was absurd . . .

"The contrast was stimulating. I saw one world in terms of the other, always slightly aloof: black against white and white against black."

Stein, like Sampson, had to cut across the barrier of *apartheid* constantly just in order to do his job. "At every turning are the signs '*Slegs vir Blankes* (For Whites Only),' '*Nie Blankes* (Non-Whites),' sorting the two races like an infallible machine, and sending them separate ways."

Once, Anthony Sampson and Henry Nxumalo stopped near a sign along the road reading "Natives Cross Here." Nxumalo turned to Sampson. "That sign is incomplete," he said. "It should read: 'Natives *Very* Cross Here.' "

*Apartheid* was more than a wall of color separation. It was a declaration of white ownership and control. It meant that the Africans, who constituted 90 per cent of the population, were limited to ownership of 15 per cent of the land. It meant that Johannesburg was out of bounds to blacks except for daytime working purposes. In order to commute to work from the locations, the Africans often had to spend four or five hours a day, much of it waiting for buses.

*Apartheid* meant a license just to live. An African needed a license to identify himself. He needed a license to work. He needed a license to be out of work. A passbook sometimes contained as many as six or seven separate items, each of which had to be countersigned regularly and kept up to date. Irregularities in passbooks—indeed, just leaving a passbook at home—could mean a prison sentence. It was in this way that Henry Nxumalo got himself arrested in order that he could write about the wretched conditions at the local prisons.

"I could never forget about *apartheid*," Sampson wrote. "It cut across nearly everything I tried to do. It made the job of a white editor on a black paper awkward. I could never travel with Henry in the same train, taxi, bus, or lift. We could not be together in a restaurant, a bar, a theatre, or a park . . .

"Even in the contents of the magazine, *apartheid* intervened. We were ticked off for showing a picture of Eleanor Roosevelt shaking hands with Mrs. Edith Sampson, a Negro woman. We could not print photographs of a black boxer pummeling a white boxer. Mixed boxing is forbidden in South Africa, and photographs of mixed fights were frequently held up by the South Africans as constituting 'incitement.' "

Sampson had also written about the time that Nxumalo saw a white woman fall down in the street.

"I was just going up to help her," Nxumalo told Sampson, "and then I stopped and thought: what will the whites think? They'll think I'm trying to rape her. If I pick her up it means I'll actually have to touch her. A native touching a European woman! Oooh! Terrible! I couldn't risk it, so I walked on."

Ironically, it was the fearful reluctance of the Africans to become involved in difficult situations that may have cost Henry Nxumalo his life.

For his body lay on the sidewalk five hours before the police were summoned. It is possible that he bled to death during that time. Across the street was a hospital.

If Henry Nxumalo's own doubts about helping a fallen woman seem unjustified, it may be helpful to consider a statement made by Alan Paton, who for millions of white people around the world has served as their conscience in South Africa. Paton was speaking at a public meeting about both the ostracism and physical danger involved in direct contact with Africans under the existing strained circumstances.

"Who is there who would not hesitate to come to the aid of an African who stumbled in the street?" he asked. "And if you say that no one would hesitate, I must tell you that there is at least one, and he is speaking to you from this platform now."

In talking to Sylvester Stein, I learned that when he took over the editor's job from Sampson he was to go through the same personal difficulties and challenges—and also the same rewards. Things happened as the result of articles such as Nxumalo liked to write. And the magazine continued to grow. I had a strong respect for the problems involved in building a magazine; but I knew that such difficulties as I had experienced were minor alongside the challenges that faced Anthony Sampson and Sylvester Stein. And when the evidence came that the magazine was hitting its mark, the satisfaction was bound to be rich indeed.

"Yes," Sylvester Stein said, "I can't think of a more varied or exciting job than this. But every once in a while the ceiling falls in on you. That's the way I felt yesterday when I heard that Henry had been murdered."

I thanked Mr. Stein for his time, then rejoined Zeke Mphahlele and Bill Modisane. Zeke took the wheel of an old Chevrolet and we started on our tour. Our first stop was a narrow shopping lane which served as the supply headquarters for fetishers, or witch doctors. Store after store offered an endless variety of goods that are part of the witch doctor's trade—masks, ceremonial objects, magic devices, herbs, potions, processed foods.

"Just how important is the witch doctor in a place like Johannesburg?" I asked my escorts.

"Not as important as he used to be, but still a surprisingly large factor in the lives of the people," Zeke said. "Many non-Europeans still go to them when they are ill—and sometimes when they aren't ill."

The gap between the Africans who went to fetishers and those who didn't was a wide one, I was told. But the conflicting opinions held about fetishers were not the only major point of separation within the African community.

"An educated African has a hard time inside many of the locations," Bill said. "He is resented by his fellow Africans because of his knowledge and because he is believed to have an advantage over most of the other non-European people. The same would apply to any well-dressed African. And by well-dressed I mean any man who wears a tie. According to this standard, it is generally easy to pick out an educated African, for they are better dressed than the others. It is not unusual for educated Africans to be beaten up."

I asked whether it was possible that this was what had happened to Henry Nxumalo.

It was possible, they agreed, that it could have been something as simple as a random act of violence directed against a black man with a tie. But in Nxumalo's case, there were too many other factors to be taken into account. He took too many chances in the things he wrote about. Nor were his crusades limited to abuses by whites against blacks. Violence by black against black deeply concerned him.

"Henry was opposed to the gangs and everything about them," Zeke said. "He hated to see young boys drawn into these different gangs, each with its own names and dress and habits. One particularly strong gang is known as the Russians. There is no political significance to the name. They just happened to fasten onto the name; another is called the Berliners; and another the Americans, etc. The Russians like to walk around wearing brightly colored blankets. The Americans like to affect zoot suits.

"It makes no sense, but that is the way it is. And each gang makes its own laws. No one informs on crimes committed by fellow members of the gang—at least, not if he expects to stay healthy. Most of the gangs are tough, especially the Russians. They carry small arms, generally long switch knives."

The favorite reading of many gang members, I learned, was the American terror comics. And the favorite entertainment was the American gangster film. We were not without influence in that part of the world.

Then Nxumalo might have been a victim of one of the gangs, I asked.

Yes, it was possible; in fact, anything was possible, they said. But, once again, no one knew enough to be sure.

We were now passing through one of the locations where the "Russians" lived. I observed a group of ten or twelve young men wearing somewhat faded blankets in cape style. Zeke identified them and said it would be a good idea to stay away from this place after dark.

Soon we were in the famous black location or quarter known as Sophiatown. Many novels written about South Africa had their central settings here. Sophiatown was the largest of the locations. But it was now being closed down. Few reasons were given publicly. It was believed by some that the government felt Sophiatown was too close to Johannesburg, too likely to become the center from which mass violence against the whites might spring. Others believed that the government was genuinely concerned about the squalor and wanted to resettle the "natives" in the new housing developments that were now being built far outside the city.

In any event, the decision to condemn Sophiatown had been strongly resisted by the people who had to move. But the government was proceeding with its plan. Much of Sophiatown would be torn down, rebuilt, and opened up for settlement by whites.

Zeke discussed the situation in a matter-of-fact way. There was no bitterness in his voice.

I recalled what he had said earlier about Henry Nxumalo's ability to think and write without bitterness; and I told Zeke he had no reason to be envious, for it seemed to me that he had succeeded in that respect.

"I try," he said. "I try very hard. I am now writing a novel. Some chapters I have had to write over and over again, maybe as many as six times, because I am afraid that they sound as though they were written by a sour old man. There's so much around you that's hard to swallow that you've got to fight with yourself to get it down and keep it down."

We arrived at Bill Modisane's home. It fronted on a small courtyard. Some poorly dressed elderly people sat on the narrow stoop. They watched with faint amusement as a toddling infant tried to embrace a dog.

Bill invited me into his quarters. It was a single room made to serve all the purposes of a small family. Mrs. Modisane was not home at the moment. The room was about twelve by ten. It contained a large day bed, table, electric refrigerator, small stove, several chairs, and a bookcase. There wasn't much open wall space but it was adorned with several attractive modern paintings. What could have been a dingy cluttered thimble of a room had been neatly and imaginatively decorated.

As I mentioned earlier, Bill Modisane was a music critic. Noticing a small phonograph player, I asked Bill about his recordings. He said he didn't have much of a collection but he was hopeful that, circumstances permitting, he might be able to build it up and perhaps even obtain a high fidelity playing unit.

He served some refreshments. Africans by law are not permitted to possess or serve alcoholic drinks under penalty of arrest. On occasions the law is observed, more the result of economic limitations than determination to comply. The law, of course, does not apply to whites.

Bill discussed his hope that some day he might be able to bring his family to Canada or the United States. He was hopeful that he could get a job on a newspaper as reporter or music critic. He also wanted to study.

It was now late in the afternoon and I was due back in Johannesburg for dinner. We drove back under a deepening sky. The colors of the landscape responded to the warm orange of the western horizon. Man-made mountains of slag heaps from the gold mines, as characteristic of Johannesburg as skyscrapers are of New York, took on the color of gold itself. But the gold mines were also the center of many of the most deep-seated economic and racial problems of the country. Once again, I could reflect on the contrast of the beauty and the squalor, and the incestuous relationship that almost seemed to exist between the two.

When we came into Johannesburg, I saw long lines of Africans stretching almost endlessly. There was one large open area where the line wound in and out and around and was so long that I found it difficult to see the end of it. In some places the line would swell out until it was six or seven people thick.

The people were waiting for the buses to take them home. Sometimes they would have to wait two or three hours. The trip itself might take an hour. Allowing the same amount of time to get in to work early in the morning, Africans sometimes had to spend many hours or more each day coping with the ordeal of transport. And when the bus company announced an increase in the fares, it touched off a series of riots.

Africans are not agreed among themselves about the methods to be used in combating inadequate bus service or the fare increase. When a movement for a boycott against the bus lines developed in the Evaton location, thirty miles from Johannesburg, it was opposed by a large number of people. The boycott went into effect. On the next day, a clash took place involving about four thousand people armed with sticks and clubs. The boycotters were in the majority by a ratio of three to one. Two persons were killed. No one bothered to count the injured. Eight houses were wrecked. Not much attention was paid to it in the outside world, but a number of whites spent hours each day ferrying the Africans to and from their homes. The significance of these acts was not lost upon the black community.

Henry Nxumalo was especially concerned about the fact that it was illegal to engage in organized protest. The government had declared the protest activities were acts of communism which were outlawed under the Suppression of Communism Act.

I asked my escorts whether this didn't mean that anyone who tried to get people interested in working together to bring about essential reform—even though the problems involved were of a non-political nature—could be brought to trial as a Communist.

This was exactly it, they said. The government had charged one hundred and fifty-three people with treason under the Suppression of Communism Act. No one knew exactly what the charges were, but no one expected that the government would attempt to prove that the accused were *members* of the Communist party. If incitement to protest could be proved, then it was tantamount to communism, and communism was treason.

Johannesburg bus stop. "Africans sometimes had to spend many hours or more each day coping with the ordeal of transport. And when the bus company announced an increase in the fares, it touched off a series of riots."

"We drove out along the 'Main Reef,' as Johannesburg's principal artery is called. We passed some of the largest gold mines and electric power stations, their mammoth vase-shaped water-pressure towers standing like giant sentries over the rolling countryside."

Some of the most distinguished Africans were now involved in the trials. Only a week earlier, preliminary hearings had been held. And now the entire nation held its breath, wondering what would happen next.

Once before, the government had been successful, through severe measures, in throttling a movement which challenged the ideas underlying *apartheid*. Manilal Gandhi in 1953 had attempted to organize a passive resistance movement along the lines made famous by his father in India. But the government used the jails and the lash with such resolute effect that the movement collapsed.

Bill Modisane and Zeke Mphahlele dropped me off at the private residence where I was staying in Johannesburg. I thanked them and told them of my desire to keep in touch with them. I knew that nothing would have pleased me more, on my return to the States, than to be able to help Bill Modisane to find a newspaper job in Canada or the United States, and to try to find a publisher for Zeke Mphahlele's new book.

That evening, after dinner, when we discussed the plans for the next day, I asked Clara whether it would be in order for me to pay my respects to Henry Nxumalo's widow. She said she was certain Mrs. Nxumalo would welcome my call. There were, however, some technicalities. Whites were not allowed to visit African locations without specific government authorization.

Clara said she would arrange with Dr. Ellen Hellman, a "European" and a distinguished anthropologist, to take me out to Orlando. Dr. Hellman, like Anthony Sampson and Sylvester Stein, moved in and out of both worlds in Johannesburg in the pursuit of her work and concerns. She had helped to organize and maintain joint councils for improving conditions between black and white. When prominent Africans got into trouble with the government for one reason or another, they could turn to her for advice. Her prestige and influence were considerable. She had helped to arrange bail for many of the accused in the treason trials.

Dr. Hellman picked me up in the morning. Our first stop was the old post office building where Dr. Hellman obtained the pass for the locations without difficulty. We then drove out along the "Main Reef" as Johannesburg's principal artery is called. We passed some of the largest gold mines and electric power stations, their mammoth water-pressure vase-shaped towers standing like giant sentries over the rolling countryside. The gold mines were having their troubles; prices had not kept pace with the inflationary spiral, and salaries were low.

The vast majority of the Africans who work in the gold mines are migrant laborers. They live in compounds furnished by the mines and usually patronize company stores. Food is furnished by the mine companies. The diet is well rounded and high in protein value, unlike the average diet of most Africans. The physical condition of the men who work in the mines is considered good.

Most of the migrant mine workers are signed up by recruiting agents for a minimum of a year. The big lure is city life and a chance to accumulate a modest amount of capital, something that is rare in village life.

At the time the young married males of the village "hold the pen" over the contracts to work in the mines, no doubt there is every intention to return to their wives and children at the end of their service. But it doesn't always work out that way. Some get caught up in city life or become involved in new domestic situations and never go back. The result is a permanent disruption in the home lives of thousands of families in the villages.

On the way out to Orlando we passed several of the older locations—vast sprawling collections of shacks and crowded alleys that lay on the land like a giant fungus. But we also passed new locations for Africans that were reminiscent of the Levittown type of housing development in the United States. The houses were small and repeated themselves endlessly, but they were neat, attractive, and sturdy. Rent was modest and convenient terms had been worked out for purchase by the tenants.

Some of these new developments near Johannesburg were spurred into being by the leadership of Sir Ernest Oppenheimer, the widely respected philanthropist and civic leader. Sir Ernest had made per-

On Sundays, the mine workers who belonged to different tribes would put on their dances before large crowds of visitors. The musical instruments were almost exclusively percussion and were home-made.

sonal contributions of millions of dollars to help get the housing projects started.

I asked Dr. Hellman whether we might stop and visit with some of the people who had just moved into a new house. She agreed; we stopped outside a home that couldn't have been completed more than a few weeks earlier. She went in first to explain the situation and find out if we would be welcome. I entered the home of a man who turned out to be a coal dealer. The rooms were small but pleasant. There was a good balance between window and wall space. There were four rooms, altogether, including the kitchen. I was impressed with the paintings in at least two of the rooms. The people couldn't have been more cordial or responsive.

We resumed our journey. At Orlando, we were delayed momentarily by the system of numbering houses not according to street but according to a general area. In any event, we found the Nxumalo home. Several people were standing outside. Three youngsters who turned out to be his children were sitting on the stoop. We introduced ourselves and learned that Mrs. Nxumalo had been called to the police station in connection with the killing. Henry Nxumalo's brother suggested that we wait. Then he said that everyone in the family was still completely mystified by what had happened. There had been some threats but Henry had convinced everyone that they were not to be taken seriously.

After about half an hour, Mrs. Nxumalo returned. She was a gracious, attractive young woman and she carried her grief with great dignity. After we were introduced she said the police had nothing new to report. They had asked her some routine questions which she had answered to the best of her ability.

I told Mrs. Nxumalo that I had heard that Henry had been writing a manuscript about South Africa for an American publisher. She replied that so far as she knew, the book was almost finished. She had the impression that her husband had asked someone a week or so earlier to read it, and she had no way of knowing whether the manuscript had been returned. In any event, she said she would be glad to look for it in the study.

Very methodically, starting with the top of a heavily cluttered table that had all the signs of being used for intensive research purposes, she began her search for the manuscript. While she was thus engaged, I glanced at some of the titles of books on the adjoining shelves—many of them having to do with literary criticism or collections of essays on writing. Among the authors represented were Somerset Maugham, Thomas Wolfe, Charles Morgan, Desmond MacCarthy, Cyril Connolly, John Dos Passos, John Steinbeck, Ruskin, and Proust. There were also a few books about writing for radio and television.

In the corner of the table was Henry Nxumalo's typewriter. In it was a sheet of copy paper on which he had started to write a story. In the upper right-hand corner were the identifying initials "nx"; in the center the slug line for the story—"pass"—and the number of the page. There were just a half-dozen lines before he broke off:

Last month a "Post" reporter was robbed of his pass in a Johannesburg township. He made a report to the police. But since then he has been arrested twice and paid a total of £2 in fines before he was issued with a duplicate pass. He got his new pass after weeks of going from one office to another—weeks of hardship, sweat, frayed tempers and wasted time, and it cost him 12/—.

When I read this I remembered something Bill Modisane had told me when he showed me his own passbook.

"If you want the story of hell, it's written right in this book. You see a policeman and instinctively you reach in your back pocket to make sure your passbook is there. When you're in a crowd you keep your hand pressed against it lest it be stolen. Sometimes, in the morning, you change your clothes and rush out of the house in too much of a hurry. Then after you arrive at your job you reach in your back pocket and you discover you left the book at home. And all day long you wonder whether you are going to make it, whether you are going to be able to get home that night. When you see a policeman you are so terrified you hardly know what to do.

"A curious thing, this dread we have of the police. I suppose in your country when you see a cop you feel reassured. In the U.S., if you are walking through a bad neighborhood at night and you see a policeman, a great deal of the fear goes out of the dark.

"But here the policeman is not the image of security and reassurance to us. If we are in trouble, the last thing we think of is the police. We look at a policeman and say, 'My God, I wonder if I'm going to be stopped. I wonder if I've got my passbook with me.'

"It's no fun being arrested and being put in jail for a passbook violation, not even under the reforms that have been put in as the result of the articles Henry wrote during his brief term in prison for being without his passbook."

It was ironic that the last thing Henry Nxumalo should have written was about the injustice of the passbook.

Mrs. Nxumalo interrupted my thoughts and said that the search for the manuscript was unavailing. She wasn't certain of the name of the American publisher but seemed to remember having received letters from Doubleday and Knopf asking whether her husband would be interested in writing about South Africa. She asked me if, when I returned to America, I might locate the publisher for whom Henry Nxumalo was writing his book and explain what had happened. She said she would continue her search and also attempt to find out if the book was still in the hands of a friend. It was arranged that when the manuscript was found she would mail it to me and I would undertake the necessary dealings with the publisher.

Before I left, Mrs. Nxumalo handed me some photographs of her husband that had turned up while she was looking through his papers for the book. They showed a young man with a strong, alert face. He was sitting at his typewriter. Another photograph showed him with three of his children in the doorway of his home

On the way back to Johannesburg, I asked Dr. Hellman about Mrs. Nxumalo's circumstances and who would take care of her.

"She's a registered nurse," Dr. Hellman said, "and no doubt she will be able to keep busy. But she has five children. I'm sure Henry's

One of the photographs taken secretly for the magazine *Drum*, in the series written by Henry Nxumalo to expose prison injustices. "There was still some responsiveness to honest and dramatic fact-finding in the community at large."

In addition to the dog, several humans occupied this particular abode. The government was undertaking a resettlement program under which Africans would be moved into new housing projects far outside Johannesburg proper.

friends, and there are many, will make up a purse for her or run some benefits. She has a lot of spunk. Even so, it won't be easy. It won't be easy at all."

Dr. Hellman was right about Henry Nxumalo's friends. Several days later more than one thousand people attended the funeral, and it lasted seven hours. All morning long at the Communal Mall people stood up to speak their tributes. Dr. Ray Phillips, one of his friends, said that if Henry Nxumalo's life and death were to have any meaning at all, it was that each individual had a fixed obligation to put an end to the terror that was disfiguring their society.

"Have we been shocked enough," he asked, "to grab the devil of hooliganism by the neck and say, 'This is as far as you go!'?"

The evening before we were to leave Johannesburg for Lambaréné, I checked notes with Clara Urquhart. I told her what had happened since I first began to find out about the questing, restless, wonderful Henry Nxumalo.

"No conclusions," I said, "just a lot of unassorted impressions. The Union is far more complicated than I had ever supposed."

"In that case, you're lucky to be leaving when you are," she said. "The longer you stay, the more complicated it gets."

"Sometimes it seems as though the trouble is actually a myth," I said. "Yesterday I sat on an attractive portico talking to reasonable people. I looked out over a rolling lawn toward a bed of tulips. Right then, all the turmoil and controversy seemed like a second-hand tale or a nightmare. Then I remembered the long lines of people waiting for the buses and I knew it was real enough. How long can it go on before it explodes?"

"The Africans are a remarkably patient people," Clara said.

"Even patient people have a sense of justice."

"Yes, but it's amazing how large the capacity of the Africans is to live with the impossible. And they're not easily stampeded into political movements of one sort or another. The Communists have tried hard to exploit the situation here but they've made hardly a dent."

"And yet the government feels sufficiently concerned about it to enact a Suppression of Communism Act," I said.

"Which has very little to do with communism. How many Communists are among the one hundred and fifty-three who are being accused of treason? If being aware of injustice makes a man a Communist, then we've just made Communists out of every great man who ever lived. The fact that the government uses the word doesn't change its real meaning."

I said I wondered whether the government's interpretation of protest as a Communist activity would stand up in court.

"That's the question almost everyone is talking about," she said. "I just have to believe that our courts have not lost their good sense or their independence. What other impressions have you had?"

I said I was struck by the fact that almost no one close to what was happening in the Union of South Africa had any clear answers. Everyone seemed to be groping. Another thing that impressed me was that the middle ground in South Africa seemed to be disappearing. Day by day, there was a growing accumulation of the forces at the extremes—the kind of extremes that made for mighty collisions and explosions.

"Most troubling of all," I added, "is the feeling I get that more and more people are losing interest in any moderate approach. I suppose there is nothing strange about this when we consider that the government itself has taken up its own position at one of those extremes."

"Isn't this another example of the hard time that moderates have had to face whenever the social problem becomes acute?" Clara asked.

"True," I replied, "but it bothers you just the same to see men and women of good will of both colors become increasingly lonesome in or near the center. Just in the process of existing, whether you are black or white, you find yourself being pushed toward the extremes.

"If you are black, you are confronted with the arguments of

those who have only to point to the policy of the government to prove that it intends to keep the African permanently disenfranchised—no citizenship, little personal dignity, even less land. According to this argument, the war has already been declared and it is only a matter of time before the fighting begins.

"If you are white, you are confronted with the arguments of those who say that whatever should or could have been done in the past, it is now too late to do anything except to hold fast. For these whites, the issue seems to be simple: survival."

"One thing that makes the position of the South African whites so tragic," she said, "is that some of them have been here for centuries. This land and way of life they know and love. Where else can they go? What would they do? In the other parts of Africa, the white people have basic ties to the European countries from which they come or from which they may be only a generation or two away. And they feel they can always go back if they have to. Not so the white people of South Africa. This is their homeland. Their ancestors came here—many of them—about the time Columbus discovered America."

Yet the remarkable thing, I said, was that while almost every white person knew the explosion was coming, none was willing to say that it would come during his own lifetime. Thus a man of about sixty would say that matters could probably be kept in hand for another five or ten years. A man of fifty thought that the Union had about fifteen years of grace. A man of forty guessed that the big fireworks might not come perhaps for a generation. In the meanwhile, life goes on. And it is a gracious and congenial way of life, despite the occasional violence and subsurface tension.

"Even so, you have no idea," Clara said, "how much more tense it is than it was only a few years ago. That was when a middle ground seemed not only possible but inevitable. Now everything is so tight and uneasy. To be decent now requires martyrdom and I have insufficient courage."

She was talking very slowly, very deliberately.

"I've decided to give up my home in Johannesburg. I've been

fighting against that decision for perhaps three, four years. But how can I make my home here if I don't feel at home here? Everything is changing, and I can't make the changes in myself that have to go with it."

MUCH OF THIS conversation, and many of the things I learned when I tried to find out about Henry Nxumalo, came to mind during the flight over the jungle from Brazzaville to Lambaréné, when Clara said that the mood of Africa was changing and that there was a tightening in the air.

"In three or four minutes we ought to be able to see the Hospital from the plane," said Dr. Frank Catchpool from across the aisle in the Air France DC-3. "These pilots are very thoughtful fellows. When they know that some of their passengers are going to the Schweitzer Hospital, they generally make a run over the place."

Dr. Catchpool was right. We flew over the Ogowe River, which connects the South Atlantic with the interior of the middle Congo. From the air, the Ogowe was a light muddy brown. Here and there along the river, we could see small African villages. Then there was a cluster of buildings, with a church tower in the center. Dr. Catchpool identified this as the town of Lambaréné.

Then, suddenly, Dr. Catchpool called out and pointed to the Hospital. It consisted of a series of long, narrow buildings close to the river. Red roofs interrupted the jungle only briefly. Immediately beyond was the deep and endless green.

Approaching for a landing at the Hospital dock. Dr. Schweitzer calls out instructions to the leper oarsmen in the pirogue. At the extreme right is the head of Dr. Frank Catchpool.

Same general scene, one minute later.

# III

THE LAMBARÉNÉ AIRSTRIP, like all the others we had seen en route from Brazzaville, was just a dirt clearing in the jungle. The "terminal" was a large lean-to in which waiting passengers could shelter themselves from the sun. There were, of course, no mechanical installations or gasoline trucks for servicing the plane.

The reception committee at the airport consisted of Dr. Jan van Stolk and Mme. Oberman. Clara introduced us. Then, when they turned to Dr. Catchpool to inquire eagerly about his experience in the Brazzaville hospital for his dogbite, Clara told me that Dr. van Stolk was now the senior staff doctor at the Schweitzer Hospital. He was a native Hollander who had gone to medical school in the Union of South Africa and who had left a growing practice to come to Lambaréné. He was about thirty-two.

Mme. Oberman had worked with Dr. Schweitzer some months each year for about four years. She, too, came from Holland. (Many of the members of the staff, I later learned, came from Holland.) She worked in a general supervisory capacity, taking care of the needs of the Hospital personnel.

The entire party got into the back of a truck and sat on benches along the sides. I was thankful for the overhead canvas. It was noon and the equatorial sun was living up to its reputation.

"The path opened out on a courtyard, with low-lying wooden structures." Relatives of hospital patients share in the daily work of servicing a community.

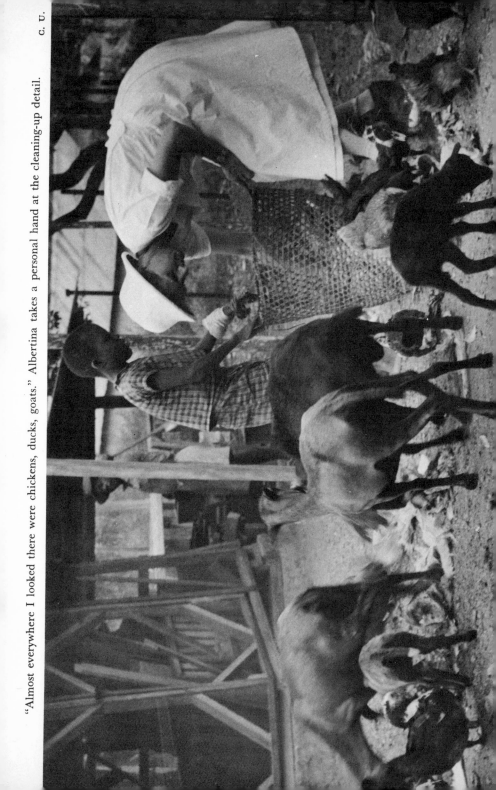

"Almost everywhere I looked there were chickens, ducks, goats." Albertina takes a personal hand at the cleaning-up detail.

We drove for perhaps a mile and a half over a bumpy dirt road, alongside which were scattered African dwellings. Then we came to a clearing, just beyond which was the river. At the narrow wooden dock was a long pirogue, the sturdy and graceful native canoe specially built to withstand the powerful currents of the Ogowe.

Our bags were loaded into the pirogue by a half-dozen young men who comprised our crew. I sat up front near Mme. Oberman. Behind us were Dr. van Stolk and Clara and the young men who sang in rhythm to their strokes with the paddles. The trip to the Hospital is against the current, and so we stayed close to the shoreline. I marveled at the stamina, power, and good spirits of the young Africans as they paddled us upstream. And I recalled something that Dr. Schweitzer had written in one of his early books, *On the Edge of the Primeval Forest*. Shortly after he founded his Hospital in Lambaréné, it became necessary to take an emergency canoe journey into the interior. For hour after hour, the African natives insisted on staying at their paddles. It was an endurance feat that made a profound impression on the Doctor, and he resolved to remember it every time he was tempted to regard the Africans as shiftless or lacking in energy.

I recalled, too, that it was on his canoe trips that Dr. Schweitzer felt that conditions were most congenial for the exercise of the moral imagination. This I could readily understand. There is a total awareness of nature, if only because the contrasts are so compelling. The stillness over the water is made dramatic by the cries of the birds in the jungle. The sky is a silver sheath sparkling in the sun in contrast with the soft filtered light of the forest. The power of the current in the center of the stream contrasts with the easy play of the waters near the shore. I could understand why Dr. Schweitzer wrote that he could never take a canoe up or down the river without reflecting on the importance of reverence for life.

"The Doctor loves the river," Mme. Oberman said as if reading my thoughts. "Perhaps you will have a chance to take a ride with him in a canoe. When you do you will marvel at his expression, at his concentration."

I asked Mme. Oberman if the Doctor was in good health; some reports I had heard recently were disturbing.

"You will see for yourself," she said. "He is in fine health. His energy is high and he is in good spirits. Watch now; soon we will see the Hospital."

The Hospital is around a bend in the river and you do not have a good view of it until you swing around and come toward it downstream. In order to do this your canoe continues perhaps a third of a mile or more on the opposite shore beyond the Hospital so that when you cross the river the current will not carry you beyond the dock.

As the canoe swung into midstream, I discovered some figures dressed in white walking down the hill from the Hospital toward the small dock. Clara waved toward the shore. When we were about three hundred feet from the dock, I recognized the Doctor. He was at the edge of the dock now, waving to us. Then, when the canoe was within perhaps fifty feet, he began to call out directions. It was like a ferry being eased into her slip by commands from the bridge.

"*À gauche! À gauche!*" the Doctor cried out. "*Lentement!*"

Then, sharply,

"*Arrêtez!*"

He stooped and grabbed the prow, then eased the canoe alongside the dock. The Africans held the pirogue firm, and the Doctor reached over to help us out, one at a time. As he took my arm, he introduced himself, then went over to greet Clara warmly. He turned to Dr. Catchpool and inquired both about the health of the doctor and the health of the dog. Dr. Catchpool replied that he was well and that he had brought the dog back apparently none the worse for the experience.

The Doctor then took my arm and escorted me up the hill to the Hospital. The lane was narrow and we threaded our way past some small shacks and enclosures on the hillside. The ground underneath was moist and slippery and had the consistency of a chicken-yard. The reason was readily apparent. Almost everywhere I looked

Hospital Row and clinic. Arriving patients check in at the bungalow at the right.

Same place, late in the afternoon.

there were chickens, ducks, goats. Then the path opened out on a courtyard, with low-lying wooden structures. The building on the left was mounted on concrete piles about six or seven feet above the ground. This was where the doctor and the immediate members of his staff lived. Directly opposite were some utility and storage buildings, also set on concrete piles a few feet off the ground.

At the foot of the steps leading to the Doctor's quarters was Mrs. Schweitzer. I had been told she was not well and was able to get about the hospital grounds only with the greatest difficulty. The Doctor introduced me. Mrs. Schweitzer spoke in English; she was most gracious, apologizing for the fact that she was unable to accompany me to my room, and saying she hoped I would drop by for a chat after I got settled. We resumed our walk, the Doctor leading the way past several other frame buildings, each with its dark-red corrugated iron rooftop. One could hardly see the sky because of the thick benevolent overhead shelter from the trees. In a moment we were walking along the porch of a long single-story bungalow consisting of about twelve rooms for members of the staff.

The Doctor opened the door to my room, bid me rest a while, then come to the dining room for lunch. He apologized in French for the fact that he didn't speak English and said that the only place in the world where he would dare to speak English was Edinburgh, for the people there had a habit of speaking very slowly.

I thanked the Doctor and began to tell him how privileged I felt in being able to be with him at his Hospital. He cut me short with a wave of his hand, saying with a smile, *"Pas des compliments."*

The room was far nicer than I had expected. Walls and furniture were painted white. The room was only six or seven feet wide but it had everything one might need: small writing table and oil lamp, bookshelf, wooden cabinet for clothes, a stand for water basin, pitcher, and toilet articles. The bed was an iron four-poster, fairly narrow, with thin mattress. It was firm, just the way I like it, and did not sag.

One end of the room was screened in and opened out on the slope going down to the clinic and the Hospital wards. Beyond was the river Ogowe, shimmering in the midday sun. Some Africans in their pirogues were drifting downstream. It was warm, but not uncomfortably so. I was delighted with my first fifteen minutes in Lambaréné.

Clara came by to escort me to the dining room. There I met several members of the staff: Dr. Margaret van der Kreek, a lovely young lady of about thirty from the Netherlands, who was chief surgeon at the Hospital; Dr. Richard Friedmann, a Czech who had been imprisoned in German concentration camps during the war and who now performed the full range of medical duties required of a doctor at the Hospital; Mlle. Mathilde Kottmann, who had been the first to join Dr. and Mrs. Schweitzer at Lambaréné and who had served alternately as nurse, administrative assistant, housing supervisor, etc.; and Albertina van Beek Vollenhoven, a nurse in charge of one of the wards. Dr. Schweitzer explained that the rest of the staff was at work but that I would have a chance to meet them during the afternoon or at dinner.

I looked closely at the Doctor as he chatted with Clara across the luncheon table. His skin was pink and firm. His eyes were clear, his manner was alert. He was in excellent health. It seemed to me unbelievable that this vigorous, fully functioning man was about to mark his eighty-second birthday. When he had led me up the walk from the docks to my bungalow I had to step quickly to keep pace.

He asked about his various friends in the United States by name and I was happy to be able to tell him that they were all in good health. When he asked about Erica Anderson and Jerome Hill, who had just produced the motion picture about his life, I told him that I had seen a preview of the film and felt certain that it would be well received. I added that he was well on his way to becoming a movie idol in the United States.

He smiled. "Well, who knows, I may be famous yet."

One of the attendants came into the room with a baby gorilla

C. U

Dr. Schweitzer's bungalow in the central compound. The Doctor is chatting with a member of his staff at the steps of the porch.

"The Doctor's eyes were clear, his manner was alert; he was in excellent health."

One of the non-working **and perennially** relaxed members of the Hospital community.

Dr. Richard Friedmann chats with "Joseph," who received his training as a medical attendant at the Hospital.

clinging to her neck. She sat down on a bench in front of the wide window. The Doctor beamed every time he looked over at the gorilla. For the next few minutes he discussed gorillas and their high order of intelligence, saying that the gorilla was much closer to man in the scale of evolution than he was to the chimpanzee. Then he got up and walked over to the bench and took the baby gorilla in his lap. He put his head down so the gorilla could play with his massive head of unruly gray hair. Now and then the gorilla would tug fairly hard and the Doctor would wince, but say nothing. He had the look of an adoring grandfather.

As we left the dining room, the Doctor advised me to rest during the afternoon and to stay out of the sun.

The last thing in the world I wanted to do right then was rest. I went back to the bungalow, got my floppy rain hat to protect me against the sun, and then took off on an unescorted tour. I visited the central compound, where Dr. Schweitzer had his bungalow, and where the dining room was located. I walked down the slope to the clinic and the Hospital wards. Dr. Friedmann and Dr. Margaret were on duty at the clinic, handling the last few patients in the afternoon line.

Dr. Friedmann invited me to sit alongside him as he explained the kind of cases that turned up at a jungle clinic: venereal diseases, leprosy, malaria, hernias, framboesia, sleeping sickness, ulcerous sores, abscesses, malnutrition, miscarriages, toothaches. Not infrequently, patients would wait until their illnesses were advanced or their suffering acute before they came to the Hospital. Hence, the clinic more nearly approached an emergency ward than the outpatient department of a hospital.

An African of about fifty, his face betraying no pain, was next on line. As Dr. Friedmann examined the man, it was obvious he was suffering from a massive hernia. It seemed inconceivable that the man could get about.

"This is one of the astonishing things you come to accept as a matter of course in this part of the world," Dr. Friedmann said. "The people have an extraordinary fortitude. They don't usually give in to sickness unless it is so serious that it hobbles them. It is not at all unusual to find men like this, with hernias that would hospitalize white men at the very start, continuing at their work for many months or years before the weight and size of the sac make movement literally impossible. That accounts for the fact that so many of the hernias are strangulated. Much of the surgery done by Dr. Margaret here has to do with strangulated hernias. By this time she probably knows as much about strangulated hernia operations as any doctor in the world."

I looked over at Dr. Margaret—*"La Doctoresse"* as she was affectionately known to everyone at Lambaréné. She was busy filling out some prescriptions. I could readily understand why Clara had called her one of the most beautiful young women in the world. There was nothing mechanical about her appearance. Light-colored hair was combed straight back and held in place by a simple ribbon. She wore no lipstick or other make-up. Yet she possessed an unmistakable quality of classic loveliness both of feature and expression. I wondered about this attractive creature—how she happened to go into medicine and surgery, why she came to Lambaréné. Here was the raw material from which legends were fashioned.

A young African mother, her face heavy with apprehension, carried in a five- or six-year-old girl and set her down next to *La Doctoresse.* The child was wide-eyed and fearful. She clung to her mother's faded skirt with both hands.

*La Doctoresse* knelt down alongside the child and spoke to her reassuringly, then stroked her head as, speaking through an interpreter, she asked the mother the trouble. The mother said the child had persistent fever. Still kneeling and speaking softly, *La Doctoresse* took her stethoscope and applied it to the child's chest. Then, very deftly, she put a tongue depressor in the child's mouth. The little girl coughed involuntarily, then cried out. *La Doctoresse* reached behind her and took a little wooden doll from one of the drawers in her desk. She made small clucking sounds as she held the doll alluringly in front of the little girl. While the child scrutinized the doll, *La Doctoresse* lifted the child's dress and examined the lower part of her body.

Then she smoothed the little girl's dress and told her that it was all over and that she would soon be well again. The child smiled shyly, still holding the doll. *La Doctoresse* told the mother the child had malaria but that it could be brought under control. The child was assigned to a place in the wards and *La Doctoresse* wrote out instructions for her care and a prescription to be filled by the Hospital apothecary. Then she asked the African interpreter to have the woman repeat everything said to her, just to be sure she understood.

In keeping with the custom at the Hospital, the woman was asked whether she could afford to pay anything. Dr. Schweitzer believed that people are more respectful of advice, especially of a medical nature, if they have to pay for it.

The woman was prepared for the question. She opened her hand and offered *La Doctoresse* one hundred francs (equivalent of about twenty-five cents), an amount that was slightly above the average. *La Doctoresse* thanked her, then carefully made a notation of the revenue in the accounts book in which the daily fees from patients are entered.

I could see that the woman had something to say to *La Doctoresse* but was choked up. The heavy apprehension in her face was gone; now there was measureless relief and gratitude. I wondered what went through her mind as she looked at this white goddess with the golden hair who had such knowledge and skill as only the most gifted of men were supposed to possess. I could tell the woman was struggling for a correct way to make known her feelings. Then she reached out and lightly touched Dr. Margaret's arm. It was a simple gesture but profound in intent. Dr. Margaret responded with the smile not of a doctor but of one woman communicating with another in a universal language.

IT WAS TIME now to close the clinic for the afternoon. Dr. Margaret marked some notations in the accounts book, then closed her desk. We walked up the short hill to the bungalow where the staff lived.

"You must see my garden," Dr. Margaret said. "The jungle flowers are just now coming up."

In front of the porch leading to her room, Dr. Margaret had built a small wire enclosure of perhaps no more than six by eight feet. Inside, protected from the goats and other wandering animals, were the young shoots of jungle flowers. I couldn't identify them nor do I remember their names; but Dr. Margaret named them for me one by one as she ran a finger around the tender buds. What I do remember vividly is the joy and pride that this lovely girl doctor took in creating a tiny sanctuary for a few flowers.

It was late afternoon. The sun had lost its fever and a small wind came up from the river and eased the heaviness in the air.

Dr. Margaret sat down on one of the porch steps. It had been an exhausting day, and now she rested her head on her hands and breathed in the coolness.

"Do you like Lambaréné?" she asked. "It has all the things that are difficult to find outside. A chance to concentrate on your work; quiet when you need it; and most of all, freedom from all the non-essential things that fill one's life."

"Non-essentials?"

She looked at me sternly.

"Surely you must know," she said. "The non-essentials of life in Europe and America. The endless running around in circles to do things that seem a matter of life and death at the time but that you can't remember two days later. The business of struggling with a checking account at the end of the month to make sure there's enough to cover all the things that we bought but that we don't even know where to put. And the desperate way we try to entertain ourselves.

"Here at Lambaréné, we do nicely without the frills. We have a purpose and we apply ourselves to it. We never have to ask ourselves whether we are really needed. We are never at wits' ends for what to do with our time. When our work is over for the day we can sit down and rest or we can make our tea and we talk among ourselves or we can read and we can think. It is very good. Do you find this strange?"

I told Dr. Margaret that I had nothing but admiration for the people at Lambaréné and for their ability to come to terms with life.

"It may take you a little time to understand Lambaréné," she said. "So many people come here just for overnight and go away appalled. You know, there's a reason for everything at Lambaréné, but it takes a little time to find it out."

I said that I had been carefully indoctrinated by Clara and that I would do my best to get to know the real Lambaréné.

"That is good," she said. "Maybe you will not make the mistake of judging this as you would a modern hospital. It is a jungle village with a clinic. If Dr. Schweitzer had put up a fully equipped modern hospital of the kind you see in large cities, I am not sure the natives would come to it. They would probably be afraid of it. They must understand something before they give themselves to it. The hospital here they understand. It is very simple. If a person gets sick and the local remedies are of no use and the sickness stays on, the entire family gets into a pirogue and paddles—sometimes many, many miles—to the clinic here at Lambaréné. When they arrive, they find an African village very much like the one they left. If the patient has to be hospitalized, we assign the entire family to a cubicle in one of the shelters. The people go into the woods for their toilet and take care of their own refuse. They get their water from the wells. They cook their own food. They can get fish from the river. We give them bananas and some rice. They get the rest from the trees. We do the diagnosis and supply the medicines and check up on the progress of the patients. When they get better they go home."

"Do you manage to establish any real contact with your patients?" I asked. "Do you find yourself getting caught up in the lives of the people you treat so that you have a real emotional stake in what happens to them?"

Some babies are king-sized . . .

. . . while others fit comfortably into a hat.

C. U.

Three photographs of Dr. Margaret van der Kreek—*La Doctoresse*—chief surgeon at the Hospital. "She possessed an unmistakable quality of classic loveliness."

Dr. Margaret looked up. "It is true that there are many patients. So very many patients. Each day they come. It is hard to keep track of them. And many of them we see just once. But at the time they stand before you and tell you their story you make the contact. I think this has nothing to do with Lambaréné necessarily. It comes with being a doctor. A person comes to you and describes his ailment. You observe him carefully. You watch closely for a little sign that will tell you what you must know. And the patient will not have confidence in you if he senses that the contact has not been made."

I asked the question I had wanted to ask her since I first saw her in the dining room for the staff. I asked how it was she came to Lambaréné.

Dr. Margaret reached over the wire fence and ran her hand lightly along one of the young shoots. She said she had wanted to be a doctor ever since she was old enough to think about what she wanted to do with her life. Her father was an artist, her mother a poet. She grew up in an atmosphere of kindness, graciousness, and intelligence. She was aware that the great happiness enjoyed by her parents in their relationship to each other and her own resultant happiness were not the lot of all people. Her family suffered no deprivations, not even during the depression; and her father felt the obligation to do what he could to help others when he could.

All this created a determination in her to serve. Medicine seemed an effective way. Then she read about Dr. Schweitzer and his work in Africa. She doubted that the Doctor would accept her, but she decided she would get the most comprehensive medical and surgical training available and then volunteer. She knew that many of the persons on the Lambaréné staff came from Holland; at least she could get one of them to write a letter of introduction.

After she completed her internship the introductory letter was written to Dr. Schweitzer.

"Then very soon after that came his reply," she said, reliving the experience in her brightened expression. "He would take me. His letter was amazing. Dozens of details. He put each one down.

How I was to travel, what the timetable was, where I was to change trains and so forth, what my work would consist of in Lambaréné, how I was to go about making arrangements for passports, and the kind of clothes I would probably need. And there was an air of great kindness in everything he said.

"It is now two years since I arrived. At first, like most of the others who came, I was puzzled by many things, sometimes even disappointed. But the more I stay the more I understand and the stronger is my admiration for the Doctor. I know now why things are done the way they are.

"People know the Doctor as a great philosopher and theologian. I am lucky to know him as a human being. It is fantastic all the things he does every day and the things he manages to keep in mind. He has been after me for several months to take a holiday. I have been putting it off but finally I will leave. Next month I go to South Africa. A surgeon, Dr. Jack Penn, visited the Hospital here recently. There are new techniques in reconstructive surgery I want to learn. Dr. Penn invited me to work at his hospital. After I accepted, Dr. Schweitzer took it on himself to make all the arrangements for passport and visas. Only two days ago, he went in the pirogue to the village and filled out the forms. He brought back the forms for me to sign. Then, because there was some hurry in the matter, he took the pirogue again to the village and didn't return until long after dinner. Maybe he spent six or seven hours that day to save me the trouble. No detail escapes him. Everything has to be just so; and whenever he comes across something that he knows will be especially burdensome, he never asks someone else to do it, though we pray that he will, but goes ahead quietly and does it himself."

"Who will do your work while you are in South Africa?" I asked.

"Right now, we are very fortunate at the Hospital. We have enough doctors. You have met Dr. von Stolk. He comes from Holland, too. He is very young but he is very talented and he is very precise. Dr. Schweitzer leans on him heavily. Dr. Friedmann is a completely dedicated man. He has a fine background of experience. We have teamed up in the clinic and there is nothing about the operation of the clinic he does not know. You probably have heard that Dr. Friedmann lost all his relatives in the concentration camps of Germany during the war. He is not embittered. He asks only to be allowed to serve because he himself was spared. Then there is Dr. Cyril Coulon. He, too, is young. His wife is with him. Soon they will have a baby. They will have it here at the Hospital. Mrs. Coulon is such a fine young woman. The Coulons have worked in the jungle before. They are Dutch. They have gone out by themselves into the jungle villages to help supply medical help. Now they are here to help Dr. Schweitzer. Dr. Coulon knows the Africans well; he is a great asset to our staff. Dr. Catchpool you have already spoken to on the plane. He has not been here very long. He has done no medical work so far but I can tell that he has had excellent training. He is a sensitive man. There will be no shortage of doctors while I am gone. When I come back, Dr. van Stolk will go on leave for six months. There is always a great deal of rotation going on."

Dr. Friedmann came up the path and greeted us. He was soft spoken and there was shyness in his manner. He had an enormous jet-black mustache; it made me think of some of the photographs of Schweitzer at the age of thirty. When Dr. Friedmann sat down on the stoop and crossed his arms on his knees I could see, on the inside forearm, the number tattooed on him at the concentration camp. He would carry the number as long as he lived. He noticed that I was staring at it.

"Just a souvenir; you're welcome to look at it," he said, holding up his arm.

I told Dr. Friedmann that *La Doctoresse* had been telling me about her approaching holiday and about the things that led to her decision to come to Lambaréné.

"Actually, it is not much different from the rest of us," he said. "Some of us may have come here because we were in good circumstances and didn't feel quite right about it; others because they were in difficult circumstances yet managed somehow to survive, and they wanted to find some way of acknowledging their debt. But always it is the debt. And always you will find that somewhere we happened to read something by Albert Schweitzer that opened up a big door in our mind and made us know we had to come."

Just then, we heard explosions of joy from the far end of the porch. Three or four of the nurses were holding bright-colored dresses to themselves, their arms clasping the garments to their waists and their chins pressing against the necks of the dresses. They swirled around and squealed ecstatically.

"Dr. Margaret, come quick," one of them shouted. "Clara brought each of us a dress."

Clara was leaning against the porch railing near the end of the bungalow. Even at a distance I could see that she was deeply pleased.

Dr. Margaret jumped to her feet with the bright alertness of a child being offered a surprise gift. She ran down the porch. Clara reached into a box and took out a long blue cotton dress and handed it to Dr. Margaret who exclaimed her thanks and then rushed into her room to try it on. Three minutes later she emerged, smiling and radiant. The dress fit perfectly.

"Look at Margaret!" exclaimed Trudi Bochsler, the nurse in charge of the lepers. "She looks like a movie star. Only more beautiful."

Then the other girls rushed into their rooms to try on their own new dresses. When they returned, it almost seemed as though the place had been touched with magic. Each girl was proclaiming her delight with the appearance of the others. And indeed, each seemed uncommonly attractive. But more appealing than anything else were their expressions of satisfied wonder as they danced up and down the porch, their dresses swooshing and swirling.

"How does it feel to be Mrs. Santa Claus?" I asked Clara. She was beaming.

"You have no idea how much they hunger for the chance to dress up and do the attractive things most young girls take for granted," she said. "It's so easy to forget that even dedicated girls like this—girls of serious purpose and high intelligence—it's so easy to forget that they are still girls in every way—warm and wonderful and full of eagerness and feminine charm."

Albertina spun around and clapped her hands. "Tonight we will have a party," she exulted. "We will surprise Dr. Schweitzer. We will wear our dresses to dinner. And we will fix our hair."

Albertina was auburn-haired. Even in the early evening light I could see that her face was flushed with excitement. Yet there was an essential quality of composure about Albertina that seemed to cause her to draw back as soon as she saw she had become the center of attention.

But Albertina never had the chance to develop any second thoughts about her suggestion, for the others seized upon it with whoops of agreement. They would wear their new dresses to dinner, they would put ribbons in their hair—and, some of them would use lipstick. They would all gather in the dining room several minutes early and would be in their places when the doctor walked in.

The plan was a complete success. When Dr. Schweitzer walked into the dining room, the girls—scrubbed, shining, bright-eyed— were all wearing their new dresses. They were sitting in their customary places, trying to make it appear that nothing special had happened.

The Doctor, in a split second, took it all in. His eyes danced behind the craglike brows. I could see he had a vast delight.

"Thank you for letting me come to your banquet," he said in the manner of a man who had just arrived at the Queen's ball. Then, when he sat down and the customary silence occurred so that he could say grace, he said that he was so overwhelmed by all the beauty around him that he had forgotten all the grace prayers he ever knew.

The girls were charmed and showed it. Then the Doctor leaned forward in the manner he adopts when saying grace, and the entire table became silent. I couldn't help noticing that the prayer of thankfulness had a special meaning for everyone on the staff.

C. U.

Dinnertime at Lambaréné.

The Doctor finished saying grace and looked up.

"I, too, have a contribution to make to the festive banquet," he announced. "A case of wine that was sent here some time ago and has been waiting for just such an occasion as this. It is a very fine burgundy. We will all dine superbly tonight and drink to each other's health. And tonight—there will be butter!"

The mood of the staff that evening, like the wine itself, had a delicious sparkle to it. Only a few hours earlier, nothing out of the ordinary had been anticipated. Then, suddenly, the evening had been transformed into a surprise party—and each person was the guest of honor.

I looked around the long table. The men were obviously pleased with the sudden assertion of femininity by Dr. Margaret and the nurses. Three of the four male doctors were young and unmarried. While the men were certainly not unaware of the attractiveness of the young women on the staff at any time, their appreciation under ordinary circumstances was subordinated to the unending demands of the Hospital. For the main part, their relationship was professional. But now, for the moment at least, the circumstances had changed. The dining room was transformed into a small banquet hall and young ladies, gay and lovely, adorned the table.

It was interesting to see the way the occasion affected the conversation. Usually the people on the staff spoke about the work of the day or about matters related to the Hospital. But now, appealing nonsense was in order.

"Dr. Schweitzer, we want you to settle a bet," Dr. Margaret said. "Trudi and I have just been talking about champagne and how it is made. She says it is made just like wine and I said it can't be because of the fizz and the sparkle in it."

Dr. Schweitzer was equal to the occasion. For almost five minutes he expounded on champagne—how it was developed historically, what goes into it, why it is expensive, what kind of water is necessary, how it is bottled in order to keep the carbonation, etc. It was an astounding *tour de force* on the one subject in the world that he might be expected to know least about.

Neither Dr. Margaret nor Trudi had been completely correct and the bet was declared a stand-off.

Then the conversation somehow veered off into playwriting and acting, and from there into exploration and geology, and finally into furniture-making. On each subject the Doctor would listen carefully and then come forward with a surprising wealth of observation backed with historical information, dates, and intricate detail. With respect to furniture, the Doctor identified the soft woods and the hard woods, spoke vividly of their uses, the relative expense in their manufacture, and the competitive problems in the world market for fine woods.

Whenever the conversation seemed on the verge of getting too heavy, the Doctor restored the mood of gaiety with an amusing anecdote, which invariably had a point to make. One of these anecdotes grew out of a question on the declining powers of observation of older people.

"Naturally, it all depends on the person you are talking about," he said with mock seriousness. "When I was a boy of sixteen, I was very much under my grandfather's thumb. One day a cousin of my age of whom I was very fond came to visit me. We wanted to leave the house for a certain purpose but feared Grandfather might not give us permission. And so we told him we wanted to visit our uncle some blocks distant, and he said we might go.

"When we were out of viewing range from the house, we turned sharply and went in the direction of our real destination—a beer tavern. After we were there about ten minutes a man sat down at our table. It was our grandfather.

" 'An old man isn't as blind as you might think,' my grandfather said. 'And sometimes he is just as thirsty as younger men. Why didn't you invite me to come with you in the first place? Now pour me a drink.' "

Then Dr. Schweitzer looked at me with a twinkle in his eye.

"I'd better be careful," he said, "or Mr. Cousins will think I do nothing except tell funny stories."

The meal came to an end. The Doctor reached up and took his

napkin out from his open collar, folded it carefully, and put it in his holder. The hymnbooks were passed out and the Doctor announced the number of the hymn to be sung that night.

It was then that I experienced the shock of watching him sit down to play the dilapidated old upright piano. But all the others were long accustomed to both the sight and sound, and it did not diminish the general festive air of the party. And so tonight they sang with added spirit, still flushed with the brightness of the occasion. The Doctor finished the hymn, returned to his place and read the Lord's Prayer in German. Then, the dinner over, the staff went to the small side tables, carrying an extra cup of coffee and tea, so the mood would not be broken, and they would chat and relax in the cool of the evening.

The Doctor said good night and went back to his room to submit himself to the inevitable tyranny of his correspondence.

I walked back to my room, turned on the oil lamp, and wrote home to Ellen and the little girls. I was anxious for them to know about a place called Lambaréné and the people who worked there. For nothing is more essential to young people than to have their natural idealism nourished—and this I felt I could give them through what I had seen in my first day at Lambaréné.

Long after I turned off the small light from the kerosene lamp, I lay in bed listening to the sounds from the open wards a short distance away. There was a hacking cough, and then a child's cry. They were contrasting sounds to the ones I had heard a short time earlier in the dining room, but it was part of the human mixture and it was real.

# IV

THE NEXT MORNING, I rose early in order to see the Hospital starting up on a new day. African women, wearing their blankets or faded colored cloth in wrap-around style, were carrying water jars on their heads and were on their way to and from the wells. Some of them had babies riding on their hips in a sort of side-saddle arrangement. In front of many of the rooms or cubicles in the open wards, women were cooking over homemade burners. I noticed one woman squatting close to an improvised stove consisting of a large pit in which she was making a banana milk stew for her family. The milk was her own. She deftly worked each breast close to the pan, sending streams of her milk into the stew. Here was the eternal woman in the oldest drama in the world—giving her totality to the cause of life near her, infinitely resourceful, inventive, responsible. The banana milk stew sent up its steam and the woman sang softly.

I continued down the path past the long rows of open rooms. African women were helping to wash the ill members of their families. A father of about thirty was playing with his little girl, throwing her into the air and catching her just above the ground. My homesickness was too much; in no time at all I managed to join the party and both of us entertained the youngster.

One of the main occupations at the Hospital is ministering to new life.

C.U.

"Some of them had babies riding on their hips in a sort of side-saddle arrangement."

To my great relief, I was still able to perform the trick I used to play on my own youngsters—reaching behind their ears and finding all sorts of strange objects—pennies, pencils, rubber bands, and buttons. Then the three of us walked down to the dock, the child riding between us on our outstretched arms as we skimmed her over the ground.

The African father's name was George Malthen. I spoke to him in French and learned that his wife had just suffered a miscarriage. He had brought her and their child to the hospital just one day earlier. There were no complications, and she was now resting comfortably. Dr. Margaret had told him that in another two or three days she would be well enough to go home.

I asked where home was. He said it was up the river perhaps eight miles from the Hospital, where he worked at a lumber camp. He had had enough schooling to learn to read and write French, but that was all. He lived with his wife and child in a compound built by his wife's grandfather who was still alive.

Two years earlier his wife's sister had died during childbirth. The baby had been thought to be overdue and the local midwife tried to induce labor. Then, when the woman's condition became serious, the witch doctor or fetisher was brought in. The woman was given potions and made to do strange things. Then external force was used on her abdomen. The woman went into labor and died.

George Malthen said he resolved then that he would never let anything like this happen to his own wife. And when, several days ago, his wife began to bleed, her mother summoned the fetisher who told his wife to do strange things if she wanted to save her baby. When George said he wanted his wife to go to the Schweitzer Hospital, the fetisher warned him not to interfere or he would cause him to turn into a vampire at night and kill his own child.

"He told me he would cause me to kill again and again and then I would be killed myself."

"Were you frightened?"

"Yes, I had much fear. The fetisher, he is a powerful one. But I could think only about my wife and what had happened to her sister. And very early in the morning, before the sun came up, I carried my wife and child down to the dock, put them in my brother's pirogue, and paddled as fast as I could to the Hospital. I am glad, very glad, I have done so."

"And what will happen when you return to your village?"

George Malthen said nothing for a moment or two. His eyes were fixed on a pirogue not far away.

"I have not thought much about that yet. I don't know what the fetisher will try to do to me. I will try to be strong. I have seen people under the spell of the fetisher do terrible things. Other people who have disobeyed him have lost their minds or have died. It is a strange power he has. I must be strong."

As he spoke, I thought back on something Clara had told me about the witch doctors at Lambaréné. They would hover on the perimeter of the Hospital, weaving their spell on the patients or members of their families as they came within range. Generally, the witch doctors wore Western clothes, complete with white shirt and tie, though the outfit would be somewhat frayed and unpressed. Occasionally, they would affect the long diplomat's frock, wing collar, and black tie—again of a distinctly secondhand origin. They would warn the patients to leave the Hospital and place themselves in the fetisher's hands. The usual threat has to do with visions of human beings transformed into vampires in the middle of the night and sucking the blood of loved ones, sometimes until death.

Once, the entire leper village was tied up when a witch doctor of obvious persuasive ability caused virtually the entire African population to refuse any treatment. Patients went on a no-treatment system. Dr. Schweitzer was angry and he was all for finding the fetisher and settling the matter once and for all.

Even before the Doctor got to him, however, Clara had managed to persuade the fetisher to call off the strike. Clara is about five feet tall and looks as though she barely weighs one hundred pounds. I had an image in my mind of this lovely, dainty little lady confronting the African witch doctor in his stained diplomat's frock, trying to convince him to rescind his authority and abdicate. Later,

Clara refused to tell exactly what transpired at the meeting or what superior magic she herself had brewed in order to persuade the African to cease and desist. Whatever it was, it worked, and the patients submitted to treatment and diligently took their medicines.

When I asked Clara at least to tell me how she felt during her parley with the witch doctor, she knitted her brows:

"I was terrified," she said, "that what the Africans said about him might be true. I expected at any moment that he might get a big hex out of his black bag and turn me into a vampire right on the spot."

She shivered, then smiled: "Now let's not talk about it any more."

The rational-minded Westerner finds it easy to scoff at the hold of the fetisher over many Africans. But before we give ourselves too much credit, we ought to take into account the countless millions spent each year in America and Europe on mediums, bogus doctors, tea-leaf readers, numerologists, astrologists, to say nothing of snake oil, cure-all drugs, and quack potions for revitalizing the blood. Both the quacks and the witch doctors can point to people they have treated who have become well again. The human body has an amazing capacity for overcoming both natural illness and mistreatment by those who profess to cure. Indeed, even in the West, doctors today are amazed that people should have been able to undergo without too much apparent harm the kind of medical treatment that not so long ago was considered routine. We need also to remind ourselves that George Washington's physician tried to treat his patient for a cold by bleeding him. It may or may not be a coincidence that Washington died shortly thereafter.

In any event, the witch doctor has had—and still has—vast power in many parts of Africa. And his exploits are sufficiently dramatic for him to maintain the myth of his magic. While standing on the dock talking to George Malthen about his own experience, I recalled an incident that had been told me by Laurens van der Post of the Union of South Africa, one of the most sensitive and skilled writers to deal with the terrifying complex subject of Africa. His *The Dark Eye in Africa* is a poetic and rich presentation of the human situation in contemporary Africa.

Colonel van der Post had been in New York several weeks before my departure for Lambaréné, and discussed some of his own experiences with witch doctors in the southern part of Africa. He liked to visit African villages in the remote interior. Knowing the ways of the tribes, he would wait with his African guide and companion just outside each village until, eventually, one of the elders would come out to look him over, and, if he were lucky, to bid him enter.

On this particular occasion, van der Post and his guide had to wait several hours. Finally, they were confronted by the witch doctor, who was apparently torn between the natural curiosity and friendliness of the villagers, which led them to want van der Post to come in, and his own need for a sense of superiority, which led him to try to keep van der Post out.

The witch doctor casually said he would permit van der Post and his guide to enter on condition that they would agree to a test of comparative magic which the fetisher would name.

Van der Post readily accepted.

"The fetisher smiled in triumph," van der Post told me, "then bid me follow him. Inside the village a large crowd quickly surrounded us. The fetisher announced that the white man had accepted his challenge. Then the fetisher handed me his staff; it was about four feet long. He told me to instruct my African guide to keep the staff planted on the ground.

"This was to be a test of power. The fetisher said he would cause the staff to rise despite anything my African guide could do to keep it on the ground. My job was to exert all the influence I could on my African guide to keep the staff from rising.

"I had known Joseph, my African companion, for a long time. He was with me all the way. He understood exactly what he was to do. He stripped to the waist, revealing a gleaming and powerful torso. Then he took the staff and plunged it into the earth as far as it would go. He spread his legs apart to assure his balance, grasped the top of the staff with both hands, and anchored it firmly to the ground.

"I announced we were ready for the challenge.

"The witch doctor seemed unimpressed with the sight of my companion leaning on the staff with his full two hundred pounds. Then the witch doctor began his magic incantations. He made large swirling motions with his arms above the straining, sweating body of my guide, clinging to the staff.

"Then the witch doctor, in African, began to chant: 'Let the staff rise! Let the staff rise!' Some of the villagers had their drums with them and took up the beat. Within a minute or two, hundreds of people joined the chant. The drums beat louder.

"I watched my companion. Joseph was sweating profusely. His eyes seemed somewhat glazed and they were fixed on the witch doctor who continued his swooshing and swooping motions. And all the time the witch doctor kept chanting, 'Let the staff rise! Let the staff rise!'

"I began to feel uncomfortably warm. My shirt suddenly seemed to have the weight of a winter overcoat. And all the time the drums got louder and louder, merging with the chanting of the crowd. The volume of the sound was unbelievable.

"Joseph kept the staff pinned to the ground. I thrust my wet face next to his and shouted to him above his din to hold on. His muscles were taut and knotted with strain. His eyes were bulging in their sockets. The staff seemed secure under his weight.

"The chant approached its crescendo. Its beat was irresistible. Strange words began to form in my throat and on my lips. Almost before I realized it, I found myself intoning hoarsely with the others, first in a husky whisper, then in full voice: 'Let the staff rise! Let the staff rise!'

"The cords were standing out on Joseph's neck and shoulders. His arms and hands were quivering. There was a moment of agony on his face, then the staff began to waver in his hands. The witch doctor swooped even closer. The staff in Joseph's hands began to rise.

"He held on to it with all the remaining strength he could muster, but it was no use. The staff rose perhaps eight inches above the ground, Joseph clinging to it all the way. Then the witch doctor clapped his hands. The drums stopped, and the spell was broken. Joseph relinquished his grip and the staff fell to the ground.

"The people began to disperse. There was neither elation nor triumph in their manner. What had happened had been entirely expected and predictable. The only excitement would have been if the witch doctor had failed.

"As for the witch doctor himself, he couldn't have been more friendly. Now that he had demonstrated the inferiority of the white man, there was nothing more to prove; and he bid me welcome. We couldn't have enjoyed greater hospitality at that village that night if we had owned it.

"The next day, several hours after we had left the village, I decided to speak my mind to my African guide and companion.

" 'Joseph,' I said, picking my words carefully, 'how did you ever let yourself fall for that old hypnotism trick? That staff rose from the ground only because you lifted it. You were hoodwinked by one of the oldest witch doctor tricks in Africa.'

" 'Colonel,' he replied, 'I'll be glad to tell you why I couldn't hold that stick to the ground if you tell me why you kept screaming in my ear to let it rise.' "

WALKING AROUND the Hospital with George Malthen as my guide, I could understand why some visitors came away with negative impressions.

The idea of a hospital creates instant images in the mind of immaculate corridors, white sheets, total sanitation. These images were badly jolted when one saw the Hospital at Lambaréné for the first time. Countless numbers of goats wandered at will all over the place; even when they were not visible their presence was perceptible. The ground was made moist and slippery by an equally large number of chickens. Hanging heavily in the dank air was the smoke from the dozens of crude burners used by the Africans for their cooking. There was also an inexplicably sweet and somewhat sticky smell—perhaps from the cooking or from fallen and fermented fruit.

The sanitary facilities were at an absolute minimum. There were only two outhouses, one for each sex. The sewer underneath was open and sometimes the wind blew from the wrong direction.

There were no bedsheets. The Africans brought their own blankets. There were no "wards" as the term is used elsewhere. There were long, bungalowlike affairs with small cubicles. When a patient came to the Hospital, he was generally accompanied by his entire family. The mother did the cooking, as she would at home. The children were usually on their own.

The difficulty, of course, was with the term "hospital" as applied to the Schweitzer colony. It created false images and expectations by outsiders. The proper term should be "jungle clinic," as Dr. Margaret had explained. Dr. Schweitzer did not come to Africa for the purpose of building a towering medical center. He came in order to meet the Africans on their own terms. What he built was an African village attached to a functional medical and surgical clinic. The Africans were attracted to Schweitzer because of the man himself and because this was a village and a way of life familiar to them rather than a forbidding building where they would be cut off from their families and frightened by a world of total whiteness, of people and walls and machines. Modern medicine has come to accept the emotional security of the patient as a vital part of any therapy. Dr. Schweitzer knew this almost a half-century earlier when he made his plans to serve in Africa.

Most visitors who stayed long enough became aware of these things. While they might never be able to accept completely all the crudeness, at least they developed a working perspective. Some visitors, however, could hardly wait to get back to Europe or America in order to make known their discoveries. I had read at least four articles by disillusioned visitors to Lambaréné who misunderstood and misjudged Dr. Schweitzer and what he was trying to do in Africa.

In addition to exposing the lack of sanitation, the articles would invariably talk about the gruffness of Dr. Schweitzer, especially toward the Africans. They would be disturbed especially by his references to the "noble savage." These were some of the things that Clara Urquhart had cautioned me about before I left for Lambaréné. I, too, was surprised when I first noticed it, but after a while I realized it was more apparent than real.

The Doctor would bark out his orders to the Africans and scold them when they were doing something wrong. The impression he gave was that he was dealing with children. There did not seem to be sufficient respect in his manner toward the Africans. But this was not the complete story. To get the full picture, one must realize that Schweitzer also treated most whites as his "small brothers" and one had to find out how the Africans themselves interpreted his manner.

I watched the Africans closely as they worked under Dr. Schweitzer's orders, pushing back the jungle or gathering up stray pieces of lumber or moving crates of medicines. When he appeared to be arbitrary or gruff in what he told them to do they would smile broadly and carry out his instructions. Sometimes when he called out sharply, he would have a glint in his eye which they would catch and it would amuse them.

In talking to one of the African leper workers at some length, I learned that the ones who had been with the Doctor for any length of time had no trouble in understanding him. They knew he was somewhat short-tempered when things did not go just right; but they knew something, too, about the pressures under which he worked. And what was most important to them was that they knew the stern manner did not reflect any displeasure by Dr. Schweitzer.

Even when the Doctor seemed to lose his temper, it was only for the moment, the leper said. Sometimes, if he had been too severe, he would go out of his way later to make amends. Once he scolded the wife of one of the patients. Fifteen minutes later he beckoned to her when no one was looking, said he was sorry and gave her thirty francs.

"We do not become angry," the leper said. "How could we? Could a man become angry at his own father for telling him what to do?"

The fact that Dr. Schweitzer's role at Lambaréné was that of father—with respect to patients, their families, the workers, the white doctors and nurses, and even the visitors—is vital to any understanding of his manner. He had a sense of total personal responsibility for everyone and everything at Lambaréné. Time was his most precious commodity and he was no longer able to expend it in lengthy and cordial explanations for what he would like to see done. When, for example, he ordered the staff and visitors to wear pith helmets, he did not have time to explain that when he first came to Lambaréné he had to deal with serious cases of sunstroke suffered by white people who had insufficient respect for the striking power of the equatorial sun. Once he had to take an overnight trip by canoe to attend the wife of a French planter who became seriously ill because she thought it was unnecessary to wear a helmet even though the sky was overcast. She hadn't understood that even the diffused rays of a hidden sun can cause trouble. Dr. Schweitzer did not intend to use all his time in Africa treating white people for sunstroke; his purpose was to provide medical treatment for Africans. And it became a little wearisome having to go through detailed explanations to each new visitor. Hence his "take-my-word-for-it" approach, whether with respect to sun helmets or other matters, each of which had its reasons.

At Lambaréné I realized that the criticism of Dr. Schweitzer's relationship with the Africans missed an important point. The somewhat arbitrary or patriarchal manner was not reserved for blacks only. Once, while Dr. Schweitzer was superintending a jungle-clearing operation, he ordered the blacks to rest. Then he turned to three white members of the staff and to me and said, "Now it's your turn." We obediently took up the work, pulling stubborn weeds from near the trunks of young trees. After about ten minutes we looked as though we had been working ten hours. Our white shirts and khaki pants were drenched. All the while the Africans stood by, looking on us with boundless compassion and appearing desperately eager to spare us further effort. Then the Doctor said we could stop; he just wanted us to have some respect for the requirements of physical labor in Lambaréné. He had made his point.

Not infrequently, his seeming brusqueness was leavened with humor. When Adlai Stevenson visited Lambaréné he was escorted on a tour around the Hospital by the Doctor. The former presidential candidate noticed a large mosquito alighting on Dr. Schweitzer's arm and promptly swatted it.

"You shouldn't have done that," the Doctor said sharply. "That was my mosquito. Besides, it wasn't necessary to call out the Sixth Fleet to deal with him."

Clara gave me another illustration of the fact that his sternness knew no color lines. Once, he became particularly exasperated at an African who was putting boards of lumber in the wrong place. He mumbled that he could almost slap the man. Clara, who was standing nearby, was shocked and said so to the Doctor.

"Well, Clara," he said, "I don't think I am going to slap him. But if I should do so, I want you to close your eyes and imagine that I am slapping a white man. In that case, it will probably be all right with you."

# V

BY SEVEN A.M., the sun had claimed the sky and the moisture was heavy in the air. I was halfway down the path leading to the dining room when Dr. Catchpool fell in alongside me.

"Better not let Dr. Schweitzer catch you without a helmet," he said. "He's very sensitive on this subject. He's had some experience with other people who saw no reason for wearing a helmet when they were going to be outside for only a few minutes. He's had to hospitalize them, and if there's one thing that annoys the Doctor, it's taking care of people who have no right to be sick."

In the dining room at breakfast, a few minutes later, Dr. Schweitzer referred to the matter. He instructed Clara to be sure to get a helmet to fit me from one of the extras in Mme. Oberman's possession. Clara nodded knowingly, then whispered to me that the Doctor meant business.

Breakfast consisted of thick homemade bread, jam, warm milk, coffee, and bananas. The Doctor ate somewhat more heartily than the rest. He supplemented the regular fare with an avocado pear and an egg.

Toward the end of the meal, I told Clara that I hoped we might have a chance to spend a few minutes with the Doctor. I didn't want to impose on his time, but I did have a few commissions that I felt ought to be performed as soon after my arrival as possible. During the course of a correspondence with President Eisenhower on various subjects related to the peace, I had mentioned that I hoped to be able to visit Dr. Schweitzer in Lambaréné at about the time of his eighty-second birthday. The President proceeded to write a letter of birthday greetings to Dr. Schweitzer which he was good enough to ask me to deliver. In that letter he spoke of his high admiration for the Doctor, saying that he had derived much inspiration from his work and thought, and felt the world was greatly in need of the kind of contribution he had been making. He ended by wishing the Doctor many more years of effective service to the human community. I had another message for the Doctor from Prime Minister Jawaharlal Nehru, whom I had seen on his visit to the United States in the fall of 1956 and who was most eager to have conveyed to the Doctor his deep admiration and affection.

I had a third commission. A fifteen-year-old boy by the name of Marc Chalufour in Concord, New Hampshire, was engaged in a crusade to save an old organ in his church from being replaced with an electronic instrument. He felt the old organ could be repaired and that it was sacrilegious to let it die—especially since he did not feel the new machine was really an organ. When, through one of his teachers, who was a friend of mine, he happened to learn of my coming visit to Lambaréné, he asked if I might deliver a letter from him to the Doctor. He felt that if he could enlist the Doctor in his crusade, he might win his fight.

I knew I would feel better when these various commissions were out of the way. And so I asked Clara what she thought would be a convenient time to act. When the Doctor got up from the break-fast table, Clara followed and spoke to him on his way out. Then she nodded to me, indicating the matter was arranged.

As I emerged from the dining room, Clara handed me a pith helmet that had just been given her by Mme. Oberman. It was a little large and I had the feeling I was walking around inside an inverted laundry basket, but it served the purpose.

In the compound not far from where we stood, the Doctor was giving working instructions for the day to perhaps two dozen Africans. First there had been a roll call and now the doctor was dividing the men into different groups, one of which was to collect stray lumber around the place, another of which was to repair the porch on one of the long bungalows occupied by the staff, and still another of which was to carry on the war against the jungle, cutting, pruning, weeding, pushing back. I could readily see the truth of the remark that had been made in connection with Lambaréné, that if you left the jungle to itself for two months, it would close in over you and you would have to tunnel your way out.

Dr. Schweitzer completed his instructions and motioned to Clara and me to follow. He went up the several steps to his quarters, which were at one end of the long bungalow facing the compound. Just to the side of it was a fenced-in area for several antelopes.

The Doctor's room, like all the others, was open at both ends, covered by a wire screen and supporting woodwork. On the far side he had his desk, part of which was under a mound of papers and books. Opposite it was a medium-sized bookcase. The Doctor asked us to be seated, then sat on a stool at least as old as the hospital. I made a mental note of the fact that nowhere at Lambaréné had I seen an upholstered chair or anything even resembling a sofa or couch.

Dr. Schweitzer apologized for the fact that we did not have a talk the previous afternoon. He pointed to a pile of forms on his desk.

"All this is to be filled out," he said. "Now the French government has asked us to prepare complicated forms for each patient at the Hospital. Miserable paper work. Also now we have to fill out workmen's compensation forms for the working people who come to the Hospital. Dozens of items for each patient. And I hardly know what to do with these."

He lifted his head in the direction of one end of his desk, indicating a large bundle of new mail.

"My paper work is killing me," he said slowly. "Week by week the mail gets larger. Mlle. Ali and Mlle. Mathilde help me as much as they can. Even so, we keep falling farther and farther behind. Many of them are important letters which must be answered. Some of them are from theologians who raise significant questions in connection with things they may have read that I had written at one time or another. The least I can do is to try to answer them. I am maybe fourteen months behind with part of my correspondence. Some of the letters involve the work of the Hospital— people who volunteer to work here.

"While the turnover is not excessively large, we do have to bring in new people now and then. The nurses and doctors here have to take a leave of absence after a year or two just to rest up after the exhausting work in this climate. And so new people keep coming. Before a new nurse arrives at the hospital, some twenty-seven letters have to be written. It is important, so very important, to be sure about these things. It is not like a girl coming to the city from the country to get a job. It is a long way from Europe and America to Lambaréné. And the only way I have of finding out what I must know about each person is through letters—most of all the letters the person herself writes, but also what I can learn from other people.

"After it is decided that a nurse or doctor is to come here, then the real work begins. I must be very clear and very complete in the instructions. Passports, visas, en-route accommodations.

"But we will not talk any more about this. Are you comfortable in your quarters?"

I said I was very comfortable indeed. I told the Doctor I was eager to take care of some matters, and handed him the letter from President Eisenhower. He opened the letter carefully, then read it without translation.

"A letter from the President of the United States," he said slowly, then read it again. I could see that he was moved by what the President had written. Then he looked up and smiled.

"Do you know," he said, "that if, eighty-two years ago, my dear mother had been told by someone that her little baby would some-day receive a letter from the President of the United States, she would have had the surprise of her life. This is a very kind letter from the President. I will take great pride in returning the greeting."

Then I conveyed the cordial good wishes of Prime Minister Nehru. At the Doctor's request, I told him about the Prime Minister's recent visit to the United States, and the importance attached to it by President Eisenhower. I referred to the fact that in the past few years some misunderstanding had developed between the United States and India. This was not merely a matter of differing foreign policies. Public opinion in both countries had been growing apart. In America, the mistaken notion existed that India had ranged itself on the side of the Soviet against the United States, especially as it concerned the decisions made in the United Nations. In India, a large segment of public opinion took the view that the United States was insensitive to the independence movement in Asia and Africa, and that we were determined to preserve the interests of the colonial powers in that part of the world.

The result of both these views was that the two peoples of the world whose national historical experience had so much in common and who, by standing together, could contribute so mightily in the building of essential bridges between East and West, were drifting apart. Hence the significance of Mr. Nehru's trip to the United States. Both the President and the Prime Minister had developed an instant mutual admiration and respect.

In talking to the Prime Minister before he returned to India, I learned that the exchange of views with the President had cleared up many matters in his mind. At the same time, he was able to impress on the President the reasons for the positions taken by India with respect to affairs in Asia and the Middle East.

The Prime Minister attached the utmost importance to the uprisings in Poland and Hungary. What happened in Hungary,

C. U.

"If, eighty-two years ago, my dear mother had been told by someone that her little baby would some day receive a letter from the President of the United States, she would have had the surprise of her life . . ."

he said, would have a profound effect everywhere, especially among young people who would now realize that communism could no longer be the headquarters for their natural idealism. And even though the uprisings were suppressed, there would be profound internal changes in the Communist world, he felt, changes that would inevitably move away from the old dictatorial shape of things. People were demanding something better than what they had known; they were demanding greater liberty and a better life. And there was no way for this demand to be resisted. He was therefore hopeful that, given world peace and a little time, there would be long steps forward for a large part of the world community.

Dr. Schweitzer listened carefully. He recalled the time he had met Nehru in Lausanne in 1936. He was impressed by the interesting combination in Nehru of the contemplative man and the man of action. And he was fascinated by the way Nehru complemented Gandhi.

But enough of politics, he said. There was work to be done around the Hospital and he invited me to tag along if I wished.

There was still the matter of Marc Chalufour and his crusade to save the church organ. I decided to bring it up another time. As for the main reasons I had come to Lambaréné—making duplicate copies of the Doctor's unpublished and unfinished manuscripts, and the possibility of direct action by him on the matter of world peace —these would have to wait for an hour when there might be some easing of the pressures of the Hospital. But the more I learned about the Hospital and about Dr. Schweitzer, the more pessimistic I became that such an hour might be found.

In any event, I accepted the Doctor's invitation to accompany him as he discharged his morning chores. The first job we had was to move planks of lumber that had been stored in several places to the porch of a bungalow that was now under repair. The wood was the finest mahogany in the world. Its use as floor boards was dictated by two factors: first, the abundance of mahogany in the general area of Lambaréné, making it inexpensive; second, the need for the hardest woods as protection against termites. Much of the Hospital, in fact, had been built with the kind of hardwoods that in Europe and the United States were reserved for the most costly cabinets or luxury paneling.

The Doctor worked alongside the Africans as the wood was transported by hand and placed in neat piles under the bungalow porch being repaired.

Our next job was to get rid of crates of medicines that had spoiled because they were improperly packed before being shipped. Shippers in Europe and the United States, apparently, have little experience in preparing medicines for storage under equatorial conditions. In addition to the heat and the moisture, medicines have to contend with raiding ants. Large crates would arrive with labels saying they had been specially sealed to guard against heat and moisture; but within three weeks after arriving at Lambaréné they had to be repacked. Even sealed metal drums sometimes failed to do the job. I watched two such drums containing millions of cathartic molasses tablets emptied into wheelbarrows for dumping into the river because they had spoiled or turned into tar.

Then several large packages of aspirin had to be dumped. There was some evidence indicating successful ant raids into the aspirin stores. Judging by the size and activity of the ants at the Hospital, I would have supposed that they were in no need of pain killers. In any event, they kept coming back for more. The ants, who have an appetite for paper, made it necessary to do double labeling on each bottle. In fact, almost everything connected with the medicines required two or three times as much work as it would in a temperate zone. Every bottle, for example, had to be resealed with paraffin after it arrived.

Ideally, of course, a hospital should have large refrigeration facilities for its medical supplies. This is especially true with respect to the new antibiotics. But there was very limited refrigeration equipment at the Schweitzer Hospital. For the most part, the medicines had to be stored in the empty space underneath buildings that were so situated on the hillside that they had relative protection against the sun. Sometimes the Doctor had to get down on his hands and knees and crawl through the dirt in order to check on the medicines and help move them as might be required.

After two hours of working alongside the Doctor I was ready to throw in the towel. The Doctor's shirt was wringing wet. His hair lay in moist gray clumps on his forehead.

Again I had to keep reminding myself that the Doctor was entering his eighty-third year. Just trying to keep up with his stride as he hurried from place to place was a feat of endurance. I could hardly wait for the afternoon siesta.

When I entered the dining room, I felt fully at home for the first time. For I knew that the look of severe midday fatigue on all the faces of the staff was reflected on my own. The Doctor, too, showed the effects of his exertions. Yet it was astounding and wonderful to see the way even a brief respite and a good meal enabled him to recoup his energies. After he had had his soup and was halfway through his cheese and noodles the color began to come back to his face. By the end of the meal he was sitting straight in his chair, his eyes were twinkling, and he had some stories to tell the staff that lifted their spirits.

"I have an announcement to make," he said. "Civilization in all its glory has finally come to Lambaréné. Less than a mile from the Hospital today there was an automobile accident. There are probably only two cars within miles of the place; today they inevitably met in a crash and we treated the drivers for some minor injuries. If anyone here has reverence for automobiles he is welcome to treat the cars."

Then he turned to Albertina.

"How is your pet monkey doing these days?" he asked.

"Not very well," Albertina replied. "I set him free in the morning but he has a terrible habit of wandering too far and has increasing difficulty in finding his way home."

The Doctor smiled. "Maybe he drinks too much," he said.

On my way out of the dining room, I asked Clara when she thought we might discuss with the Doctor the two main purposes that brought us to Lambaréné. She said we might have to wait another day or two, perhaps longer; when the right time came, the Doctor would come to us.

Walking toward my room, I passed the Doctor's quarters. Through the wire screen I could make out his silhouetted form perched over his correspondence. For the rest of the afternoon I stayed in my room, taking care of office matters that had been air-mailed to me from *The Saturday Review* in New York. Like everything else in the world, my job had been revolutionized by the air age. There is hardly a place in the world that cannot be reached by person or mail in forty-eight hours. I have had airmail bundles of work from New York only two and a half days old delivered to me in a Moslem village fifty miles from Dacca in East Pakistan. On another occasion, a postman on a bicycle met me coming out from a small Japanese inn near Nagasaki and handed me page proofs of the forthcoming issue of *The Saturday Review*. And now in a jungle hospital in Africa, I was no farther away from my office than I would have been twenty years earlier in Colorado.

The airplane has made it not only possible but necessary for an editor to move his desk from place to place in the world. Indeed, the airplane itself has become an efficient and productive office. Nowhere outside of a plane have I found such ideal working and thinking conditions. When at my job in New York, my main business each day is to preside over interruptions. I have little time for reading, less time for writing. Sequence is annihilated. One takes things as they come, and they come in short, uneven bursts. Once, I became sufficiently objective about my job to keep track of everything that happened in the course of an average day. I discovered that the telephone rang on the average of once every six minutes; that there were at least six callers each day, three of whom came by previous arrangement; that the important business that had to be transacted with the staff had to be squeezed into less than an hour and a half. In between the telephone and the appointments, I would work on my unanswered-mail folder, which weighed heavily on me even when out of sight. I would type my replies on the back of each letter for retyping by my secretary on office stationery.

On a plane, however, especially when we are high above the clouds, the fragmentation disappears. Sequence comes to life again and the mind has a chance to reorganize itself for consecutive thought. The meal tray serves as an excellent stand for my typewriter and the seat beside me becomes a side table for working papers. I estimate that one hour in the sky office is the equivalent of about four hours on land in terms of actual output.

There are, of course, distractions. The sky itself. Even so, it is the kind of distraction that nourishes thought. For nowhere else in the world is there grandeur like this. Sometimes, when flying above a storm, vast, clearly defined cloud masses catch the light of the sun; the result is a Grand Canyon of color multiplied by infinity. Once, flying from Seattle to San Francisco, I saw three distinct cloud levels, with clear sky in between. Each level had its own character. The first was a massive purple floor, swelling and bulging when we were close to it. The second layer was full of tunnels and canyons, exploiting every gradation of gray and black. The third level abandoned itself to color, with gentle formations suggesting lakes of gleaming silver, or long lemon-colored slopes meeting a light green-blue sky.

Another time, flying from Beirut to Karachi, we darted in and out of vast thunderheads, each of which looked as though it was made up of the combined masses of all the mountains in the world. The thunderheads each ran up to forty thousand feet, growing fatter and more menacing on their way up, and then, suddenly, they flattened out on the top, connecting with each other to make a giant white roadbed. Underneath were the long caverns, deep gray on one side, streaked blue on the other. For more than an hour, we rode through the caverns until we finally hit a long open clearing.

I need not even mention the sunsets. The combination of the setting sun, the winds, and the swift movement of the plane changes the landscape so rapidly that no two minutes are the same. It is a developing wonder and makes a moving picture that stays in the mind. And, if you catch the late sun just right as you fly west, it will cause thin golden threads to spin off the propellers.

As I say, these are nourishing distractions, and I welcome as many of them as I can find. Indeed, I have made a hobby of collecting skyscapes, and now have some two dozen in my memory box to think back upon. It is the kind of hobby that goes naturally with the kind of job that has the world for its locale.

# VI

ON MY THIRD MORNING at Lambaréné, after breakfast, I crossed the small compound from the dining room to the porch of the Schweitzer quarters in order to pay my respects to Mrs. Schweitzer. She was seated in a dilapidated beach chair. In her hand was a long bamboo rod which she used to fend off some of the animals that would otherwise disturb her rest. In particular, there was one bird she had to guard against. He looked like a cross between a raven and a parrot. He would come swooping down and alight on your shoulder or the back of your neck. If you weren't used to this sort of thing, it could be somewhat unnerving, for it would happen very suddenly and you hardly knew what hit you. Generally, it would be accompanied by a sudden flapping of wings and occasionally a nip of the ear. This bird was totally gregarious and it was hard to persuade him that his attentions were not always the ultimate in human enjoyment. Mrs. Schweitzer liked to doze in her chair. So long as she held onto the bamboo rod, the bird respected her solitude. But when the rod was absent, the bird invariably interpreted it as an invitation to a shoulder.

The first time I saw Mrs. Schweitzer I could see she was not well. The blue veins stood out in her forehead and seemed stark against the pure whiteness of her skin. She had lovely gray-brown eyes but they seemed to look at you through a mist. When she spoke it was with considerable effort. Her breathing was labored. Despite her difficulties, she would not allow anyone to treat her as an invalid. She insisted on coming to the dining room for lunch and frequently for dinner. It was easy to see how much of a struggle it was for her, even with the aid of a cane, to negotiate the two dozen or so steps across the compound and climb the short stairs to the dining room. Once, I saw Mrs. Schweitzer start out across the compound, her weight bent forward on the cane and her whole being struggling for breath. I rushed to her side and took her arm. She looked up at me, somewhat puzzled, as though I did not know the rules of the game at Lambaréné. Then she smiled and thanked me but said she was in the habit of getting around by herself. She expressed the hope that I might come to visit her on the porch and talk to her.

"You know," she said, "I am so interested, so very much interested, in what is happening in the world, and here one does not often get a chance to talk about it. The Doctor is very busy with the Hospital and all his other work and I do not want to tire him by asking him about things. So there is very little talk. And I am hungry for talk, especially about the world. It would be very nice if you could come to visit with me; that is, if you are not too busy."

I accepted the invitation gratefully. And now, the next day, I visited Mrs. Schweitzer on the porch. She bid me draw up a chair and told me she would do her best to protect me against the bird now making short circles above my head in preparation for a landing.

She apologized for the fact that she was unable to escort me to such parts of the grounds as I might like to see. Many years earlier, she had had a skiing accident that had broken her spine; in recent months she found it increasingly difficult to get around. Hence the cane and her seat on the porch.

"It makes me feel so foolish," she said, "this being so helpless. I ought to be working with the Doctor. He is an amazing man. I really think he is working harder now than he did twenty years ago. And twenty years ago I was afraid he was killing himself with work. He has always said that he has a favorite prescription for anyone over sixty who does not feel well—hard work and more hard work. As you can see, it is a prescription he follows himself.

"I am only sorry I cannot work hard, too. We have been working here a long time . . . more than forty years. I used to be able to help the Doctor; but in the last few years it has no longer been possible for me to do so."

I asked Mrs. Schweitzer which place she considered her real home—Gunsbach or Lambaréné.

"Actually, there are three homes," she said. "When I was young, shortly after I was married, I contracted tuberculosis, so we went to live in the Black Forest of Germany, where the climate was considered helpful for people in that condition. We liked it very much, so very much that we kept a home there even after I was cured.

"Then there is the home in Gunsbach—in Alsace, as you know. At first it was very much a home, but we spent less and less time there. And whenever we returned, there would be so many people who wanted to see my husband that I came to regard it more as an office than a home.

"And so, of the three places, this place seems most like home to me. It is very hot here in Lambaréné; it is very moist and there isn't much more than a bedroom to our own quarters. But this place to me has been our main home for more than forty years."

When I asked about their daughter, Mrs. Schweitzer said that she wasn't sure that Rhena had the same feeling about Lambaréné being a home for her.

"You see," she said, "when Rhena was small, the Doctor did not feel that Lambaréné was a good place in which to bring up a child. And so we sent her to boarding school. In later years, when she was fully grown, she came here for a visit—for a few weeks. She is now married, to Jean Eckert of Zurich, an organ builder, and she of course has her own home."

Mrs. Schweitzer then got up from her chair and went inside.

After a moment she reappeared, holding the family album. One of the early photographs showed the Albert Schweitzers with their little girl. Albert Schweitzer was young and sturdy, with black wavy hair, a thick mustache, and an unwrinkled face. Helena Schweitzer was pretty and vivacious; her features were thin and sensitive but there was great strength and directness in the face. Rhena had large dark eyes and held her mother's hand. She wore the expression so familiar to parents of little girls—part shyness, part curiosity, part uncertainty, part feminine delight in being asked to pose.

Another photograph showed Dr. Schweitzer at the age of thirty-seven or -eight. This was about the time he was getting ready to come to Lambaréné. He was in a reflective mood. His wide-set eyes had caught a distant light and seemed to carry him outside the frame of the photograph. His chin was resting on his closed right hand. There was a quizzical look around the mouth. There was no trace of whiteness in the hair.

Then there was a picture taken of Rhena and her husband just after they were married. Mr. Eckert was alert and well groomed. Right after this were other family pictures of the Eckerts, the most recent of which showed their sixteen-year-old boy, Philippe, who had recently been very ill but who was fast recuperating. Mrs. Schweitzer said the Doctor was very fond of Philippe.

"Here," Mrs. Schweitzer said, "you were interested in our home in Gunsbach. This is a picture of the street and this is a picture of the house itself. It is a nice house, not pretentious, but very sturdy and comfortable.

"I like these pictures. I turn to them very often. Those early pictures were taken so many years ago that it is hard to remember exactly how many. Now we do not take pictures any more.

"I know I am talking like an old woman. But it is true. I am old. I am very old. And now I have very little to do except to look at these pictures and old letters and think back upon our life together.

"You will forgive me? Thank you. I am so eager to hear about your own family. Clara has told me about your wonderful wife and your children—four or five girls, which?"

I took out a small group photograph showing Ellen and three of the four girls. I did not happen to have a picture of the youngest girl.

Mrs. Schweitzer lingered a long time over the photograph, asking me to describe each child. This I did with relish and supporting anecdotal material. Then, at her request, I told her about our home and the openness and sense of freedom it afforded the children. She eagerly devoured the information about my family and made me promise to send additional photographs from time to time.

I feared I might be intruding on her rest, so I got up to leave. She made me sit down again, saying that it had been a long time since she had had the opportunity to talk to anyone so fully. Besides, we hadn't even mentioned the world situation. What about the political situation in the United States? What about the struggle between East and West? What about communism? What about atomic energy?

"You must understand," she said, "that there aren't many visitors from the United States. And those who come do not stay for very long. They have urgent business, or so it seems, with the Doctor, and there isn't much opportunity to talk to them."

For the next hour I answered Mrs. Schweitzer's questions as best I could—and she had comments of her own to make now and then.

"It is so terrible," she said, "the world gets rid of one monster like Hitler but there are always others waiting in the wings to take his place. Isn't it strange? Human beings allow themselves to become all twisted under the influence of these men. I have seen it happen to many people I knew in Germany. I have seen the change that came over these human beings. I have seen decent people turn into killers and sadists."

She was of Jewish origin; she knew what she talked about.

"Anyway, we do the best we can and we work for better things. I don't know why I say 'we.' My time is over. I am too old. But I can hope. No one is ever so old that he cannot hope—even if his hopes are for others. I have done much such hoping in my life—and some of the hopes have come true. In recent years for me there has been not much to do except to hope—and to look at the pictures and think back."

Mrs. Schweitzer did not come to dinner that night. I feared that our conversation had been a drain on her energies, but Clara told me that she had been coming to dinner with decreasing frequency in recent weeks.

I next saw Mrs. Schweitzer at lunch the following day. She was at least fifteen minutes late in coming to her place. For the longest time after she sat down she had difficulty in getting her breath. Her lips were a stark blue. But she smiled graciously and did her best to engage in conversation. All she had for the noonday meal was a cup of tea. When she got up to leave she beckoned to me. Once we were outside she said she hoped I would visit her again on her porch. It meant so much to her, she repeated, to talk to people who have just come from the United States.

I had several visits with Mrs. Schweitzer while I was at Lambaréné. I came to admire her pride, her resourcefulness, her tenacity, her continuing interest in the outside world.

Two months after I returned from the Hospital at Lambaréné, I picked up a newspaper and learned that Mrs. Schweitzer had died. Her life had not been an easy one, but it had known purpose and hope and grace.

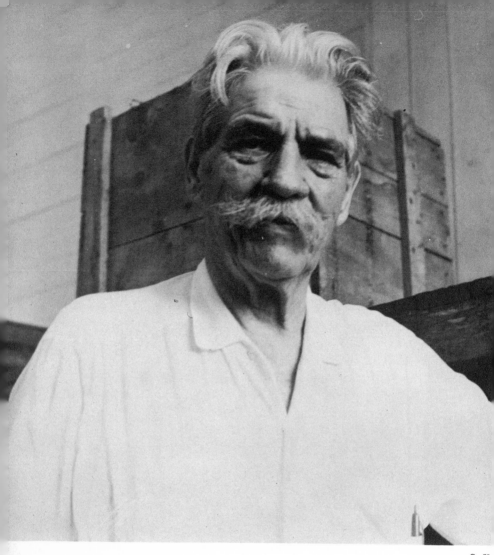

C. U.

"There is such a thing as being too detached."

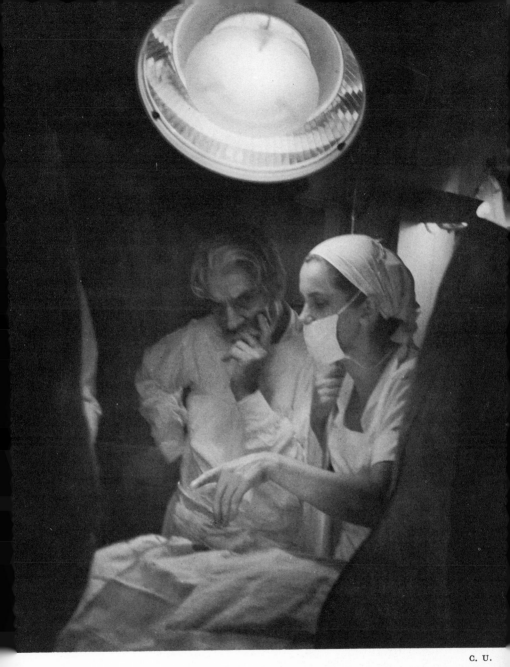

"One can't expect philosophers to be romanticists . . . the philosopher must deal not only with the techniques of reason or with matter and space and stars, but with people."

C. U.

"Goethe became a prisoner of his own promises. I don't want that to happen to me."

# VII

ON THE AFTERNOON of my fourth day at Lambaréné, Dr. Schweitzer came to my room and said he hoped it might be convenient for us to talk. He said it would be nice if we could be incommunicado from the rest of the Hospital for about two hours. We found Clara who suggested that we meet in her room which was out of sight from the main Hospital paths.

Dr. Schweitzer seated himself on the bed, his arm resting on the white metal tubing on the lower end. I don't recall how we happened to get on the subject of philosophy, but for at least two hours he discussed his debt to the great thinkers in history.

At one point, he stopped short, saying that one did a disservice to philosophy in general and certain philosophers in particular if one had nothing to offer them except praise. He pointed out, as Bacon and others have done, that the uncritical acceptance of Aristotle for many centuries actually had the effect of retarding speculative and systematic thought. There was a tendency to view Aristotle as the be-all and end-all of philosophy, with the result that hardly anyone felt there was any point in further exploration or development.

"Philosophy will never be complete and can never be complete, by the very nature of philosophy. The human mind is capable of infinite growth. There are endless adventures in creative thought ahead of us. It is only when men bow low before great thinkers and proclaim them to have said the last word that philosophical growth becomes arrested.

"Aristotle wasn't the only one. There was Kant. He was such a giant that men who should be working on the philosophical frontiers drew back and confined themselves to endless interpretations and theorizing about Kant. It was important, of course, for qualified thinkers to analyze Kant, but they had a tendency to make this analysis an end in itself. And since Kant was regarded as the last word, the new words were delayed in coming."

Then Dr. Schweitzer ran his fingers through his long, shaggy hair—one of the most characteristic of his gestures.

"I can't blame all this on Kant, of course," he said. "But I can blame him for being so much system and so little compassion. One can't expect philosophers to be romanticists, but it is important to remember that the philosopher must deal not only with the techniques of reason or with matter and space and stars, but with people. After all, it is the relationship of man to the universe, and not solely the relationship of one galaxy to another, or one fact to another, that should occupy such an important part of the philosopher's quest. There is such a thing as being too detached. I fear this may have been true of Kant."

There were other things in philosophy that bothered Dr. Schweitzer. Philosophical catchwords, for example. The way a phrase would be picked up and used almost automatically whenever a man's name would be mentioned, as though the phrase described the sum total of his life and thought.

He spoke of Descartes.

"One would think that Descartes lived just to emit a line of staggering profundity: 'I think, therefore I am.' How rare are the full-bodied examinations of his work. There has been too much genuflecting before *Cogito, ergo sum*; too many philosophical monuments have been erected in its behalf.

"I find it difficult to be impressed by 'I think, therefore I am.'

One might as well say, 'I have a toothache, therefore I exist.' These catchwords are tricky things. I don't think they serve the cause of creative thought in philosophy."

Then he looked up and said, "I have been negative enough. Too negative. There are philosophers whom I like and who have exercised strong influence on my own thought. Hegel, most of all. A man of reason. But also a man with a deep respect for the possibilities of the human being, especially the capacity to embrace important new concepts. Hegel is a philosopher who deserves well the mind concerned with the problems of its own growth.

"As for a school of philosophical thought, I acknowledge my great debt to the Stoics. To my mind it is the greatest formal philosophy in human history. To the extent that I can be identified with any one school, I should be proud to be related to the Stoics. There have been other influences, of course. I have a high admiration for the English philosophers of the late seventeenth and eighteenth centuries. They were imaginative but precise. They were concerned with man's estate.

"I have also found myself influenced by the early Chinese philosophers. There is an intense human quality in their thought. They never allow themselves to get too far away from their speculations about the nature of man or the purpose of man. And it is only logical that out of all this thought should come such a creative naturalism."

Dr. Schweitzer stared at the opposite wall. He said that these real influences in his philosophical thought were sometimes overlooked in the appraisals by others. It would be said, for example, that the Buddhist influence in his work was pre-eminent. This was a mistake.

"The mistake is natural enough," he said. "There is a disposition to think that because I am so deeply concerned with the need for reverence for life that my philosophy must be Buddhist, especially in connection with the Buddhist emphasis on the importance of animal life. But there is much more to Buddhism than that; and I hope there may be more to my own philosophy than that.

"A moment ago I was talking about the need to be careful of philosophical catch phrases. It is true that no single phase of my own philosophy is more representative of my thinking than 'reverence for life.' But the phrase is related to a whole line of thought. Rather than have people speculate on whether this phrase connects me to this philosophy or that, I would have them look into the purpose and the meaning of my work as a whole, such as it is.

"Christian theology has found it difficult to come to terms with my thought, though Christians have not."

He paused as he made this distinction.

"I have the feeling that the Christian theologians are reluctant to come in through the door I have tried to open. I have tried to relate Christianity to the sacredness of all life. It seems to me this is a vital part of Christianity as I understand it. But the Christian theologians, many of them, confine Christianity to the human form of life. It does not seem to me to be correct. It lacks the essential universalization that I associate with Jesus. Why limit reverence for life to the human form? As I say, I have tried to open the door; I hope the Christian theologians will come in."

I asked Dr. Schweitzer whether these philosophical ideas were reflected in the two books I understood were now close to completion.

He replied, yes, these ideas were reflected in the two books, but they were not necessarily the main theme. And he proceeded to describe the two unpublished works. He had published two of four volumes which together he calls *The Philosophy of Civilization*. The last two volumes were not quite completed. They were a continuation of the work which had appeared more than a quarter of a century ago. Another book now in manuscript form would be called *The Kingdom of God*. It would be a fairly short book. It would contain his ideas about Christianity and the reality of the human spirit in general.

"In the early 1930's I was determined to finish the two last volumes of *The Philosophy of Civilization*. But I kept being diverted. We were in a period of considerable building at the Hospital in Lambaréné and I found myself putting off my writing from month to month.

"Then, in 1938, I decided to leave Lambaréné in order to work for at least a year on the books. But even before we arrived in Europe—in fact, while we were crossing the Bay of Biscay by boat—I learned that Adolf Hitler had just made a speech in which he tried to reassure the world that his sole aim in everything he did was peace. The speech was such a patent cover-up that I realized the war was not far off. I put all thoughts of my book aside and concentrated on planning for the Hospital's needs in the event of war. I obtained a large stock of medical and other supplies, knowing that we would be largely cut off from Europe. Then I returned to Lambaréné in order to prepare for the war."

He knitted his craggy brows and I could tell that he was reliving the anxieties of the late thirties, when the heavy clank of the iron heel grew louder and nearer with each passing day.

The war came and only occasionally, late at night, was he able to work on the books. And then, at war's end and in the years that followed, a whole new crop of problems sprang up at the Hospital, resulting in yet other postponements for his serious writing. First of all, he said, additional buildings were needed. Then there was the biggest single new project since the Hospital started—his leper village, built with the money he received from the Nobel Peace Prize. And, of course, his visit to the United States in 1949 in connection with the bicentenary celebration at Aspen, Colorado, of Goethe's birth.

"Interruptions—important interruptions—but interruptions all the same," he said. "I worked for months on the Goethe lectures to be given in America. I worked for almost half a year on the Nobel acceptance speech."

I asked if the original Nobel acceptance manuscript was still available; I felt that any number of book publishers would feel privileged to issue it. Perhaps some of the matter might be adapted for article purposes. In that case *The Saturday Review* would like to put in a bid for the material right now.

Dr. Schweitzer said he had not looked at the material for some time, but would, at a later date, send me a copy. He feared, however, that it might need a great deal of work. And he shuddered when he thought of all the other work that claimed priority. Here he was telling me of the difficulty he was having in finding time to finish his two major manuscripts and now, just in the act of talking about it another book was being mentioned.

"Perhaps I had better do what I can, when I can and if I can, on the *Philosophy* and *The Kingdom*. Even though it has been some years now since I have been able to do any sustained writing on these books, my mind has kept ticking away like an old clock a long time after the key had been lost. The central ideas have been developing in my mind; I have a fairly good idea of what it is I should like to say; now all I need is time to say it."

His reference to time jolted me into the awareness of my own responsibility for having taken up so much of his afternoon. I told the Doctor that there was much I wanted to discuss with him, especially as it concerned his manuscripts; but perhaps we should leave it for another time.

He said he had understood from Emory Ross that I had some mission or other in connection with his manuscripts but the whole thing was very mysterious. He smiled and got up to leave.

"Tomorrow, at four o'clock, I will come here again to Clara's room and we will talk. And the mystery will be no more."

As he walked out the door he turned around and cautioned me to wear my helmet at all times. He said it would take him too long to give me all the reasons why it was necessary for me to do so but he hoped I would take his word for it. I told him that by now I was completely indoctrinated.

After he left, Clara said she had a feeling we would get what we came for. She was deeply pleased to see the Doctor so relaxed.

"It's been a long, long time since he has talked this fully about philosophical matters with anybody. He needs it. It is a busy but lonely life here for him at Lambaréné. Always it is the pressure of the Hospital. And he does not have the chance to exchange ideas in the fields that mean so much to him. He will come tomorrow and we will put the two propositions to him."

BUT AT FOUR o'clock the next afternoon, Dr. Schweitzer did not come. Fifteen minutes later we were in the middle of an equatorial squall. I had a large umbrella and I asked Clara whether I ought to take it to him. Clara said that even though the Doctor had no umbrella and seldom used a raincoat he would not be stopped by the heavy rain. This troubled me and when, at four-thirty, the Doctor had failed to arrive, I put on my raincoat and set off after him with the umbrella.

He was not in his bungalow. I learned that he was out in the rain, looking after a new shipment of medicines, making sure they were in a dry place. When he came to his bungalow and saw me standing there with outstretched umbrella, he shook his head slowly and said he hoped I had not thought he had forgotten our engagement.

We walked together to Clara's room. He took off his helmet, hung it on the bedpost, then seated himself on the bed as he had done the day before, resting his arm on the metal tubing. He started off by lightly scolding the both of us for having thought it necessary to come after him. I exonerated Clara and said I had come to fetch him because of the storm. He put me at my ease at this, saying that even old ducks can shed water.

"Now then," he said, "there are some matters you wish to take up with me?"

There were two such matters, I replied. One of them concerned the manuscripts. It was this that Emory Ross had in mind when he had told him of my coming. The other matter was of a more general nature. I would like to defer it until we had had a chance to discuss the first.

"Very well, then," he said, "let us proceed with the first matter. Emory Ross has asked me to give it my most careful and favorable attention. Judging by the way he mentioned the matter, I almost have the feeling of approaching doom."

His face lighted up with a big smile. This put me at ease. I told Dr. Schweitzer that his friends in America and throughout the world, in fact, were deeply concerned about his unpublished manuscripts. They felt it a matter of the most urgent importance that these works be completed and issued. In this latter connection, I said there was serious apprehension over the physical safety of the manuscripts.

"All right," he said, "we will discuss the manuscripts. So that you will know fully all there is to know about them, I will review the situation as it stands at this moment. As I told you yesterday, there are two unpublished manuscripts. Having mentioned *The Philosophy of Civilization* at our meeting yesterday, we will now discuss *The Kingdom of God*.

"No one has seen this book and I have not talked about it. This book is practically complete. The thesis is that Christianity has veered away from Christ. Christianity has constructed an elaborate dogma but it has not really comprehended that the mission of Jesus was to enable every man to discover the Kingdom of God in himself. Jesus wanted to prepare man for the Kingdom of God; it was his dominant concern. But Christianity, as it has developed, has been more concerned about the forgiveness of sins and the resurrection than it has been about the thing that was closest to Jesus—the fact that mankind must understand the meaning of the Kingdom of God. Jesus did not claim to be the Messiah. He claimed to be none of the things that have been claimed for him. He claimed only to know the reality of the coming of the Kingdom of God.

"There is not much more to say about it. An author is really a poor person to talk about the central ideas in his book."

I asked where the manuscript of *The Kingdom of God* was.

"Right here in a trunk at the hospital," he said.

"When will *The Kingdom of God* be ready for publication?"

"As well ask a hen exactly when she expects to produce her egg. I do not know. There are so many things for me to do here at the Hospital. Everything is changing. Now that the Africans have all become French citizens, our whole relationship to the patients is different. Now there are elaborate forms to be filled out for the administration. Each African who wants workman's compensation has to have detailed information supplied about him.

"Formerly, I would just dash off a short note, sign my name to it, and that was that. Now it's endless. And if I say that a patient has been treated and is ready to go back to work, he may become outraged for he would like to stay away from his job yet get paid for it. Little by little, all the joy is going out of our work."

(I recalled reading that when he first came to Lambaréné, Dr. Schweitzer wrote to his friends in Europe, saying that he wished they could come to see his joy when a patient left for home after being treated at the Hospital—well and smiling.)

"And this is not all," he continued. "There is so much work to be done. Little things that have to be done that no one else will do. Look at that curtain. It should have a rod at the bottom, too, so it won't blow into the room when the wind is strong. I must really get to it.

"A few years ago, I thought that my long-range problems at the Hospital were in the process of being solved. I had a doctor here who could take over and keep the Hospital going after I died. And I felt that I would soon have the time I needed to finish my books. But it became necessary for him to leave—the reasons are long and complicated—and then I had to take the entire burden of the Hospital again. The doctors who are with me now are very good but they are very young and so I cannot relinquish the supervision of the running of the Hospital.

"It takes two years before a doctor is really on his own here at Lambaréné. There is so much to learn that can only be learned by experience. When a doctor first comes out, we have to go through certain stages. They all have the same ideas for bringing in electricity and for doing this thing and that—things that many years ago we had decided against for specific reasons—and one has to be patient."

I remarked that the wonder to me was not that he had been able to write so little but that, under the circumstances, he should have been able to write at all.

"The published writings have turned out to be helpful friends," he said. "As I may have told you earlier, the first buildings at this Hospital were paid for by Johann Sebastian Bach. The royalties from my book on Bach, that is. Who knows, perhaps these other works, if I am able to complete them—remote as this may now appear to be—may be similarly helpful."

I asked if we could revert to his *Philosophy of Civilization*. What led him to conceive of it and write it?

"Just before the First World War," he said, "I received an invitation from the London representative of Harper & Brothers to do a book on my philosophy. I started work on it, finishing it shortly after the end of the war. Then I sent it to Harper's, which apparently lost its old enthusiasm for the project, saying they were very sorry, and they returned it. I guess the idea of a book by anyone with ,a German background—even a French-speaking German from the Alsace—was not too popular in those days.

"And then I sent the book to my publisher in Berlin. And he, too, had no taste for the manuscript. Finally, I didn't know what to do with it, so I asked my friend Emmy Martin, who was going to Munich, to dispose of it any way she wished—on almost any terms.

"Emmy took the manuscript under her arm one day to a publishing house named Beck. I don't think she knew that Beck was primarily interested in legal books, but she went in anyhow and asked to see Mr. Beck. She was told that Mr. Beck was out but that the bookkeeper would see her. He was Herr Albers. Emmy introduced herself. Her hands trembled as she held the manuscript, and with an air of defeat she said she supposed that there would be no interest in a book on philosophy.

"Herr Albers took the manuscript, leafed through some of the pages and said nothing for ten minutes. Then he said that Emmy couldn't have the manuscript back. He wanted to read the entire thing.

"When Emmy Martin wrote to me that Beck would take my book, I was overjoyed. On my next trip, I went in to meet Herr Albers who was to become my close friend for many years—until his death by suicide during the Hitler regime."

At this point Dr. Schweitzer interrupted himself.

"Are you sure that you are interested in this sort of thing?" he asked.

I assured Dr. Schweitzer that nothing could interest me more.

Dr. Schweitzer looked at me closely, smiled slightly, and said he did not want to seem like a garrulous old man. He continued:

"It was just about that time that Albers had persuaded his associates to publish another very long book, this one by a writer named Oswald Spengler, who gave his work the ominous title, *Decline of the West.* And so now Beck had two non-legal heavy philosophical works on his current list.

"One day, the three of us—Albers in the center, Spengler on one side, myself on the other—were going out to lunch together. And I burst out laughing, and said he reminded me of a butcher out walking with his prime oxen.

"There was much trouble before my book was published. All the money Beck made on Spengler was wiped out by the runaway inflation. And my own book was held up because the printing press had been requisitioned by the government in order to turn out more paper money. Finally the book appeared and I was pleased when I learned that it would be translated into other languages.

"But the *Decline and Restoration of Civilization,* as the book was called, was actually only part of a much larger whole. And it is the last part of this larger work that I have had such difficulty in finding time to work on. Most of my time was spent in building up the Hospital. Then one day, during the early thirties, two doctors who were here at the time came to me and insisted that I work at the Hospital during the mornings only and that the afternoons and evenings I give to my manuscript. It was wonderful. I was able to accomplish a great deal. In fact, I even had time to undertake a collection of various autobiographical papers. And so, *Out of My Life and Thought* was born. It was this book, I am told, that called my ideas to the attention of people who had not heard of me before."

That was how I happened to hear about him, I said. Because of my interest in the organ a friend had sent me a copy of the book in 1938, with a slip marking the chapter on organ building. And I remembered what a pleasant discovery it was to find among the moderns someone who still used the Aristotelian system for exploring a subject. The way Aristotle explained plant life was the way Dr. Schweitzer's chapter on the organ treated the art and science of organ building—what kind of wood to use in the pipes, where to place the organ, where to place the choir, what kind of nave was most suitable—dozens of factors that had to be correlated before an organ was built and installed.

While we were talking about *Out of My Life and Thought* one of the nurses came in to ask a question about some drugs that were needed. He disposed of the matter, then resumed:

"I was saying that after *Out of My Life and Thought* appeared, I began to make interesting connections in many places. I received an invitation to give the Gifford Lectures in Edinburgh, which I accepted. I was glad that I did so, for I worked on the lectures for almost two years and derived much intellectual stimulation from it.

"Then came World War II. You will recall that I told you yesterday that I happened to be on a boat passing through the Bay of Biscay, when I became convinced that the war was near, upon hearing that Adolf Hitler had just assured the world that all he wanted was peace. I then returned to Lambaréné in order to prepare the Hospital for carrying on its work even though it might be cut off from a large part of the world in the war years. I was able during that time to give little sustained attention to my writings. Now I don't know whether I shall ever finish *The Philosophy*."

"How much more work is there?" I asked.

"Once again, the mother hen: I have no idea. There is more

work to do here at the Hospital than ever before. More patients. More new buildings. More letters. And the forms to fill out, of course."

"Where is the manuscript of your third volume of *The Philosophy of Civilization?*"

"It is also here at the Hospital—in my trunk."

It was at this point that I told him that his friends were worried about the condition of these manuscripts. I understood that there were no duplicate copies. There were hazards of fire, flood, decay —any one of a number of things that could happen to his work. We believed that it would be wise to have photograph copies made of the manuscripts. And it was for this purpose that I had come with Clara with special camera equipment that would enable us to have film negatives made of all his unpublished work.

He looked somewhat startled at this.

"It would take many, many hours," he said.

I assured him that his own presence would not be necessary, that Clara and I would team up to carry out the project, and that all that would be needed would be the manuscripts.

He said he would consider this and give his answer the following day.

I moved on to the next point and told him that a publisher in New York had authorized me to offer him an advance against royalties of twenty-five thousand dollars for his manuscripts.

"Two things about this," he said.

"The first is that Goethe taught me a lesson that has become a stern rule. Goethe made the mistake of committing himself to deliver books by a certain date without being absolutely certain that he could fulfill the commitment. It tore him apart. He became the prisoner of his own promises. It affected him deeply. I don't want that to happen to me.

"The second thing is that I have learned from the Niebelungen to keep one's loyalty to one's associates. I already have several publishers in the United States who have worked with me."

This matter was discussed rather fully, and I told Dr. Schweitzer that I felt that at whatever time the manuscript would be ready, we could go to the other publishers and inform them of the particular proposal I had just relayed to him, and receive from them some idea of their own capacities and desires in the matter.

"This, too, I will think about," he said. "But now tell me what the other matter is that you wish to discuss."

I took a deep breath, then plunged in. I approached the question of world peace at its largest, discussing existing world tensions in the context of an age that had available to itself total destructive power. No people and no nation were really secure. And the race for security, inevitable under conditions of world anarchy, had the paradoxical effect of intensifying the insecurity. For the new weapons placed a high premium on surprise. Yet the nations felt they had no choice except the pursuit of military strength in a lawless world.

But far more serious was the fact that man was now able to tamper with his genetic integrity. Through radioactivity he was in a position to pursue and punish the unborn generations. He now had the power to change and cheapen the nature of human life. This was power of an ominous order indeed.

I watched Dr. Schweitzer carefully as Clara, speaking very slowly in German, relayed my message. His eyes were closed and his brows were knit. He sat with his head bent forward.

I asked Clara to emphasize that I realized I was not saying anything that he did not know far better than I. My reason for bringing up the subject was that I felt there was no one in the world whose voice would have greater carrying power than his own.

When he asked what it was that we would have him do, I said we had nothing specific to suggest other than that he might feel free to express his concern to the world.

At this moment, Dr. Margaret rapped on the door. She apologized for interrupting but there was an emergency case—a woman with an extrauterine pregnancy who had just been brought to the hospital. An immediate operation was necessary but the woman was already too weak to undergo surgery.

Dr. Schweitzer stood up to leave.

"*La Doctoresse* has come at just the right time," he said. "It is good to be reminded now and then that even in a world struggling with the momentous issue of war and peace the individual has problems.

"We will talk further."

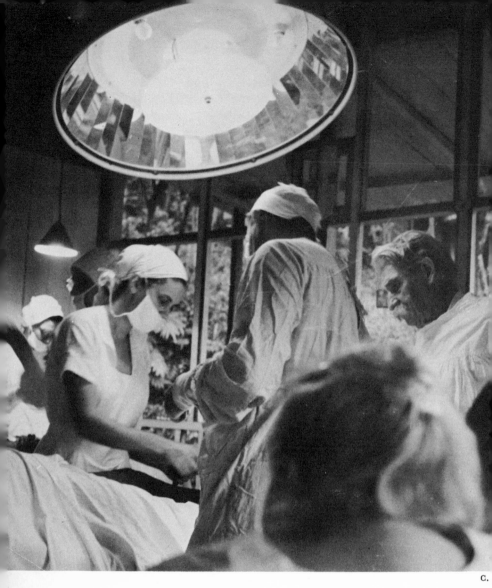

C.

The operating room at the Schweitzer Hospital. Power for the electricity is furnished by a generator. The rest of the Hospital is illuminated with oil lamps.

C. U.

A young African mother in a pensive mood while waiting for her baby to be examined.

Dr. Jack Penn, of Johannesburg, on a working visit to the Hospital, consults with Dr. Margaret van der Kreek—*La Doctoresse*.

c. u.

# VIII

THERE IS ONLY one place at the Schweitzer Hospital where electricity is available. This is the operating room, where there are modern overhead surgical theater lights. The electricity is furnished by a generator. When the motor is turned on, the throbbing noise alerts everyone that the operating room is going into action. During the night, an African medical orderly is assigned to the river landing. When an emergency case arrives and it appears that surgery may be necessary, he switches on the noisy generator. The effect is that of a gong in a fire station. Doctors and nurses swing out of their beds like firemen, jump into their clothes, and do every-thing except slide down a brass pole.

Not long after midnight on the day Dr. Margaret came to Clara's room to tell Dr. Schweitzer about the woman with the extrauterine pregnancy, the generator started up. Instantly, I could hear activity in some of the rooms. There were hurried footsteps on the porch. The steady throb of the generator continued for perhaps an hour and a half, or two. Then the motor stopped and after a while I could hear the shuffling footsteps on the porch of the doctors and nurses trudging back to their rooms. I didn't need the sounds to tell me how exhausted they must have been. It was 4:30 A.M. In another two hours the Hospital would be starting on another day. I wondered how many on the staff were too tired even to sleep. The air was heavy and the act of breathing was almost a conscious one. Little wonder that Dr. Margaret, for all her loveliness, showed darkness under her eyes and at times seemed to walk with a stoop. She was no different from the others; fatigue was their constant companion.

About a half hour after the doctors and nurses returned to their quarters, I heard strange voices coming up the hill. At first, the sound was a low rumble. Then, all at once, the voices, now sharp and insistent, seemed to be collected at the end of the porch. These were African voices and they were angry. I got out of bed quickly, put on my khaki pants, and went out on the porch. Not far away in the dim light I could make out some two dozen people in a cluster.

I walked toward the group and perceived the small, slight form of a person who was apparently saying something to the Africans in subdued tones. It was Clara. One of the Africans had stepped forward and said that the group wanted *La Doctoresse* to produce the body of the woman who had just died on the operating table. They had been instructed by the witch doctor to obtain the body in accordance with their customs.

In a voice that indicated understanding but also strength, Clara said that *La Doctoresse* was sleeping; she was exhausted from her night's work. Under no circumstances should she be awakened. Then Clara stepped down from the porch and approached the tall African who was spokesman for the group. For a second I feared for her safety—but only for a second. Clara was saying something to the tall African that caused him to nod, then turn around and speak to the group in a native tongue. There was a brief exchange, but the voices became increasingly moderate, and then the tall African said something to Clara, politely and almost in a whisper, and turned away and started down the hill. The others followed him.

Clara sat down on the steps. I could tell from the way she did it that she had spent herself emotionally. I know she didn't want to talk about it, so I said nothing and went back to my room.

Later, at breakfast, sitting next to Dr. Coulon, I learned what had happened.

The woman had been brought into the Hospital late the previous afternoon. The case had instantly been diagnosed as an extrauterine pregnancy. That is, fertilization had taken place outside the uterus. The embryo had been allowed to develop long past the danger point. Apparently the woman's parents had insisted on witch-doctor treatment, but when the pain became fiendish, the woman's husband took the matter into his own hands, put his wife into a pirogue, and paddled thirty miles to the hospital.

Dr. Coulon said that all the doctors had agreed on the need for an emergency operation to save the young woman's life. But the woman had lost so much blood by the time she had arrived at the Hospital that *La Doctoresse* felt any immediate surgery would almost certainly kill her. Under the circumstances, they decided on immediate transfusions in the hope that the patient might regain sufficient strength to endure an operation.

At about 2:00 A.M. it became clear that the decision to operate could be put off no longer. The embryo had ruptured and an immediate operation became mandatory, however small the chance for success.

*La Doctoresse* did the operation assisted by the others. The patient died during surgery. It had affected *La Doctoresse* deeply.

"After something like this, it is natural that a doctor should torture himself with doubts," Dr. Coulon said. "Would the patient be alive if we had operated earlier? Perhaps we should not have operated at all? Was there something that should have been done that hadn't been done? I know that *Doctoresse* did everything possible, but I know how she feels."

This sort of experience was not new to Dr. Coulon. He had studied graduate medicine at the Union of South Africa. Then he became an itinerant jungle doctor. He worked in the African villages. The only apothecary available to him was what he carried in his bag. Surgery would generally be performed in the open, on a flat board.

When he arrived at Dr. Schweitzer's Hospital, he marveled at the facilities.

"People come here from New York and are appalled at the crudeness and the startling lack of modern sanitation and instrumentation," he said. "When Mrs. Coulon and I came to the Hospital after working in the open jungle, it was as though we had come into a fairyland. It was wonderful. There was a ward in which patients could be kept under observation. There was a good supply of modern medicine, including the newest antibiotics. There was an operating theater, with overhead electric spot. There was a wide variety of operating instruments. I couldn't have been happier."

I asked Dr. Coulon about the mortality statistics at the Hospital. He said that the mortality might be high by Western standards, but that one had to keep in mind that a larger number of the serious cases had been brought to the Hospital too late for effective medical or surgical care.

All this time we had been conducting our conversation at the breakfast table in very low voices. The death of the young woman after such a long fight to save her had darkened the spirits of the staff. Even Dr. Schweitzer, who ordinarily could be counted on to keep a mood from becoming too heavy, remained silent through the meal. Dr. van Stolk and Dr. Friedmann came in late for breakfast. Neither had shaved. The Doctor nodded to them; they ate quickly, saying little.

After breakfast, on the way back to our quarters, Clara spoke to me about the incident earlier in the morning on the end of the porch. She had been awakened by the voices and she had intercepted the group before they could get to *La Doctoresse's* room. The Africans lived in the constant presence of death. Even so, its mysteries gave birth to many dictates and rituals. And one never knew at the hospital when the fact of sudden death would cause a group to act in an inexplicable and apparently hostile way.

I asked what she had said to the group to cause it to turn back. She said she had tried, as simply and sincerely as she could, to give them some idea how Dr. Margaret and the others had fought to save the young woman's life.

"I told them that only *La Doctoresse* had the authority to release the body," she said, "and that when she did so she would turn it over to a member of the family as was the custom, and not to a crowd. For if the body were turned over to a crowd every time it asked for it, people would lose their confidence in the Hospital staff. I said I was certain the tall African could see this.

"These are basically very kind people and they respond to kindness," she continued. "Somehow they had gained the notion that the young woman had been sacrificed in some way or another. I knew if I could make them see how Dr. Margaret had given of herself to save the young woman, and how unkind it was to make demands on her now that she had finally earned some rest, they would change their minds. The leader of the group must be a very fine man; just before he left he said he was sorry that they had awakened me and sorry that the group had wanted to disturb *La Doctoresse*. As I say, these are wonderful people."

Just before we came to our bungalow, we met George Malthen, with whom I had had a talk some days earlier and who had brought his wife to the Hospital against the orders of the local fetisher. He said his wife was now well again, thanks to Dr. Margaret and the other doctors. They would be leaving for their village directly and he wanted to say good-by.

I thanked George Malthen and we made our farewells.

When we came to the end of the porch, Dr. Margaret was working in her little flower enclosure. She looked up at us and smiled and said how happy she was that two more of her jungle flowers had just bloomed.

Neither Clara nor I referred to the events of the night and early morning. I think that both of us, without saying anything about it, were marveling at Dr. Margaret's ability to become absorbed in the miracle of new life and growth and to have it nourish her at a trying time. She had apparently rested well, brief though it was, and now she was ready to face the Hospital again.

Albertina van Beek Vollenhoven, nurse and ministerial assistant to Dr. Schweitzer. "My doubts no longer seem so terrible."

"A litter on which an object of human size was wrapped in ferns and large leaves."

A young African woman carrying a baby walked up to Dr. Margaret. There was a slight hesitation in her manner but it was obvious she had something to say.

Dr. Margaret put her at ease. Then the mother exclaimed that her baby was all well again and she was eager to have *La Doctoresse* share her joy.

Dr. Margaret took the child and cradled him in her arms and held him close. He couldn't have been more than six or seven months old.

I had no way of knowing what had been wrong with the baby; but I was glad that the incident had happened just then and in just this way. I also thought of George Malthen's wife. The scales were being balanced; it couldn't have happened at a better time.

Dr. Margaret handed the little boy back to his mother, then turned to us and said she was going down the hill to release the woman's body to the family, so that it could be brought back to the village for burial in accordance with tribal customs.

I went back to my room to take care of some more work that had been mailed to me by the office. But I couldn't put my mind to it. I was thinking of the hospital and of the young people who had come there to help Dr. Schweitzer. I wondered whether they had been prepared for the drama that was waiting for them—for the pain of occasional defeat and the constant challenge of unpredictable human response. Life at the Schweitzer Hospital involved much more than the glamour of being associated with one of the great figures in history. It meant hard work, fatigue, heartbreak. All this came with the decision to be a doctor, of course; but there was an extra dimension at Lambaréné that was almost beyond anticipation.

And what about the Africans themselves? How did they feel about the young men and women in white who had a method of treating them and their families far different from the treatment their own customs sanctioned or dictated. Why, when death occurred, was there such a powerful tug of reversion? Did they regard this inability of a white doctor to save a black life as proof of the fallibility of the white man, and as punishment for the Africans for having wandered outside that which was part of them?

I looked out the screened door and saw a small procession in single file led by Albertina going up the hill. Mme. Oberman was walking behind Albertina, followed by two Africans carrying picks and shovels. Behind them two Africans were carrying a litter on which an object of human size was wrapped in ferns and large leaves.

I put down my work and followed the procession along the narrow path leading up the hill toward the leper village. A little more than halfway toward the village, Albertina led the procession on a path off to the right. Very shortly we came to a clearing; it was a burial ground for those persons who had died at the hospital and had no relatives or friends to claim them. In accordance with the custom of the Africans who lived in this part of the Continent, the dead are carefully wrapped in long ferns before they are laid to rest.

Two Africans were at work preparing the grave. Their picks and shovels cut easily into the soft rich ground. Albertina and Mme. Oberman stood off to one side. When the grave was ready, Albertina read from the Bible while the body was put to rest. She continued the prayers for several moments and closed the book. Then the Africans filled in the grave again, working steadily. When they were through, they looked up at Albertina who said a final prayer. Then we walked in single file down the narrow path again.

As we neared the bungalow and the path widened, Albertina waited for me to catch up with her. She told me that the man who had just been buried had been deranged and had been put in an isolation cell. For days he had refused to eat, yet there was something about him that Albertina liked and admired. So far as anyone knew, he was alone in the world. He had come to the Hospital on his own, unusual for a mental case. As his condition became progressively more acute, he seemed to know that he had come to the right place. He would have periods of agony and violence, which necessitated his isolation, but he would also have moments of apparent clarity.

Only the previous evening, when Albertina came to look after him, she found him in a rational and quiet mood. He beckoned to her and told her that he thought he was going to die. He reached out and took her hand and said he was grateful to her for taking such good care of him. Then he lapsed into silence for a long time. Albertina could tell he was slipping. After a few moments he said in a low voice that he had always feared that he would be alone when he died. But now he was not alone because of Albertina, and he was content and grateful.

He asked Albertina to pray for him, which she did. Before she was through, he passed away.

And now, on the way down the hill again, Albertina was telling me about the man.

"Sometimes I wonder," she said, "whether my coming here makes sense—whether I am doing anything that is really useful. Then something like this happens, and my doubts no longer seem so terrible."

In the days that followed, I was to learn that Albertina fulfilled a dual role at Lambaréné—nurse in charge of the psychiatric ward and ministerial assistant to Dr. Schweitzer.

There was no chapel at Lambaréné. Sunday services were held in the open air. The Sunday I was there I was directed to the long, narrow compound in front of the Hospital clinic. The Africans sat on the steps or in the doorways of the small wooden structures in order to get the protection of the shade. Albertina stood in the hot sun; she wore her customary white pith helmet. During most of the service she read from the Bible. There were a few hymns and the voices of the small African choral group off to one side were heard above the rest.

Like everything else in Lambaréné, the services were simple but effective. After Albertina gave the final benediction, the Africans chatted with each other. A number of them were dressed in their

Sunday best: several men wore long pants and clean shirts, instead of the customary shorts and frayed undershirt. The women, for the most part, wore colored cotton dresses and their hair was neatly done.

Albertina exchanged pleasantries with the people who went up to her after the services. Then she walked slowly up the hill to her room. I watched her until she disappeared from view. Here was a young woman of about twenty-eight or thirty, tall, slender, attractive, intelligent, compassionate. She had defined a mission for herself in life. It was not an easy one. It involved a constant drain on her energies—physical, mental, emotional. Her satisfactions were centered in her exertions. Most of what little time she had to herself she spent in study—reading intensively in the field of mental disease and psychiatric treatment.

And the more I thought about Albertina and the other young people on Dr. Schweitzer's staff, the more I realized their importance and the greater was my admiration. For it took purpose of a very high order to work at Lambaréné. The glamour of the work, such as it is, eventually fades. Those who came expecting to spend most of their time communing with Dr. Schweitzer, or working valiantly alongside him, soon discovered that Lambaréné was no playground for the human spirit. The contact with the Doctor, except at mealtime, was necessarily brief. For the most part a staff member was on his own—that is, if he did a competent job. If not, the Doctor stepped in and attempted to set him straight. Then, if his work was still substandard, the Doctor might switch him to something else. He didn't waste time on idle flattery or trying to keep up appearances; there was too much that had to be done, too little time to do it. More than a few people never made the grade at Lambaréné and came limping home.

That is why Dr. Schweitzer tried to be so careful in screening applicants—why it required a prolonged exchange of letters and detailed checking before a person joined the staff. As a result, an extraordinarily high percentage of the people who came to Lambaréné stayed for years.

Some of the staff members, for example, had served for a quarter-century or more. To be sure, not all of them made their acquaintance with Dr. Schweitzer by mail. A few knew him in Europe and followed him to Africa to serve in any way he wished them to. Mlle. Mathilde Kottman, general supervising assistant who has a total dedication to the Doctor, came to Lambaréné from Alsace about 1923. Mlle. Ali Silver, from Holland, general secretary and personal assistant to the Doctor, and who helped to translate letters from and into English, had served the Doctor for over ten years. Mme. Oberman, who was in charge of billeting, had worked with Dr. Schweitzer on and off for forty years. Mme. Emma Hauss-knecht, who died in 1956, was the second person to join the Schweitzers at Lambaréné.

While I was at the Hospital, considerable excitement attended the departure of one of the nurses, Maria Langendyk; she had served at the Hospital for sixteen years. And now she was leaving for a rest in Europe. She had completed the term of service she had set for herself.

When I walked into the dining room on the evening before Maria's departure, I could tell that something special was happening. Dr. Schweitzer was wearing his black clip-on tie; some of the younger nurses were wearing lipstick; and there was butter on the table. Clara told me about Maria, a tall woman of about forty-five, with dark hair, dark eyes, and a face of quiet strength. Maria had come to the Hospital, as did the others, asking nothing except the privilege to serve. But her service was now coming to an end. She had spent almost one-third her life with Dr. Schweitzer. This dinner was now in her honor.

Halfway through the meal, just outside the window on the far end of the room, a choral group began to sing. It was a passage from Handel's *Messiah*. The choir was most unusual in the range of its voices. I was able to pick out perhaps a dozen children's voices and the voices of perhaps eight or ten grownups, some of which were deep and strong. The singing was superbly blended.

Clara whispered to me that the voices belonged to African lepers and that they had been trained by Trudi Bochsler, the young nurse in charge of the leper village about a third of a mile from the main body of the Hospital. They had now come to serenade Maria and wish her Godspeed.

I was profoundly impressed by the singing and told Clara so. She said she hoped she might be able to persuade Trudi to have these same lepers put on a repeat performance of the Nativity Play that had so captivated the staff when it was produced at Christmas.

The next morning, virtually the entire Hospital turned out to wave good-by to Maria as she left by pirogue. I saw Trudi, holding a small leper child by the hand, standing not far from Dr. Schweitzer, on a bank overlooking the dock. Dr. Schweitzer was standing in his most characteristic manner—his head forward and his hands folded behind his back. Dr. Friedmann, standing close to Dr. Schweitzer, was waving his helmet as the pirogue carrying Maria slid away from the dock. Perhaps a dozen others rounded out the tableau, as the pirogue, aided by the swift current, eased downstream. I remember thinking that the group on the bank would have made the ideal subject for a mural symbolizing the Schweitzer Hospital.

On the way back from the dock, Clara spoke to Trudi about the possibility of having the lepers do another performance of the Nativity Play. Trudi was delighted by the suggestion and said she was certain that the members of the cast all remembered their parts and would welcome the opportunity to play before the people who hadn't been at the Hospital during Christmas.

Trudi herself was clearly excited about the prospect. Her large gray-blue eyes sparkled as she anticipated the joy that the idea of a repeat performance would bring to the lepers. It would take no time at all, she said, to prepare the production. Perhaps it could be done as early as the following afternoon. Since our next meeting with the Doctor was to be in the evening, the suggestion seemed ideal.

As I got to know Trudi, I was to learn that nothing was more characteristic of her than her spontaneous enthusiasm and sense of immediacy for things that were worth doing. She was open and direct in her dealings with people; no one who knew her entertained any doubts about her intentions or purposes. Nor did she hesitate for a single second to put up a battle for her point of view concerning the operation of the leper village.

Two years earlier Trudi had had a disagreement with Dr. Schweitzer on an administrative matter regarding the leper village. Unable to persuade the Doctor, she left the Hospital and returned to her native Switzerland. For the next six months, neither Trudi nor Dr. Schweitzer rested easily about her departure. Then she wrote to him, asking whether she could return. Though Dr. Schweitzer had strongly espoused his own viewpoint, he was big enough to tell Trudi he was glad she would return and that she could administer the village in accordance with her own ideas.

Trudi returned to Lambaréné at once. While she had been away, she had put each minute to good use, studying every book she could find on the subject of leper colonies. In fact, there was very little that had been published on lepers—whether with respect to medical treatment or their social problems—that Trudi had not read. Her nurse's training had been of a general nature. When she had come to the Hospital for the first time, and she had started to work with the lepers, she went at it as though it would become her lifework. This combination of day-by-day experience and observation and her constant study had now made her, as Dr. Friedmann had remarked, one of the best-informed and most competent persons on the subject in the world.

When Trudi returned to the Hospital she resumed her work with the enthusiastic devotion that is basic in her personality. Increasingly, Dr. Schweitzer tended to give her autonomy in the administration of the village. She became general manager, nurse, interne, teacher, confidante, minister, and family head. And she was getting good results. It was fascinating to see Trudi at work. The lepers worshiped her as though she were their queen, which in many ways she was. She was directly concerned not only about their physical but their emotional well-being. Their trust in her was absolute. When she prescribed a certain routine of medical treatment at the leper clinic, they would accept it without question.

Trudi's only real difficulties were with the witch doctors who hovered just outside the village. Once, when Trudi discovered that some of the patients were not taking their medicines because of omens cast by a witch doctor, she tracked down the fetisher and threatened right then and there to wreak all sorts of havoc on him if he didn't remove his hex at once. The fetisher lost no time in notifying the patients that it was proper and necessary for them to resume their medication.

This young white goddess looked the part. She was twenty-four or twenty-five, flaxen-haired, slender, attractive, intense, inexhaustible. It was clear that her work was giving her the kind of fulfillment she sought in life. She was responsible for the education of the leper youngsters, and each morning she conducted classes and arranged for the purposeful use of the children's time during the afternoon when she had to attend to her medical duties.

When Trudi proposed to her lepers that they put on a performance of the Nativity Play for the visitors at the Hospital, they responded with gleeful anticipation, as she knew they would. Getting up the production for the following day was not an easy undertaking—the costumes had to be located and repaired; the stage props had to be put in order; each person had to brush up on his part; and there had to be a complete rehearsal. But Trudi's certainty that the project was well within their reach was shared by all those who had a part in the play.

The next afternoon, I joined Clara, Dr. Frank Catchpool, and the staff members who were eager to see the play again. We walked down to the leper village. The compound in front of the leper clinic had been transformed into an open-air theater. There were several rows of benches, one of which was already filled with visitors from the Catholic Mission across the river. Back of the benches were perhaps two dozen young African girls from the Mission who had come to pay their respects. They comprised a

The cast.

SCENES FROM THE NATIVITY PLAY, ACTED AND SUNG BY LEPERS AT THE SCHWEITZER HOSPITAL.

Mary and the infant Jesus.

The audience (from a world of total options).

The audience (lepers; no options).

choir and were to sing during the intermission and at the end of the play.

At the sides of the compound, sitting on benches or on chairs or on the ground, were the lepers. Some of them had bulky bandages on their feet. Others had crutches stretched out alongside them. Even without the bandages and the crutches, the effects of the disease were clearly visible. Toes or fingers were missing, or the feet would be stubbed. A little boy of about ten whose foot had been amputated several weeks earlier went hopping around as he helped put the props into place.

Within five minutes after the play began, a spell of magic settled over the compound. The singing of the actors was full of life and conviction. Two or three of the leper voices had excellent depth and tone. The costumes were crude, very crude, but they helped to create the necessary illusion. The baby Jesus was beautifully behaved, and did not cry until the intermission, and then only briefly. The Three Wise Men were very deliberate in their roles. The leper who took the part of Joseph was compassionate and gentle in his interpretation. Mary obviously relished her role and sang with vigor.

And all the time the play was unfolding, Trudi sat off to one side, her hands clasped and held close to her chest. Her mouth moved in the manner so well known to prompters.

If I say that the entire experience was almost beyond awareness or comprehension, what do those words suggest? Can they possibly indicate the range of emotion or the stretches of thought produced by watching condemned people give life to a spiritual concept? The play was concerned, essentially, with the triumph of hope through faith; but the brief moment of the lepers in a glittering spiritual universe was surrounded on all sides by the evidence of a closed-in world. Yet in that brief moment, they were connected to the things that meant life for most people.

There was something else. The play, in a sense, was almost a symbol of forgiveness. For the white man in Africa had not, in the main, been a friend. Historically, he had not been a liberator or a

Trudi and Dr. Frank Catchpool tend a nature-study class for leper children.

Leper children rehearse for the production of the Nativity Play, under director-producer Trudi Bochsler.

benefactor. He had used his superior knowledge and power to capture the Africans and impress them into slavery. He had brought with him venereal disease. He had caused the African women to bear children and then he had discarded the mothers and their young. He had advertised a religion of mercy and compassion but there had been little of either in his manner. While I watched the lepers at Trudi's village in the Nativity Play, I became conscious of the fact that the play was saying something about a world large enough to hold both black and white. This was close to the original purpose of the play, but original purpose in religion is not too often a remembered part of life. The actors, however, clearly seemed to reflect this original purpose, to which forgiveness belonged.

All this was possible because we were in a segment of Africa where there was a Trudi and an Albert Schweitzer and a Dr. Margaret and an Albertina and a Dr. van Stolk and a Dr. Friedmann and all the others who rescued their whiteness from the evil that the others had spawned.

To the lepers who were on the sidelines, the spotlight for them frequently moved from the actors to the white audience. Comparatively, the poorest among us were rich as kings alongside an African, for we were in a place where wealth was measured by a pair of shoes or an extra pair of pants or an unfrayed blanket. And the lepers heard that we had come from an unbelievable world where people lived in beautiful dwellings of more than one room and could turn a knob that would produce water right in the house, and another knob that would control several fires for cooking. And, miracle of miracles, they saw that education for everyone in the white world was free. There are now schools for the blacks in French Equatorial Africa, of course, but these were mainly in the large cities. And the schools didn't go as far or teach as much as every white child was able to get.

There was nothing resentful in the gazes of the African lepers as they watched us. From their vantage point, we were the main show. The imaginative play of their minds was directed to us rather than to the actors. We probably touched off all sorts of wonderment

Trudi Bochsler, nurse in charge of the lepers, makes her morning rounds of the leper village, located about one-third of a mile from the main body of the Hospital.

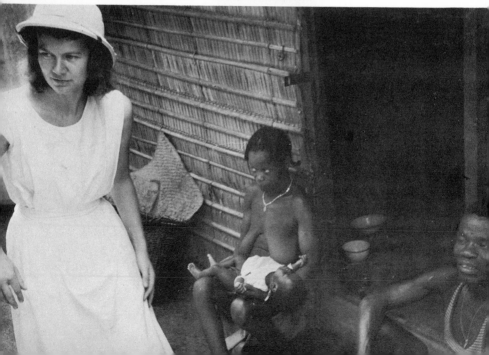

and speculation, especially for the children. Nothing is stranger to the eye of an African child than a white person when seen for the first time. The lack of color creates an aspect of ghastly pallor. It is as though layer after layer of skin had been stripped away, producing what Africans generally refer to as a "peeled" look. The thinness of the facial features of a white person gives him a pinched, closed appearance. I could almost feel the scrutiny of the African children near me.

After the play ended, we sat still for a few minutes. I had no way of knowing whether the same thoughts that had preoccupied my mind were being shared by the others; in any case, the other white visitors sat quietly. They seemed far away in their thoughts. Then the children's choir from the Catholic Mission sang for a few minutes.

We got up to leave. Trudi announced that the leper children had a surprise for us. They had made gifts for each of us—hand-carved letter openers or ship models for the men and beads or pendants or necklaces for the women. Each child had a little presentation speech to make in French; he said he hoped we had enjoyed the play and that we would come to visit the village again.

LET ME DIGRESS a moment.

I suppose people will ask: Is it safe to touch things handled or made by lepers? Is it safe to shake their hands or get close to them?

I have heard much conflicting medical testimony on the subject. Modern theory tends to support the view that leprosy is somewhat akin to tuberculosis and that there are varying stages and degrees of contagion. If the leprosy involves the muscles or nerves, it is not believed to be infectious. If it involves the skin, however, especially where open sores are concerned, then there is the possibility of contagion. Lepers who have lost fingers or toes are in this category. Inflammation of the nasal mucosa with consequent discharge can be a source of infection. Children contract the disease through constant intimacy with leprous members of their families.

Generally speaking, however, outsiders who are in good health are not considered to be in any danger as the result of casual contact with lepers.

It goes without saying that Dr. Schweitzer and his staff expose themselves constantly, regardless of the virulent condition of some lepers. But Dr. Schweitzer is careful to the point of seeming over-cautious in his policy about visitors. He advises them not to get too close to the leper patients and not to shake their hands or touch them. The reason, of course, is that he feels a total responsibility where any risk, however small, to outsiders is involved.

It isn't always easy, however, to follow the Doctor's admonitions. When the little leper boy who handed me the hand-carved letter opener also offered me his hand, I did not see how I could turn away.

On the occasion of another visit to the leper village, a leper introduced me to his wife who was carrying a baby boy of perhaps twenty months or two years. The father told me his child was leprous, too. The baby's nose was running and he had sores all over his face. As I chatted with the parents, the baby twisted in his mother's arms and reached out for me to take him, which I did. To do otherwise, under the circumstances, would have been awkward and indeed impossible.

In general, if you want to see something of Asia or Africa—which is to say, if you want to get close to the people—you would do well to leave your apprehensions behind and take the plunge. If you are in good health, you will have little to fear. Now and then, of course, you are apt to find yourself in delicate situations where a few uneasy thoughts may cross your mind. Once, in a grove not far from Calcutta, I came across a crying child of three or four who was obviously feverish. I carried the child to the nearest village compound. There I learned that the child's parents had died several days earlier, the victims of a smallpox epidemic still raging that had decimated the population of the village. I turned the child over to one of the older women. Certainly it occurred to me that I was in an exposed situation; but I tried to take some comfort at least in the fact that I had been inoculated against the disease.

This is not to say that I have been totally immune to infection or illness while abroad. I think I have sampled every species of dysentery recorded in the annals of medicine. I have had to make numberless, unscheduled, and speedy departures from receptions or social functions or public platforms all the way from Djakarta to Mexico City. There were times on the lecture platform in universities in Asia when I was certain I was capable of setting new records for the short sprint and I desperately longed for nothing so much as the chance to prove it. At such times, philosophical calm seems only vaguely attainable and, in fact, quite academic.

On occasion, I fear I may have made a washed-out impression on my hosts. At the University of Yokohama some years ago, for example, I was having dinner at the home of a professor of medicine. I was in my third week of dysentery, having just come from Indo-China. I was as diligent as possible in making it appear that I was enjoying the meal.

All this time my host was observing me closely. Then, at the end of the dinner, he asked to examine my hand. He scrutinized it carefully but dispassionately, as though it were so much disembodied flesh. Then he put his head close to mine and peered into my eyes. His manner became increasingly grave. Finally, he shook his head and said something in Japanese in a low voice to a fellow professor sitting next to him. This gentleman shook his head in turn and passed the word along to another colleague. Soon almost everyone at the table was nodding apprehensively. I felt like a prime exhibit in a surgical theater.

I asked the man at my right what it was that the professor had seen in my hand and in my eyes that had apparently made him so sad. My companion replied that the professor could tell at once that I was dying from cancer. When I inquired about the time factor, I was told that I had at least six months. That was in 1949.

In defense of the professor, it should be said that he had seen me under misleading circumstances; I had had a fairly prolonged adventure in non-retention and no doubt I looked even worse than I felt. It did not seem to him that what was bothering me could be anything short of the worst.

Five years later, I saw the professor again in Tokyo. He had come to call on me at my hotel but I was out at the time—in the public park across the street. Some students had read in the newspapers that I had once played some baseball and they had invited me to join them in a game. The hotel clerk had informed the professor where I might be found. I happened to be on second base when he caught my eye. One of my teammates hit safely and I sprinted for home plate, arriving just ahead of the throw. It occurred to me as I did so that the professor must have thought my ghost was doing the running. It was also true that this was the only time the professor had seen me run when I exercised some option in the matter.

Afterward, when we spoke, he made no reference to his earlier forecast. We discussed various matters. He was eager to come to the United States. He said he had some ideas about medical diagnosis he was eager to put before his professional colleagues in America, and wanted my advice about arranging such meetings. I gave him such help as I could, though I was not unmindful of the fact that I was something less than the perfect advertisement for his theories.

This digression, of course, is touched off by a consideration of the inevitable questions generally asked by prospective travelers about the personal health factor. Generally speaking, the risk while traveling is extremely small. Such deviations as occur are more the result of sudden changes in one's living routine than of basically unhealthy situations. Just the business of adjusting to severe time changes can be constitutionally unsettling. The pursuit of new foods is a delightful adventure; when carried to unreasonable lengths, it can produce deceptively intense but fortunately temporary discomforts.

Experiences such as these, of course, are not peculiar to Americans abroad. They apply equally to visitors to our own shores. The foreign visitor also has to adjust himself to exhaust gasses from buses, trucks, and cars on a scale far beyond anything he would experience in his own city, wherever it may be. The heavy monoxide fumes added to the industrial smokes of American cities, to a person unaccustomed to them, can produce severe headaches and nausea.

There is also the matter of crowds. When Americans think of dense population masses, they are apt to think of cities in India or Japan. But nowhere in the world are more people crowded into a smaller space than in New York City. I have seen the teeming millions in Asian cities, but never have I seen greater human congestion than in the Grand Central section of New York City during the lunch hour. The business of navigating even a short distance through the crowded sidewalks calls for a special knowledge and a certain conditioning. Little wonder that people from abroad who see New York City for the first time have the feeling of being overwhelmed and exhausted—almost to the point of illness.

I have had the good fortune to be the host in the United States to visitors from Asia. And I could tell that the noonday crowds charging through New York streets produced an effect on some of them akin to panic. They had never seen so many people under such compressed conditions before. All this, combined with the new rich diet and the accelerated American pace, caused more than a few of them to long for the comparative openness of Calcutta or Tokyo.

In any event, Americans need not fear that they are in mortal jeopardy every time they leave their own country. Nor need they believe that such minor discomforts as they may encounter are the exclusive products of foreign places.

THIS DIGRESSION OUT of the way, I should like to revert to the people at Lambaréné.

Like the Schweitzer Hospital itself, the leper village was undermanned. Trudi received some assistance from the regular staff members and she had considerable assistance from the lepers themselves. But her personal workload was very heavy indeed. One of the newcomers to the Hospital, Olga Deterding, was aware of this and volunteered to work full time as her assistant. In particular, she wanted to accept responsibility for the education of the youngsters. But Olga had been at the Schweitzer Hospital only two months at the time of our visit. Her apprenticeship was not yet complete. It was necessary for her to spend a few more additional weeks on kitchen duty and mop-up detail before she would feel justified in asking Dr. Schweitzer to assign her to the leper village.

Olga Deterding came to Lambaréné almost by accident. She had started out across the African desert in a large jeep with some friends. The jeep had broken down several times and the expedition finally broke up. Olga decided to press on by herself and see something of Africa. That was how she happened to arrive at the Schweitzer Hospital. What she saw she liked and she decided to stay on and work for Dr. Schweitzer for as long as he would have her.

Olga arrived with a secret and did her best to keep it. She was the daughter of an English multimillionaire. She said nothing about her economic and social station because it might appear to others on the staff that she was staying on at Lambaréné for a lark. Also, she didn't want the staff people to feel any disparity in their relations with her.

Unfortunately for Olga there were factors beyond her control that made it impossible for her to keep her secret. Word slipped out in London, with the inevitable result that stories began to appear in the press, some of which filtered back to Lambaréné. Olga began to get cables from magazines or news services asking for permission to send writers and photographers to the Hospital to do feature stories.

All these Olga ignored. But one day Dr. Schweitzer received a cable from a Paris picture magazine saying it wanted to send staff members to the Hospital for a story built around Olga. Dr. Schweitzer, after talking to Olga, politely declined. Despite this, two days later, there arrived at the Hospital, unheralded, a French photographer-writer and his wife and a Japanese correspondent (also carrying a camera).

Olga Deterding visited Lambaréné on a safari and stayed to serve.

Olga was working in the scullery when the photographers passed.

Dr. Schweitzer was informed by the French correspondent that his editor had discovered that his magazine's files were somewhat out of date and he wanted some fresh material about the Hospital. Dr. Schweitzer was aware that the correspondents had made a long trip and did not want to send them away empty-handed. He said he would be glad to escort them around the Hospital grounds. Then he put a direct question to them: was there anything they wished?

Yes, there was. They said they would like to meet the members of the staff and do some stories about them, too.

Was there anyone in particular they would like to meet, the Doctor asked.

Why, yes; they would like to see Miss Deterding.

Well, then, said the Doctor, this was something only Miss Deterding could decide. He would put the question to her.

The reporters waited in the dining room while Dr. Schweitzer sought out Olga. Her reply was given without hesitation. She did not under any circumstances wish to see the journalists. She felt that any publicity would be inconsistent with the character of the Hospital and that of the people who worked there. She felt, too, that her purpose in staying on at Lambaréné would be defeated if she were placed in the spotlight. Besides, she was a comparative newcomer and wanted to prove herself.

Dr. Schweitzer relayed this message to the waiting reporters. They tried to convince Dr. Schweitzer that he could persuade Olga to change her mind.

The Doctor said he did not feel free to do this. Was there anything else he might do for them?

The reporters consulted among themselves and then said they were eager to accept Dr. Schweitzer's offer to escort them around the Hospital grounds.

Dr. Schweitzer replied that he would be happy to do this on condition that they would agree not to pursue Miss Deterding.

They agreed.

The tour of the Hospital began, but every now and then one of the reporters would quitely detach himself from the party and go searching for his quarry. It was apparent that they had no idea what Miss Deterding looked like, for they confronted Dr. Margaret, Albertina, Trudi, and at least three nurses and proceeded to act as though they knew the person was Olga.

They did see Olga but failed to make the identification. Olga's assignment that week happened to be kitchen detail. The reporters peered into the kitchen. It didn't occur to them that the girl with the soiled apron involved in the messy job of cleaning the innards from fish and peeling vegetables could possibly be the glamorous multimillionairess they were looking for.

It was hot in the skullery. Olga was drenched with perspiration. Her hair clung to the sides of her face. When the reporters looked at here she smiled politely. They smiled back and passed on.

Finally, the party gave up the chase and left the Hospital. When Dr. Schweitzer met us that evening in Clara's room, he sat down hard and breathed deeply.

Well, now, he said, maybe we could see the way it was. Another day very largely lost. One day is given over to filling out elaborate forms for the government. No sooner is that finished than another day has to be devoted to parrying with the fourth estate.

Then, characteristically, the Doctor's sense of humor lifted him out of his despairing mood.

"The tour of the Hospital was most thoroughly done," he said. "Even if it wasn't what they really wanted, they saw all that there was to see. Finally"—and here he grinned widely—"just to keep them occupied, I gave them a lecture on philosophy. It was a good lecture, but I'm not sure they were in a philosophical mood. Anyway, they have now gone and we can get back to our various projects."

# IX

"WHICH OF YOUR two purposes do you consider more important?" Dr. Schweitzer asked. "You and Clara have said you had two principal objectives in coming here: one was concerned with my unpublished manuscripts, the other with the general question of world peace. Now, which one are you most eager to pursue?"

I told the Doctor that we felt the purposes were related, but that we were perfectly willing to have him decide which one was to be taken up first.

"But you are leaving in a few days," he said. "We may not have enough time to do both. Therefore, I ask you again: which one of your two purposes are you most bent on achieving before you leave?"

I said Clara and I had hoped to leave with some statement by him having to do with world peace. We also wanted to photograph such unpublished material as he had with him at Lambaréné. Our further presence was not essential for the statement. We had already put the case for the statement to him.

As for the manuscripts, however, this was something that would have to be done while we were at the Hospital. And since his presence was not required for the photographing of the manuscripts, Clara and I could work on this project by ourselves. We could devote the mornings to it; it shouldn't take more than three or four days.

The Doctor assented. Let us revert, however, he said, to the matter of the peace. Over the years he had been collecting materials on the question of nuclear energy, military and non-military. When he had visited Europe some months earlier, his concern had been considerably increased as a result of a meeting of Nobel Prize winners in Lindau, Germany. Many of the scientists there spoke with the utmost sense of urgency and gravity about the growing problem. Alongside the problem of peace, everything else seemed small.

Nothing to me was more striking than Dr. Schweitzer's face as he contemplated and spoke about the situation that confronted people in the world today. There seemed to be an infinity of detail in that face; it seemed as though every event in human history were clearly recorded there. Most of the time he sat forward in his straight chair, his eyes seemingly fixed on a distant object.

Only a few years ago, he added, the statement that this planet could be made unfit for life seemed absurdly melodramatic. But there was no longer any question that such power now existed. And even without a war, the atmosphere could become dangerously contaminated.

"After our earlier talk," he said, "I reflected that danger of this magnitude is not easily grasped by the human mind. As day after day passes, and as the sun continues to rise and set, the sheer regularity of nature seems to rule out such terrible thoughts. But what we seem to forget is that, yes, the sun will continue to rise and set and the moon will continue to move across the skies, but mankind can create a situation in which the sun and moon can look down upon an earth that has been stripped of all life.

"We must find some way of bringing about an increased awareness of the danger," he continued. "It is a serious thing that the governments have supplied so little information to their people on this subject. There is no reason why people should not know exactly where they stand. Every once in a while, the governments will reassure the people but this comes only after there has been a serious alarm. What is needed is genuine information. Nothing that a government knows about the nature of this new force is improper for its people to know."

Dr. Schweitzer asked me if I had brought any documentry data on the matter we were discussing.

I took out from my bag a number of papers, among them an abstract of a report, of which Dr. Willard F. Libby was a co-author, that had been prepared for the United States Atomic Energy Commission in August, 1954. The report was concerned with the effect of fallout of radioactive strontium on milk resulting from the uranium and plutonium nuclear explosions that had taken place up to that date.

Samples of milk taken from various places showed evidence of some contamination. The quantities of this radioactive strontium were found at the time to be well under dangerous levels. Even so, Dr. Libby's report showed evidence of growing apprehension, especially in his recommendation that the federal government undertake estimates on the cost of decontaminating milk. The decontamination would be effected by removing the calcium from the milk. Calcium has an affinity for radioactive strontium.

Three things were significant about that report. The first was that most of the radioactive fallout resulting from previous nuclear explosions had yet to come to earth at the time the survey was made. The second was that the biggest nuclear explosions were to occur after the report was published in August, 1954. The third was that no precise data are available on the tolerance limits of human beings to radioactive strontium.

In other forms of radiation, it is definitely known that there is far less safety than had earlier been supposed. Only ten or fifteen years ago, for example, the public was being assured that it had nothing to fear from regular X-ray examinations. More recently, however, it was disclosed that the tolerances were astoundingly lower than once had been so confidently claimed. Scientists had yet to perform the same kind of exhaustive researches into the tolerance limits of radioactive strontium that had been made on

X-ray radiation. If, through additional research, it developed that the effects of radioactive strontium had been underestimated, as in the case of X-rays, then colossal damage to all living creatures would have been done. And this is the kind of damage that cannot be undone.

The discussion with Dr. Schweitzer then turned to the power of the new bombs. One way of visualizing this new power would be to imagine a procession of one million trucks, each of which contained ten tons of TNT. The total tonnage would form a man-made mountain of dynamite several times the height of the Empire State Building. If this mountain were to be detonated it would represent the approximate power in a single twenty-megaton hydrogen bomb that can be carried by a single plane.

Dr. Schweitzer said that a very high order of public understanding throughout the world was necessary in order to deal with this problem.

I then told him that this was precisely what he was in a position to do. He was among the few individuals in the world who would have an almost universal audience for anything he might say.

His eyes turned from a distant object and he looked at me directly.

"All my life," he said, "I have carefully stayed away from making pronouncements on public matters. Groups would come to me for statements or I would be asked to sign joint letters or the press would ask for my views on certain political questions. And always I would feel forced to say no.

"It was not because I had no interest in world affairs or politics. My interest and my concerns in these things are great. It was just that I felt that my connection with the outside world should grow out of my work or thought in the fields of theology or philosophy or music. I have tried to relate myself to the problems of all humankind rather than to become involved in disputes between this or that group. I wanted to be one man speaking to another man."

I asked whether the Doctor felt that the matter we had been discussing was as much moral as it was scientific or political. I told him I believed there was no living person whose voice on such an issue would be more widely heard or respected.

Dr. Schweitzer thanked me for the compliment, but said that this was a problem for scientists. He believed that it would be too easy to attempt to discredit any non-scientist who spoke out on these matters.

I told him that I thought it inconceivable that this would be true in his case. Moreover, this was not solely a laboratory question. If nuclear power could have the effect of damaging the genes of human beings, then the nature of man himself was involved. Sovereign nations were now in a position to make decisions that were not properly theirs to make.

In saying this, I told him I recognized that the problem could not be considered apart from the larger uncertainty in the world today. Nuclear experimentation did not exist in an otherwise placid world. This, of course, added to the peril of mankind. For what we had most to fear was not merely the tests themselves, hazardous though they might be, but a saturation of tensions resulting in all-out nuclear war.

Dr. Schweitzer agreed, saying that anything that would be done against nuclear experimentation should not have the effect of putting the West at a disadvantage with respect to Soviet Russia.

He said, however, that the very real challenge of world communism should not be used as the reason for withholding vital information from the human race concerning the dangers of unlimited nuclear testing. It was possible that an informed and determined world public opinion could serve as a powerful force in bringing about enforceable agreements with respect to arms control and in leading to other long-range measures for peace.

In view of all this, I asked the Doctor whether he felt justified in putting aside his reticence about making a public statement.

He said that he would continue to give careful thought to the matter. He was still troubled, he said, about the form a constructive statement might take. How would it be issued? How would one go about drafting a statement that would be outside the context of the ideological struggle in the world today? He re-emphasized that he didn't want people to think that he was admonishing the United States or trying to intrude into domestic concerns. He wanted more time to think about these things.

When we resumed our discussion the next day he said he was still uncertain about the form of a statement or the method of its release.

Meanwhile, he was eager to consider an aspect of the problem that was highly significant. This was the fact that nations which were setting off nuclear explosions in the pursuit of their own security were possibly jeopardizing the health of other peoples.

On the basis of recent visits to Japan, I could report to Dr. Schweitzer that the Japanese government was confronted with a profound dilemma. It did not wish to oppose the American government, nor did it see any way of condemning Soviet Russia at the United Nations without including the United States. But Japan had increasing evidence of soil contamination as the result of the Russian tests and fish contamination as the result of the American tests. Autopsies had indicated the presence of radioactive strontium in a number of corpses. The American hydrogen bomb explosion called "Operation Castle" had not been under complete control. Japanese fishermen outside the prohibited area had been hit by radioactive ashes. The Japanese government had just issued instructions to its people about precautionary measures in the preparation of leafy vegetables and fish. But decontamination of food was a complicated laboratory process; it was doubtful whether even the most careful washing and boiling would be adequate.

As a result, Japanese public opinion was sensitive on the subject and was now becoming articulate and potent. Meanwhile, Communists were exploiting the issue of testing against the United States, making it appear that America was responsible for the failure to arrive at cessation agreements, and saying little about the fact of Soviet nuclear testing.

As we discussed the role of the bystanders with respect to nuclear testing, I could see that Dr. Schweitzer felt that this was a vital issue. As a citizen of a democratic nation, I did not feel that we had any right to take measures that were of possible danger to others without their consent. Indeed, the principal argument against Nazism and more recently against communism was that they were scornful of the rights of others and did harm to innocent people in their pursuit of military advantage. Is it any less immoral for any nation to jeopardize the health and safety of other peoples through uncontrolled air dispersal of radioactive poisons? If other peoples are involved, then they have a right to participate in the basic decisions involved in testing. There is no more basic tenet in democratic government than that people who are affected by the acts of government have a right to participate in the affairs of that government.

If it is wrong to impose a tax on a man without giving him a voice in government, is it any less wrong to deprive his soil or water of their purity without giving him a chance to be represented and heard?

There was no argument about any of this, the Doctor said. What concerned him was the propriety of his making any statement. It was something he wanted to think about carefully. Meanwhile, he asked me if I would put down in writing a summary of such facts on the question of nuclear fallout as I happened to have with me.

Then he got up to leave. He opened the door, then said, as though by afterthought, that he would put together the manuscript of *The Kingdom of God* and would turn it over to us in the next day or so.

# X

WHEN DR. SCHWEITZER came for our meeting the next afternoon, he was carrying a small bundle. It was neatly wrapped in a large napkin. He handed it to me. "This is the manuscript of *The Kingdom of God*," he said simply. "It is practically complete."

I opened the bundle. Here, for all I knew, was one of the most important books of our time. The sheets had been perforated at the top and were tied together by a string. But I gasped when I saw the kind of paper that had been used for the manuscript. There were sheets of every size and description. Dr. Schweitzer had written his book in longhand on the reverse side of miscellaneous papers. Some of them were outdated tax forms that had been donated to Lambaréné by the French colonial administration. Some were lumber requisition forms used by a lumber mill not far away on the Ogowe River. Some came from old calendars. I couldn't even begin to count the number of manuscript pages which were written on the reverse sides of letters sent to him many years earlier.

In any other man, this would have seemed quixotic and inexplicable. In Schweitzer, however, it represented a complete consistency with everything else in his life. There was the crude piano in the dining room he wouldn't replace or repair because the money could better be used elsewhere. There was the fact that he shaved without soap or lather because he considered it a luxury. There was the fact that he traveled third class only because "there was no fourth class." He could no more think of buying paper for his own literary use than he could buy an easy chair.

This, then, was *The Kingdom of God*. It was written in German longhand. The interlinear editing was prodigious. To conserve space, he had written in a very small hand. Down the right-hand margin, in even smaller longhand, were penciled notes.

Thrilled though I was with the manuscript, I was severely apprehensive over the ability of the camera to deal with such close and faded written material. I feared, in fact, that the penciled notations might be substantially lost and said so to the Doctor.

"That is nothing to worry about," he replied. "The penciled notes are for me alone. That is the way I like to write. When I sit down to my work, the first thing I do is to write my outline down the right-hand margin of each page. Then I use the main part of each sheet for the actual writing by pen."

He sat down, then bid us do likewise.

Dr. Schweitzer said that he had started the previous evening to assemble the manuscript. All the pages were in a trunk, but they had not been in the proper order. It had taken him several hours to get all the pages together and put them in sequence. I shuddered at the invasion of his time represented by this special effort.

As for *The Philosophy of Civilization,* Dr. Schweitzer feared that getting this manuscript together would take several days of uninterrupted work, at least. There were more than four hundred thousand words in the *Philosophy,* which made it several times longer than the *Kingdom.* It was doubtful, he said, that he would be able to get to this for several months. But he promised that he would make a special effort in this direction.

Then the Doctor said that if there was nothing further to discuss about the manuscript, he would like to revert to the matter of the statement about the peace.

"Even while I was sorting the manuscript papers last night and this morning I have been thinking about the declaration or statement or whatever it is you want to call it," he said. "I have no reason to believe that anything I might do or say would or should have any substantial effect. Even so, if there would be even the smallest usefulness that I or anyone else might have on this question, it would seem almost mandatory that the effort be made.

"This crisis intimately concerns the individual," he continued. "The individual must therefore establish a connection with it. The leaders of the world today have to act in an unprecedented way if the crisis is to be met. Therefore they must be strengthened in their determination to do the new and bold things that must be done. That means that the individual has a greater role to play than before. The leaders will act only as they become aware of a higher responsibility that has behind it a wall of insistence from the people themselves. I have no way of knowing whether I can help in this. Perhaps I may be justified in trying."

This was the first intimation Clara and I had had that he was giving favorable consideration to the issuing of a statement. We exchanged glances that revealed clearly how heartened each of us was by what the Doctor had said.

"I am still worried, however, about some of the special problems involved in this declaration," Dr. Schweitzer resumed. "I am not sure that I agree with you that a broad statement addressing itself to the danger of war and the consequences of war would be the most effective way of approaching it. Yes, the world needs a system of enforceable law to prevent aggression and deal with the threats to the peace, but the important thing to do is to make a start somewhere before we get into the broader questions."

He paused for a moment. Some baby goats were bleating just outside the rear of the bungalow and he went to the screen to see what was happening. The bleating stopped and he returned to his seat at the end of the bed.

"I think maybe the place to take hold is with the matter of nuclear testing," he said. "The scientific aspects of testing may be complicated but the issues involved in testing are not. A ban on testing requires no intricate system of enforcement. All peoples are

C. U.

Beach at Lambaréné: "Mankind can create a situation in which the sun and moon can look down upon an earth that has been stripped of all life."

involved, therefore the matter transcends the military interests of the testing nations. It is clearly in the human interest that the tests be stopped. Even if there is a small chance that the tests are harmful, it is important that the nations set aside the tests until they are absolutely certain what this chance involves.

"If a ban on nuclear testing can be put into effect, then perhaps the stage can be set for other and broader measures related to the peace. That is why I am inclined to a fairly limited objective. Later we will be in a better position to do the bigger things you have been talking about."

I told the Doctor that I agreed it might be easier to rally world public opinion around the need to suspend nuclear testing by all nations than it would be to deal with the basic structure of a workable peace. The main problem was war itself. But if he felt that it might be wise to confine his initial efforts to the matter of nuclear testing, I would of course respect his decision.

Dr. Schweitzer said his mind had not been finally made up on this question; he had been thinking out loud. He would continue to study the alternatives. Meanwhile, he said, there was yet another important question: what form would the statement or declaration take? How would it be issued?

Had I given any thought to these questions, he asked.

I said that I felt a direct statement, released to all the news agencies, might be effective.

He shook his head at this, saying that he had serious doubts about the news release type of story. What it gains in immediate attention it tends to lose in long-term impact. Besides, he said, one runs the risk of competing with all the other news that may be breaking on a certain day. Here he reverted to one of his favorite themes.

"I am worried about present-day journalism," he said. "The emphasis on negative happenings is much too strong. Not infrequently, news about events marking great progress is overlooked or minimized. It tends to make for a negative and discouraging atmosphere. There is a danger that people may lose faith in the forward direction of humanity if they feel that very little happens to support that faith. And real progress is related to the belief by people that it is possible.

"Well," he continued, "maybe this is the wrong time to worry how the statement is to be issued. Our first job is to bring the baby into the world. Then we can decide what to do with it. We will therefore worry first about what the statement should be; then we will study it and determine how it might be used."

Mathilde Kottman came to the room and reminded us that it was getting late. I could see she was concerned about the Doctor. I stood up but the Doctor put me at my ease.

"In a way," he said, "the two of you in coming here have broken down my resolve not to involve myself in anything remotely concerned with political matters. But as I said the other day, the problem goes beyond politics. It affects all men. All men must speak. Some way must be found to bring about an increased awareness of the danger. Anything that is done should above all be simple and direct. It should not be ponderous or academic."

Once again, as he spoke, he leaned forward in an aspect of intense concentration. His eyes were closed and he seemed to measure every word. As Clara interpreted, I could see that he listened to her carefully in order to make sure that the precise nuance or emphasis would be given to what he was saying.

It was clear, both from his words and manner, that this subject was now preoccupying most of his thoughts. He looked out through the latticed window at the scudding clouds in the late afternoon sky.

"I know the weather in this part of the world like the back of my own hand," he said. "For forty-three years I have observed its habits and variations. And this is the time of the year when unfailingly there is hardly a breeze. And yet now there are winds that I have never known before. One must always be prepared, of course, for sharp changes in natural phenomena. But it is important at the same time not to ignore new factors which may not be of nature's making. Some scientific reports I have seen raise serious questions concerning the effect on weather caused by nuclear explosions. Obviously much more study is required. But at least the question exists. If there is the slightest chance that man's crops are being jeopardized, it is the duty of the nations to find out definitely before they proceed blindly."

I told the Doctor of the report *The Saturday Review* had carried by Dr. Irving Bengelsdorf of the General Electric Company in the United States. Dr. Bengelsdorf had correlated freak weather occurring in various parts of the world with hydrogen explosions and the patterns of weather movements.

"All things are now possible," Dr. Schweitzer said. "Man can hardly even recognize the devils of his own creation."

Mathilde Kottman was still waiting in the doorway. The Doctor rose to leave.

"If you are not too tired tomorrow, we can think and talk further," he said. "Meanwhile, you now have the manuscript of the *Kingdom* for such use as you wish to make of it."

Late that evening, after most of the oil lamps at the Hospital had been turned out, I walked down toward the compound. From the direction of Dr. Schweitzer's quarters I could hear the stately progression of a Bach Toccata. The Doctor was playing on the piano in his small workroom. I had heard that this piano had an organ footboard attachment so that the Doctor could keep his feet in playing condition.

I went up on the porch and stood for perhaps five minutes near the latticed window, through which I could see Dr. Schweitzer's silhouette in the dim-lighted room. Then there was a pause in the music and the Doctor called out to me. It surprised me that he should have known I was standing outside in the dark. I entered his room and he bid me sit on the piano bench next to him while he continued the Fugue. His feet moved over the organ footboard with speed and precision. His powerful hands were in total control of the piano as he met Bach's demands for complete definition of each note—each with its own weight and value, yet all of them intimately laced together to create an ordered whole.

Sitting there in the dim light with the vibrations of the Toccata racing through me, I had a stronger sense of listening to a great console than if I had been in the world's largest cathedral. I knew there might be things about Albert Schweitzer I would never comprehend or reconcile; but this particular Albert Schweitzer I felt I knew and had complete access to. The yearning for an ordered beauty; the search for a creative abandonment—yet an abandonment inside a disciplined artistry; the desire to re-create a meaningful past; the need for outpouring and release, catharsis—all these things inside Albert Schweitzer spoke in his playing. And when he was through, he sat with his hands resting lightly on the keys, his great head bent forward as though to catch any still-lingering echoes.

He was now freed of the pressures and tensions of the Hospital, with its forms to fill out in triplicate and the mounting demands of officialdom; freed of the mounds of unanswered mail on his desk; freed of the heat and the saturating moisture and the fetishers and the ants that get into the medicines. Johann Sebastian Bach had helped make it possible for Schweitzer to come to Lambaréné in the first place, through book royalties. Now Bach was restoring him to a world of creative and ordered splendor. For perhaps half an hour we chatted on that piano bench in the thin light from the flickering oil lamp at the far side of the room. He was speaking personally now—about his hopes mostly, like a young man just starting out in a career and musing about what he would like to accomplish. First, he would like to see his Hospital in tiptop running order. Second, he would like to be able to train others to run the Hospital after he is gone. Third, he would like to have just a little time to himself—to work quietly and finish his writings.

He did not wish these longings of his to give the impression that he was unhappy in his work. Actually, he never thought much about happiness or unhappiness in terms of his own life. Generally, he thought in terms of what had to be done and the time required for doing it. Now and then something would happen that would give him a sense of fulfillment and deep reward. Only a few days earlier, for example, he received word from a professorial colleague in France about an examination paper turned in by a nineteen-year-old boy. The question that had been put was: "How would you define the best hope for the culture of Western Europe?" The answer given by the student was: "It is not in any part of Europe. It is in a small African village and it can be identified with an eighty-two-year-old man."

Dr. Schweitzer paused. He held in his hand the letter that had told of the student's conception of his role in the modern world. He was profoundly moved.

"In the morning," he said, "when the sun is up and I hear the cries of the Hospital, I do not think of these lofty ideas. But at a moment like this, when the Hospital is asleep, it means much to me that the student should believe these things, whether they are true or not."

"We must find some way to bring about an increased awareness."

"There is no reason why people should not know exactly where they stand."

"I wanted to be one man speaking to another man."

"I have tried to relate myself to the problems of all humankind rather than to become involved in disputes between this or that group."

C. U.

# XI

HOW DO YOU go about preparing to make copies of a literary treasure?

The problem was complicated by equatorial conditions—fungus that insisted on finding its way into the most delicate parts of precision machinery; moisture with knifelike powers of penetration; a sun directly overhead that cut down human efficiency.

Then there was the manuscript itself. From the handwritten letters of Dr. Schweitzer I had received, I thought we would probably be dealing with small, even writing in a light blue or green ink. But I wasn't prepared for the variations between pen and pencil, or the varieties of paper, some tinted and some not.

Before leaving for Lambaréné it seemed to me that, theoretically at least, the best method would be microfilming. This had to be ruled out, however, because of the size of the equipment and because there was no electricity at Lambaréné except in the operating room. While it might be possible to tap the power from the generator for our purposes, there were other problems of voltage and cycles that made this approach too complicated and risky. For the same reasons, we had to eliminate various other copying devices, some of them otherwise beautifully suited to our needs.

This left plain photography.

For two weeks before I left New York, I experimented with

various cameras borrowed from friends. I went all the way from the Italian "mini" cameras to the Speed Graphic, including my own Rolleiflex. I photographed the smallest specimens of handwriting I could find and used films with different speed ratings. Nothing I tried seemed adequate. Finally, I decided to turn the problem over to the people who knew best—Eastman Kodak of Rochester, New York. I took one of the executives, Robert Brown, into my confidence. He responded most favorably and put some of his experts on the job. Five days later, Charles Kenyon, representing Eastman Kodak, came to my office with a complete plan of battle.

First of all, he handed me a Retina IIIC. He annotated the various requirements it met:

1. Automatic unit correlating shutter speed with f.1.5 opening.
2. Built-in light meter.
3. Maximum speed and ease in reloading.
4. Handled films with 36 frames.
5. Had special magnifying attachment for close-up copying work.

Charles Kenyon said Eastman Kodak recommended that I use slow-speed Panatomic-X film. It had a fine grain and was especially adapted to enlargement problems. Then he patiently instructed me in the use of the camera, which the company would lend me.

Now, at Lambaréné, we had the manuscript of *The Kingdom of God* in our possession and it was time to get down to work.

The biggest single factor in our favor, of course, was that Clara Urquhart was a superb photographer. I had seen the high quality of her work the first time we met in New York when she showed me the proofs of her new book of photographs-with-text on Albert Schweitzer. What I had not known but now discovered was that she was an expert technician in the business of copying a manuscript.

Dr. Schweitzer had told us that he wanted the photography project to be carried out as unobtrusively as possible. He didn't want the undertaking to become a topic of conversation at the dinner table or anywhere else. Hence we could not do the work outdoors, which would have afforded the best light. We set up our improvised studio in Clara's room. We placed a low table up against the latticed window in order to have maximum light. The iron bedstand we used as the base for a clamp to which we attached the Retina IIIC. We then placed a two-inch board under the table in order to provide a camera range of fifteen inches. The close-up range finder was attached and the special lens for correct parallax was inserted. The first page of the manuscript was put into position, we crossed our fingers, drew a deep breath, and started to shoot. Clara moved with characteristic ease and dexterity as she arranged the manuscript, placing each page in position, numbering each page, and unpacking and repacking the film. All I had to do was to focus, press the trigger, and reload. By the end of the first roll, I knew that if I had been left to myself, it would have taken me two hours just to complete one spool. As it was, Clara's help enabled us to shoot thirty-six frames, or one spool, in from fifteen to twenty minutes.

On this basis, we calculated it would take about six hours to complete *The Kingdom of God*. Assuming *The Philosophy* was in the same physical condition, we guessed it would take us about eighteen hours to complete that, too. But Dr. Schweitzer was unable to put the manuscript of *The Philosophy of Civilization* together. He came to the room and said that he had gone to the trunk, looked over the manuscript, and calculated that it would take too many hours to put it together in the same orderly form in which he had given us *The Kingdom of God*. But this he would do in the months ahead; perhaps arrangements might be made for later copying.

That was entirely reasonable, I said. Meanwhile, we would at least have one of the two books on film.

As we had anticipated, the photographing of *The Kingdom* took about six hours, spread over two sessions. During the rest of the time I worked on a memorandum covering the nuclear situation as Dr. Schweitzer had requested. This memorandum was based on various documents I had brought with me to Lambaréné, including the Libby report on radioactive strontium in milk, the report of the National Academy of Science in the United States, the report of the British Academy, and the official findings of the Indian government.

The following morning at breakfast, Dr. Schweitzer told Clara that he wished to spend most of the day with us and asked if we might be available. He would come to us at 11:00 A.M. and stay with us until lunch; he would meet us again at four in the afternoon; then we would spend the evening together.

The morning meeting was devoted to a discussion of the nuclear situation. The Doctor had been giving his constant thought to it. Concerning the matter of a possible news release, of one thing he now seemed certain: a direct press statement would be unwise. It was too pretentious, smacking of publicity-seeking. Perhaps a small journal would be best. If the statement were of any value, it might be reprinted in various ways. It might stay in circulation for a period of time and have a cumulative value.

I recognized that such an approach would be entirely in character, I said. At the same time, it was important to face the likelihood that anything the Doctor said on this subject, even in a small journal, would be picked up instantly and made into headline news.

In any event, I added, it seemed clear that the best way for the Doctor to proceed would be in a manner that he felt was natural to him.

We come now to the statement itself, he said. He felt that he should not aim at any arbitrarily chosen length as being especially desirable but rather should concentrate on making the statement as clear and as complete as was humanly possible regardless of length.

"I must be careful to develop the facts very fully," he said. "I don't want to be criticized for leaving large gaps in the argument."

The Doctor then recalled an experience in Oslo in connection with the Nobel Prize. It was an experience from which he had learned a great deal.

"I was told I was expected to give an acceptance talk. I worked on the talk for several months, developing my theme with great care. When the message was completed, I estimated that it would take from seventy to eighty minutes in the reading.

"But when I arrived in Oslo I learned that I would be given thirty-five minutes for the talk. That meant cutting the message in half. I was most unhappy but I proceeded according to the limitations. The original message had been as closely knit as I knew how to make it; the shortened version was uneven and the main points inadequately developed.

"For a moment, just before I got up to speak, I was tempted to reach for the full message even at the risk of being stopped halfway through my speech. But I downed the temptation out of courtesy to my hosts. After all, it was up to them to decide what kind of program they wanted.

"The printed version, of course, was of the short talk as delivered. I didn't like it and had no interest in seeing it or acknowledging its existence. Even so, I was surprised to see how it kept being reprinted by various journals throughout the world and how long it seemed to stay alive."

He leaned back and drew a deep breath. Then he said the reason he brought up the matter was that he felt the question of length in the new statement should be put to one side; the only thing that should concern him was the accuracy and the relevance of what it was he had to say. I agreed.

What should be of most concern to him now and in the weeks ahead, he said, was to complete his study of the materials I had turned over to him. He would also correspond with some of the names I had given him and with various scientists in Europe who had expressed concern to him about these matters the last time he had seen them in Europe, at the meeting of the Nobel Laureates in Lindau, Austria.

At the afternoon meeting, he did not sit down on the edge of the bed, as was his usual manner. Instead he cleared the small table and arranged three chairs around it. He had brought with him his regular stub pen and a bottle of ink.

"We must clear up our remaining business," he said. "Are there any new matters you would like to discuss?"

I broached a subject that had been very much on my mind but that was awkward to bring up in direct discussion with him.

He said that if this had anything to do with his affairs in the event of his death, it would please him if I felt free to proceed.

What about the manuscripts? I asked. If anything should happen, who would have the authorization to edit them and approve them for possible publication?

Did I have a suggestion to make on that score, he asked.

I replied that I felt a small group, consisting perhaps of Emory Ross, J. D. Newth, his English publisher, and Clara might be in a good position to superintend the publication of the books.

"Yes, that would be a good idea," he said. "If anything happens to me, you have the authorization to bring such a group together and make the essential decisions about my unpublished work. I ask only that you observe the following conditions:

"1.   That you consider carefully whether the material is actually worth publication, and, if so, what form it is to take.

"2.   That nothing be added to my work of any nature.

"3.   That no one be permitted to write anything in my name."

I gave him my assurance that I would convey these conditions to the other members of this literary trustee group. As I did so, however, I could not help thinking that the Doctor would probably outlive all of us. His skin had the color and texture of a man at least forty years younger. His eyes were clear and sharply focused. His hands—perhaps the most impressive hands I have ever seen on a human being, combining strength with sensitivity—were without the slightest trace of any tremor. I could recall, too, that three days earlier I had accompanied him on his "rounds," which included everything from moving large boxes of medical supplies in a hot sun and carpentry work to an inspection tour of the Hospital, walking up hill and down. He was less fatigued at the end of four hours than I was at the end of one. In fact, I found it difficult to think of any middle-aged man who could have kept up with him.

This matter disposed of, the Doctor inquired about other items of business.

I told the Doctor that a Schweitzer Fellowship group in Darien, Connecticut, was raising money for his Hospital and had asked me to obtain his autograph on various papers that might be auctioned. This commission was quickly executed.

Next, I referred to the fifteen-year-old boy, Marc Chalufour and his fight to save the old organ in his church from being replaced with a new electronic machine. Marc, who knew something about the subject, felt that the church was giving up a superior instrument, and had given me a letter for Dr. Schweitzer. The letter identified the old organ according to make and year, and reported on its condition. Marc felt that if he could have a note of support from Dr. Schweitzer, he might win over the elders.

Twenty-five years ago, as a youth myself deeply interested in the organ, I had dreamed that it might some day be possible to talk to Albert Schweitzer about the art of organ building. No man alive knew more about the Silbermann organ—the finest church organ up through the nineteenth century. And now, a letter from a fifteen-year-old boy was the open sesame. To my great delight, he spoke at length about the wonders of the Silbermann organ—about the knowledge and craftsmanship that went into it, about its tone and its unique features, about methods for keeping it in good working condition.

Then, with obvious relish, he reached for paper and pen, and wrote a letter to Master Chalufour:

Dear Marc:

If the organ was made in 1858, it is a good one. That was an age of fine organ building in Germany, France, England, and the United States. The period of good organ construction lasted from 1850 to 1885. After that, organs were built in factories. The organs built in the 1850's had an excellent tone—sweet, not too strong, but noble. If your church organ is still in reasonable good physical condition, it is certainly worth conserving and restoring, for it is most valuable.

In Alsace, there are organs dating from 1730 that are still in use in the churches; and if your church does not want the organ, perhaps you may be able to find a place for it yourself. One day its value will be realized. The old organs are better because they were built by artisans. In those days there was not the competition in price that there is today. The organ builder was able to use the finest material and did not have to count the hours necessary to put into it, as one has to do today. He could deliver an instrument of the highest quality.

Good luck from my heart to the courageous and intelligent young man who wants to save an organ.

ALBERT SCHWEITZER

The Doctor put down his pen; the conversation about the organ and the letter had taken the better part of an hour. Despite the expenditure, I had no regrets for having opened up the subject; he had enjoyed himself thoroughly.

We went off to dinner, at the end of which several members of the staff remarked that they had seldom seen the Doctor in such a joyous mood.

After dinner, Dr. Schweitzer came to Clara's room again. The first thing that had to be done, he said, was to write a letter to President Eisenhower. In this letter, he wanted to thank the President for the cordial birthday greetings. He was also eager to express his concern about the world situation, especially with reference to the armaments race. After about twenty minutes, he showed us the draft:

Lambaréné<br>January 10, 1957

The Hon. Dwight D. Eisenhower
The White House
Washington, D. C.

Dear Mr. President:

I send you my heartfelt thanks for your friendly letter in which you send me your good wishes and those of Mrs. Eisenhower on the occasion of my eighty-second birthday. This expression of your good wishes was the first birthday greeting I received. Your generous and kind thoughts touch me deeply.

In my heart I carry the hope I may somehow be able to contribute to the peace of the world. This I know has always been your own deepest wish. We both share the conviction that humanity must find a way to control the weapons which now menace the very existence of life on earth. May it be given to us both to see the day when the world's peoples will realize that the fate of all humanity is now at stake, and that it is urgently necessary to make the bold decisions that can deal adequately with the agonizing situation in which the world now finds itself.

I was very happy to have Mr. Cousins, who will take this letter to you, here with me in Lambaréné. It was rewarding to spend time together and to see how many ideas and opinions we shared.

With assurance of my highest esteem, I am,

Yours devotedly,
ALBERT SCHWEITZER

He then wrote a final copy on thin white paper, put it into an unsealed envelope, and handed it to me to transmit to the President. He leaned back in his chair and asked whether I was glad I came to Lambaréné.

Most certainly, I replied. I hoped he didn't think it presumptuous of me if I asked him the same question.

He said he had some forty years in which to reflect on the answer to that question, so that there need be no hesitation in his reply. Yes; he was glad he came to Lambaréné, very glad. It was while he was coming up the river Ogowe one day many years ago, passing one of the luxuriant islands in the river and looking up at the scudding clouds, that the idea of reverence for life occurred to him.

Lambaréné had also made it possible for him to make his life his argument, he said.

This puzzled me and I looked up quizzically.

"As a young man, my main ambition was to be a good minister," he explained. "I completed my studies; then, after a while I started to teach. I became the principal of the seminary. All this while I had been studying and thinking about the life of Jesus and the meaning of Jesus. And the more I studied and thought, the more convinced I became that Christian theology had become over-complicated. In the early centuries after Christ, the beautiful simplicities relating to Jesus became somewhat obscured by the conflicting interpretations and the incredibly involved dogma growing out of the theological debates. For example, more than a century after Christ, there was a theological dispute growing out of questions such as these:

"Is Jesus actually God or the son of God?

"If he is God, why did he suffer? If he was the son of God, why was he made to suffer?

"What is meant by the spirit of Jesus?

"What is the true position of Mary in Christian theology?

"Elaborate theology dealing with such questions disturbed me, for it tended to lead away from the great and simple truths revealed in Jesus' own words and life. Jesus Christ did not proclaim himself to be God or the son of God; his mission was to awaken people to the Kingdom of God which he felt to be imminent.

"In my effort to get away from intricate Christian theology based on later interpretations, I developed some ideas of my own. These ideas were at variance with the ideas that had been taught me. Now, what was I to do? Was I to teach that which I myself had been taught but that I now did not believe? How could I, as the principal of a seminary, accept the responsibility for teaching young men that which I did not believe?

"But was I to teach that which I did believe? If I did so, would this not bring pain to those who had taught me?

"Faced with these two questions, I decided that I would do neither. I decided that I would leave the seminary. Instead of trying to get acceptance for my ideas, involving painful controversy, I decided I would make my life my argument. I would advocate the things I believed in terms of the life I lived and what I did. Instead of vocalizing my belief in the existence of God within each of us, I would attempt to have my life and work say what I believed."

I recalled a discussion I had had several nights earlier with Dr. Friedmann and Dr. Margaret. We agreed that it was necessary only to see Dr. Schweitzer working with his hammer or making the rounds of the Hospital to recognize a profound symbol. The symbol, of course, had to do with the carpenter and the healer.

And so I asked Dr. Schweitzer if this symbol to him was a conscious one; in short, whether he had come to Lambaréné in imitation of Christ.

When I asked the question, I couldn't be sure whether I had asked the most obvious question in the world, or whether I was pushing at a door that was meant to be kept closed.

Dr. Schweitzer looked up and said simply that the pursuit of the Christian ideal was a worthwhile aim for any man.

Then, after a moment, he said that he did not want anyone to believe what he had done was the result of hearing the voice of God or anything like that. The decision he had made was a completely rational one, consistent with everything else in his own life.

Indeed, he said, some theologians had told him that they had had direct word from God. He didn't argue. All he could say about that was that their ears were sharper than his.

He said, however, that he believed in the evolution of human spirituality, and that the higher this development in the individual, the greater his awareness of God. Therefore, if by the expression, "hearing the voice of God," one means a pure and lively and advanced development of spirituality, then the expression was correct. This is what is meant by the "dictates of the spirit."

By an advanced spiritual evolution, he emphasized that he was not thinking so much in theological terms as in ethical and moral terms. Thus he disagreed with the impression created by some of the Psalms that if people were good they would receive their reward. Goodness need not depend on rewards, or on the absence of punishments. True spiritual evolution means that there is an awareness by the individual of the natural goodness inside him; therefore he is not reaching out but actually discovering his true self when he brings the goodness to life. There is the need to do good. If one does it because he expects tangible rewards he will be mistaken.

This led to a discussion of man's expectations with respect to the Deity. If man conceived of the Deity as an omnipresent guarantor of the good he was stretching the concept of the Deity to suit his own needs and therefore he was mistaken. There is no point in expecting God to prevent injustice by man. He said that after the last great war, with all its killing and injustice, with its persecution of religious minorities and the concentration camps and gas chambers and soap made from the remains of slaughtered Jews—after all this, he did not see how it was possible to hold to the concept of a God who would intervene on the side of justice.

This, he felt, is not how God manifests himself. God manifests himself through the spiritual evolution of man and through the struggle of man to become aware of the spiritual nature of his being and then to nurture it and give it scope. The existence of evil—or the occasional triumph of the evil over the good, as in the case of persecution and concentration camps—did not mean that God was oblivious of evil or indifferent to it. It means that man had the responsibility to deal with the evil and should not sit back and expect divine intervention.

Not infrequently in history, he said, religious leaders themselves would invoke the name of God for acts that were unjust. Calvin killed his enemies; Luther failed to speak out against the persecution of Jews; the early Israelites believed at times it was their divine mission to kill; the Crusaders used the sword freely in the name of God. Modern instances were many. One example was furnished by religious Spaniards who went to Mass and then went to see creatures slaughtered in the bull ring.

To talk of the "will of God" was a presumption and often a profanation, especially when one used the term to purify ungodlike acts. Moreover, to speak of the "will of God" is to use illusion. We must accept reality. And the dominant reality, to repeat, is that God manifests himself through the human spirit. Insofar as the individual is able to discover and develop his spiritual awareness, he is at one with the Deity. Nothing is more wonderful or mysterious than the workings of the inner awareness by which man discovers his true spirituality.

I asked Dr. Schweitzer whether I detected an echo of Hegel's "only the spiritual principle, which is synonymous with God, is real." Hadn't Kant made a similar distinction—between believing God and believing *in* God?

He replied that he had always regarded Hegel as one of his most important teachers; he also reminded me of his debt to the Stoics and the early Chinese. As for Kant, he agreed with the distinction but said that so far as Kant's ideas in general were concerned, he could be impressed by them without being moved by them.

There is no reason why religion should not grow and evolve, as man himself must grow and evolve, he continued. When, for example, Apostolic belief is used as final authority for theological positions, it should be remembered that the Apostolic doctrine was not developed by the Apostles but was created by interpretation in the middle of the second century after Christ.

Theological rigidity, he feared, was hurting religion. Young people, especially, were looking to religion as a great spiritual adventure but were being disappointed because it was not searching and probing for expanding truths. Instead it was holding to fixed ideas which young people could not fully accept. There is nothing irreligious in the search for true religiosity. The more we think and the more we are aware, the greater our concentration on the development of human spirituality and the more religious we become.

A great cathedral may help to awaken the human spirit, but it cannot create it.

Each man, in a very real sense, carries his cathedral inside him.

St. Paul preached out in the open, so that all could hear. At the Hospital in Lambaréné, there is no chapel. The preaching follows the pattern of St. Paul; it is in the open.

In all these respects, the Reformation is not yet complete. It has not given adequate weight to the Christianity of the Spirit. Reformation means change; Christianity must not be afraid of change; it must not be afraid to examine and re-examine and grow. Jesus symbolized change and growth.

The ideas contained in Reverence for Life are consistent, he said, with an evolving Christianity. The door is open but Christian theology has not gone in.

To the extent that his thinking on Christian theology and religion in general have created differences of opinion, Dr. Schweitzer said that this was a matter that affected him deeply.

"I have not wished to create problems for Christianity," he said. "I have suffered deeply because some of my ideas have become problems for Christianity."

"Wasn't this what had happened to Ernest Renan?" I asked. "Didn't Renan suffer because his own thinking caused him to veer away from the tradition in which he had been brought up?"

Dr. Schweitzer replied, "It is true that Renan, who was concerned with the life and meaning of Jesus, found himself at odds with the Christian theology he himself had been taught. This caused Renan great pain. In the end it tore him to pieces because he failed to find an outlet by which he could make his own new ideas come to life. New ideas in this field of thought are powerful things. One cannot just conceive of them as mere intellectual properties and then take leave of them. My own ideas do not happen to coincide with Renan's. I am much more concerned with the actual shape of history in the life of Jesus. But I think I can understand how Renan's work affected him when he didn't allow it to redefine his life for him. This is what I mean when I say I came to Lambaréné because I wanted my life to be my argument. I didn't want my ideas to become an end in themselves. The ideas took hold of me and changed my life. Resistance to those ideas would have been impossible."

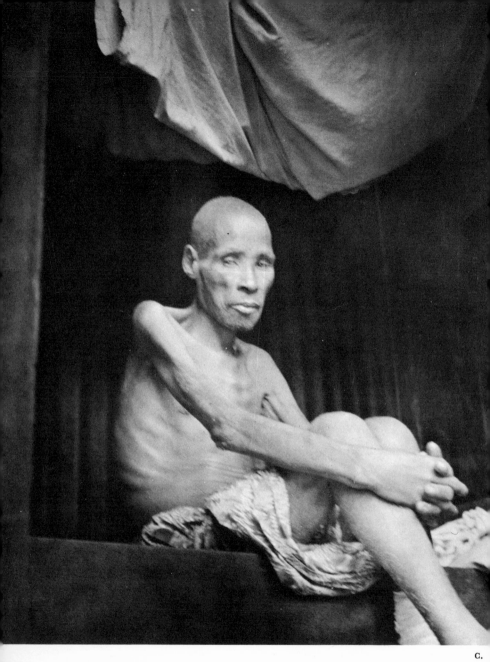

c.

Earthquake under conscience.

Same man, same situation.

Open door.

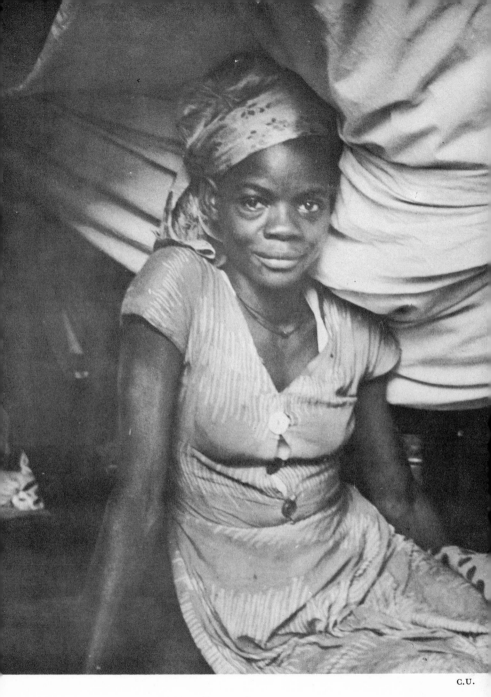

Communicator.

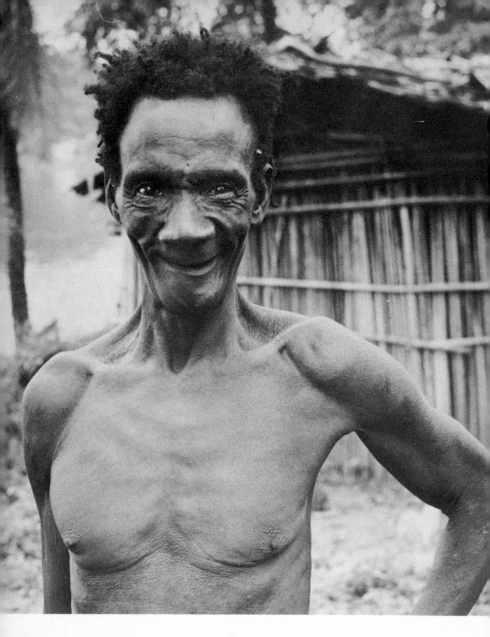

Encounter.

Special Delivery.

C. U.

Act One.

Intermission.

C.U.

C.U.

Responsibility as an art.

# XII

THE DOCTOR GOT up to leave. It was now close to 11 P.M. I was angry with myself for having caused him to use up so much of his time, but, after his earlier admonition on this score, I had avoided any mention of the hour or the need to conserve his energies. I consoled myself with the thought that even St. Paul recognized there were limits beyond which it was unreasonable to berate a weak conscience.

I looked over at Clara. She was tranquil and content. But she had been scrutinizing me and could read my concern.

"You mustn't worry about the Doctor," she said after he left. "He does exactly what he wants to do. That is one thing no one can take from him. Besides, he feels cut off from the world and relishes the chance to talk to people from the outside.

"We should be proud and happy," she continued. "Everything we wanted to accomplish has been done. The Doctor will do the statement and it will be a good one, a very good one, and he has promised to work on his manuscripts, and I believe this time he will do it.

"You know, for the past week or so I've been watching you closely and I have the feeling that this place had no strangeness for you from the start. I don't know how to explain it except to say that you made me think you had once lived here or a place like it and knew the kind of thing to expect."

She was right about one thing: I did have a special feeling about Lambaréné long before I came. This grew not only out of Schweitzer's own descriptions of life in the jungle, but out of some experiences I had had as a child. For when I read his account of the Hospital, it made a connection with part of my boyhood.

His books had been for me, as they have been for countless others, an intense reading experience. What he wrote had the effect of reawakening incidents and ideas in one's own life. The ideas may be only half-formed in your mind, and the incidents may be part of long-slumbering memories; but they spring to life full-size under the stimulation of his descriptions. When, for example, I came across a passage in Schweitzer's *On the Edge of the Primeval Forest* I had a burst of recognition.

This particular passage was concerned with the effect of illness on Dr. Schweitzer in his early forties. His future seemed uncertain. Two operations were necessary. They were a success and the recovery was complete. After it he could write:

"The misery I have seen gives me strength, and faith in my fellow man supports my confidence in the future. I do hope I shall find a sufficient number of people who, because they themselves have been saved from physical suffering, will respond to those in need."

He had coined a striking phrase: "The Fellowship of those who bear the Mark of Pain." He identified the members of this Fellowship as "those who have learned by experience what physical pain and bodily anguish mean. They belong together, all the world over; they are united by a secret bond."

It was at this point that the magic junction of Schweitzer's ideas with my own experience took place. Years earlier, as a boy of ten, I had been sent to a public sanatorium for tuberculosis. The pain I felt was not one of sickness but of loneliness. It was the pain of being detached from everything warm and meaningful and joyous in life.

There was also the pain of being separated from hope. On Sundays I would find my perch on the wall near the entrance gate to the sanatorium and would watch the healthy people, many of them young married couples, as they walked up the hill from the bus station. They had come to visit relatives and friends but they owned the miracle of being able to leave when they wanted to. I wondered about their world of health and total option and whether they rejoiced in it. I had no reason to believe I might ever be able to join that world. The boys in my age group in the ward seldom spoke about a life beyond the sanatorium. We knew that cures took place but we also knew that although people who left the sanatorium lived a careful, disciplined life, many of them came back.

In those days, antibiotics had not yet been born and tuberculosis sanatoriums existed as much for the purpose of keeping infectious people out of society as for treating them. As a result, the outside world was something of an unreality. We longed for it but it was like pressing our noses against the windowpane of a non-existent tomorrow.

The physical suffering was not severe. There would be some dizziness, a slight fever in the evenings, but we were used to it. Most of the discomfort, I suppose, came during the winter nights. Cold was considered useful therapy at that time and the ward would be converted into open-air shelters at night. Each patient had two blankets, ample for most weather, but there were a dozen or more subzero nights during the winter when it was too cold to sleep and we shivered violently until daybreak.

There was also the pain of human relationships. As in most dependent or disciplined groups, whether in a sanatorium, school, or penitentiary, there was a sharp division between the strong and the weak. Weakness was not necessarily a biological trait but was assigned to the newcomers, who were treated as outcasts. They served the old-timers and took orders; and they marked time until the shine of their newness wore off, when they were relieved of much of their underprivileged status. It had never occurred to me until then that individuals who were in the midst of suffering themselves could be cruel, but I soon realized that when people are thrown together, whether ill or healthy, some of them assert power by being bullies, just as there are some who have positive qualities of leadership. We had our share of both, especially the former.

On occasions, there would be something of a dissolution of the animosities within the sanatorium; indeed, the dissolution would extend to the barriers that separated us from the outside world. One such occasion, of course, was at Christmas, when we put on the Nativity Play for the visitors. Even the bullies became just so many good voices in the choir. The feeling we had then of being able to join the world of the healthy and the warmth of being in their favor was a feeling I have never been able to forget. And when, years later, I sat in the section for the healthy outsiders at Trudi's leper village at the Schweitzer Hospital and watched the leper children perform in the Nativity Play, I could sense their emotions in this hour of glorious connnection with the rest of the world. I could also sense their emotions when, after many rounds of applause, the connection ceased and the outsiders returned to the universe of unlimited options. The feeling the youngsters had, I was certain, was that only a miracle would enable them to become whole again.

The miracle had happened for me at the sanatorium. Six months after being admitted, I was discharged and went back to family and school. It took many years before I could comprehend the fact that I was fully cured. I didn't quite know how to go about establishing my membership in the world of the healthy. I hesitated to ask other boys whether I could play ball with them for fear they would laugh at me. And well they might: on my fourteenth birthday I weighed seventy-eight pounds. But we lived near a public park and I invested my weekly allowance of twenty-five cents in baseball practice. I hired a boy of ten to throw grounders to me and retrieve my throws from the outfield. Within two years—some six inches taller and forty pounds heavier—I became the captain of the neighborhood baseball team, alternating between first base and shortstop. Sometimes, after league games, I would stay on the field, practicing running bases and sliding, so that I would know automatically where my feet should leave the ground without taking my eye off the ball in order to hook my slide away from the bag.

My debt to baseball was great. Even when, at college, I had a head-on collision in attempting to catch a fly ball, resulting in a serious concussion and temporary loss of memory, I almost rejoiced in the fact that this was the sort of thing that sometimes happened to healthy people in the fullest use of their physical resources. What was happening, of course, was that I was enchanted with the idea of being alive. The magic of being able to run swiftly was worth almost any price. I gloried in the fatigue and perspiration that came from muscular activity, for earlier in my life the only perspiration I had known came from fever.

Even after I was able to accept fully the fact that I could live a normal life, I carried with me the feeling that I had the obligation somehow to pay back. The sense of debt was much more than an intellectual one. It lay deep in my bones and I had no way of ignoring it. Indeed, from the moment I walked out of the sanatorium and looked back at my Sunday perch on the old wall near the entrance, I knew that my life would be unbearable unless I could find some way of making good a debt I couldn't quite define but that I knew would be with me as long as I lived.

Albert Schweitzer expressed this feeling for me in *On the Edge of the Primeval Forest* when he wrote that "he who has been delivered from pain must not think he is now free again, and at liberty to take life up as it was before, entirely forgetful of the past. He is now a man whose eyes are open with regard to pain and anguish, and he must help to overcome these two enemies and bring to others the deliverance which he has himself enjoyed."

If I felt "at home" in Lambaréné, as Clara put it, perhaps it was because there had been something in my own life years earlier that enabled me to recognize the chemistry of human emotions that existed in a place like this. I could also recognize what Schweitzer meant when he spoke about the sense of summons he felt. He was not tormented by the tragic question that has stunted the growth of civilizations—the inevitable question asked by the individual: "What can one man do?" If the purpose were strong enough, the question answered itself. The fact that thousands of doctors were required to take care of the millions of people who were in need of them in Africa did not produce in Schweitzer either awe or surrender. A single doctor, he knew, could show what was possible. In the very act of accepting a responsiblity, he could make it visible to others. Besides, to deprive one man of help because many more also needed help was to design the moral paralysis of society.

"A single doctor in Africa," Dr. Schweitzer has said, "even with the most modest equipment, can mean very much for very many. The good which he can accomplish surpasses a hundredfold what he gives of his own life and the cost of material support he must have. Just with quinine and arsenic for malaria, with novarseno-benzol for the various diseases which spread through ulcerating sores, and with emetine for dysentery, and with sufficient skill and apparatus for the most necessary operations, he can in a single year free hundreds of men from the grip of suffering and death."

It may be said that only a Schweitzer has the knowledge and personal power to answer satisfactorily the question: "What can one man do?" Certainly we can't all be Schweitzers. But what should concern us is not what it takes to be a Schweitzer but what it takes to be a man. Nature has not been equally lavish with her endowments, but each man has his own potential in terms of achievement and service. The awareness of that potential is the discovery of purpose; the fulfillment of that potential is the discovery of strength.

For Albert Schweitzer, the assertion of this potential was not a matter of charity but a matter of justice. Also moral reparations. He had always been troubled by the fact that the white man, carrying with him the cross of Jesus, not infrequently also carried the means of cheapening the lives of people he sought to change or dominate.

"Ever since the world's far-off lands were discovered, what has been the relationship of the white people to the colored?" he has asked. "What is the meaning of the simple fact that this and that people have died out, and that the condition of others is getting worse because they were 'discovered' by men who professed to be followers of Jesus? Who can measure the misery produced by the fiery liquids and hideous diseases we have brought to them?

"We are burdened with a great debt. We are not free to confer or not confer benefits on these people as we please. It is our duty. Anything we give them is not benevolence but atonement. That is the foundation from which all deliberations about 'works of mercy' must begin."

And for the appropriate time to act? The time, inevitably, is now. It can only be now. "Truth has no special time of its own." When circumstances seem least propitious, that is the correct time.

These were some of the thoughts I carried to bed with me in the bungalow overlooking the wards on my last night in Lambaréné.

# XIII

The plane made a half circle as it rose from the jungle clearing. Looking out the right side I could see the fast-shrinking figure of Dr. Schweitzer slowly waving his white helmet. In a few seconds the airstrip was obscured by the jungle hills.

I observed my fellow passengers in the plane. Their faces were close to the windows. This was Lambaréné country. It had something to say to the moral imagination; its images and symbols filled the mind. I looked down at the Ogowe River and thought of the thousand-year gap between the pirogue paddled by the lepers and the man-made metal bird that had lifted us out of the jungle and into the clouds.

I thought, too, of the gap between the Hospital at Lambaréné and science in modern dress; but that was not what was important. Nor was it important that the Schweitzer Hospital should be at odds with advanced hospital design and practice. What was important was the timelessness of the Lambaréné message and the enduring nature of the teachings and the commitment and the symbol that belonged to it. The lesson was not new; in fact, it was one of the oldest lessons in the world, but it had yet to be fully understood. The lesson was concerned with the nature of human connection

Everyday is washday on the Ogowe.

One of the transportation centers at the Hospital.

and obligation, with the reality of pain, and the chemistry of human response, and the reach of the moral man.

I thought back to the long discussions I had had before coming to Lambaréné—of Clara's compassionate cautions and her concern lest I leave Lambaréné under a burden of disenchantment and hurt. I thought, too, of the talk with Frank Catchpool in the plane just before arriving at Lambaréné. And now I could dwell on these earlier cautions and anticipations in the full play of retrospect.

The biggest impression of Albert Schweitzer that emerged was of a man who had learned to use himself fully. Much of the ache and brooding unhappiness in modern life is the result of man's difficulty in using himself fully. He performs compartmentalized tasks in a compartmentalized world. He is reined in—physically, socially, spiritually. Only rarely does he have a sense of fulfilling himself through total contact with total challenge. He finds it difficult to make real connection even with those who are near to him. But there are vast yearnings inside, natural ones, demanding air and release. They have to do with his moral responses. And he has his potential, the regions of which are far broader than he can even guess at—a potential that keeps nagging at his inner self for full use. Schweitzer had never been a stranger to his potential.

This is not to say that Schweitzer achieved "happiness" in acting out that potential. He was less concerned with happiness than with purpose. What was it that had to be done? What was the best way of doing it? How did a man go about developing an awareness of important needs? How did he attach himself to those needs? Was he able to recognize the moral summons inside him? To the extent that he lived apart from these questions, he was unfulfilled and not genuinely alive.

A full life, thus defined, however, is not without the punishment of fatigue. Albert Schweitzer was supposed to be severe in his demands on the people who worked with him. Yet any demands he made on others were as nothing compared to the demands he made on himself. He was not concerned about the attainability of perfection; he *was* concerned, however, about the pursuit of perfection. He considered the desire to seek the best and work for the best as a vital part of the nature of man. When he sat down to play the piano or organ, and he was alone, he might stay with it for hours at a time. He might practice a single phrase for two hours or more. The difference between the phrase when he first played it and when he himself was satisfied with it might have been imperceptible even to a trained musical ear. But he had a stern idea of his own capacity for interpreting Bach, for example, and he felt he must stretch himself to whatever extent was necessary to achieve it. This was no mere obsession. He sought his own outermost limits as a natural part of purposeful living. If he seemed to prod and push others, it was an almost automatic carry-over of his own work habits.

There were other thoughts that occurred to me as we flew over the jungle hills of French Equatorial Africa. I considered the matter of Dr. Schweitzer's relationship with the Africans, and the many misconceptions about it that had found their way into print. When the Doctor first came to Lambaréné the life of the African had barely been touched by industrial civilization. It was difficult to get Africans to work steadily in putting up the buildings and in doing hard jobs for the Hospital.

There was the temptation at first to think that the Africans were naturally lazy. But Dr. Schweitzer very early realized that it made a difference when one lives in a climate and in an environment where the needs are few. Living close to nature, the African saw no need to work beyond that which was necessary to the immediate well-being and the minimal needs of his family. The idea of putting up extensive buildings, making concrete piles, sawing and storing woods—all this seemed to have little connection with reality as the African lived it. But the lack of incentive did not mean, as Dr. Schweitzer soon came to realize, that the Africans would not work hard under any circumstances. When they understood the reason for making a special effort, they were more than equal to the challenge.

"Watching them one day as we made an emergency trip in the canoe to save the life of a woman who was seriously ill," Dr.

Schweitzer had said, "I marveled at their stamina and their determination and I resolved never to fall into the careless habit of regarding them as shiftless."

Here we come to the real point about Schweitzer.

It is not whether he was severe in manner toward the Africans, any more than it is whether he failed to bring a gleaming modern hospital to Lambaréné.

The point about Schweitzer is that he brought the kind of spirit to Africa that the dark man hardly knew existed in the white man. Before Schweitzer, white skin meant beatings and gunpoint rule and the imposition of slavery on human flesh. If Schweitzer had done nothing else in his life than to accept the pain of these people as his own, he would have achieved eminence. And his place in history will rest on something more substantial than an argument over an unswept floor in a hospital ward in the heart of Africa. It will rest on the spotless nature of his vision and the clean sweep of his nobility.

The greatness of Schweitzer—indeed the essence of Schweitzer— is the man as symbol. It is not so much what he has done for others, but what others have done because of him and the power of his example. This is the measure of the man. What has come out of his life and thought is the kind of inspiration that can animate a generation. He has supplied a working demonstration of reverence for life. He represents enduring proof that we need not torment ourselves about the nature of human purpose. The scholar, he once wrote, must not live for science alone, nor the businessman for his business, nor the artist for his art. If affirmation for life is genuine, it will "demand from all that they should sacrifice a portion of their own lives for others."

Thus, Schweitzer's main achievement is a simple one. He has been willing to make the ultimate sacrifice for a moral principle. Like Gandhi, the power of his appeal has been in renunciation. And because he has been able to feel a supreme identification with other human beings he has exerted a greater force than millions of armed men on the march.

The Doctor and members of his staff await the arrival of a visitor.

Schweitzer to Stevenson: "That was *my* mosquito."

C. U

It is unimportant whether we call Schweitzer a great religious figure or a great moral figure or a great philosopher. It suffices that his words and works are known and that he is loved and has influence because he enables men to discover mercy in themselves. Early in his life he was accused of being an escapist. He was criticized for seeming to patronize the people he had chosen to serve. Yet the proof of his genuineness and his integrity is to be found in the response he awakens in people. He has reached countless millions who have never seen him but who have been able to identify themselves with him because of the invisible and splendid fact of his own identification with them.

"I must forgive the lies directed against myself," he wrote, "because my own life has been so many times blotted by lies. . . . I am obliged to exercise unlimited forgiveness because, if I did not, I should be untrue to myself in that I should thus act as if I were not guilty in the same way as the other has been guilty with regard to me."

Albert Schweitzer is not above criticism. Few men of our century have come closer to attaining the Greek idea of the whole man—the thinker, the leader, the man of action, the scientist, the artist. But like all great figures in history, he becomes real not despite his frailties but because of them.

Men, like history, come to life in their paradoxes. Gandhi welded a nation of four hundred million people but he couldn't hold his own family together. The cause he defined required a Congress party to fight for it, but Gandhi never gave it the power of his own name. He was the apostle of non-violence in the attainment of national freedom, but once the freedom was won he did not object to the use of military force in the Kashmir.

No man was more effective in defining and working for the liberties of the American people than Thomas Jefferson. He was permeated with the cause of human rights; he saw it in all its aspects—historically, philosophically, spiritually. His great subject in life was the anatomy of freedom. Yet he owned slaves. Like Solon and Pericles centuries before him, he made prodigious contributions to the democratic design of his nation. But, also like the Greek leaders, he did not become passionately involved in the fight against human slavery. All this is now seen in perspective. It is what Jefferson did rather than what he failed to do that inspired his generation and has given him his place in history. Moreover, the principles defined by Jefferson later became the philosophical structure for the victorious fight against slavery.

The American name most associated with the uprooting of slavery, of course, is Abraham Lincoln. Yet only a few days before he became President, Abraham Lincoln said that he did not argue against slavery where slavery existed; he would argue only against its extension to new states. He said he would not eliminate slavery in the South even if he had the power to do so. He appalled those who did not want to compromise on the issue. But when the moral summons was presented by history in its final form, Lincoln accepted magnificently.

The story of Lincoln in his relations with the Negroes would be incomplete if told only in terms of his attitudes during the early days of the Presidency. The inconsistencies and the paradoxes are neither ignored nor set aside by history; they merely yield to the consequential and to the main impact made by the man on the lives of others.

The sublimest paradox of all, of course, is represented by the fact that the most important prophecy of Jesus was proved to be historically false, yet this did not interfere with the establishment of a religion based on the total truth of his mission. Jesus prophesied the imminent end of the world. By imminent he did not mean a matter of several generations; he meant a few years. The fact that this did not eventuate was no obstacle to the creation or the growth of Christianity, based on the divinity and omniscience of Jesus. What was central and what made its impact on the spiritual nature of man were the Godlike qualities of Jesus. His example and moral teachings awakened the natural spiritual responses of people; the rest was subordinated or forgotten.

History is willing to overlook almost anything—errors, paradoxes, personal weaknesses or faults—if only a man will give enough of himself to others. The greater the ability to identify and serve, the more genuine the response. In the case of Schweitzer, later generations will not clutter their minds with petty reflections about his possible faults or inconsistencies. In his life and work will be found energy for moral imagination. This is all that will matter.

Albert Schweitzer will not be immune from attack. There may be a period of carping and intended exposure, much of it with an air of fresh discovery and all of it in a mood of disillusion. But in the long run the inconsistencies and paradoxes will be as nothing alongside the real meaning of Albert Schweitzer and his place in history. For Albert Schweitzer has done more to dramatize the reach of the moral man than anyone in contemporary Western civilization. No one in our times has taught us more about the potentiality of a human being. No one in our times has done more to liberate men of darkened skin. No one in our times has provided more inspiration.

If Albert Schweitzer is a myth, the myth is more important than the reality. For mankind needs such an image in order to exist. People need to believe that man will sacrifice for man, that he is willing to walk the wide earth in the service of man. Long after the Hospital at Lambaréné is forgotten, the symbol of Albert Schweitzer will be known and held high. It would simplify matters if Albert Schweitzer were totally without blemish, if his sense of duty toward all men carried with it an equally high sense of forbearance. But we cannot insist on the morally symmetrical. In the presence of renunciation and dedicated service such as few men are able to achieve, we can at least attempt responsible judgments and we can derive spiritual nourishment from the larger significance of his life as distinct from the fragmented reality.

There is something else we can respect: we can respect the image of Schweitzer that exists in the souls of people. This image gives them strength and purpose; it brings them closer to other people and establishes connections beyond the power of machines and explosives to alter or sever. This is what men most need for today and tomorrow but especially for today. For the making of tomorrow requires most of all a sense of connection beyond reward or compulsion. Also a sense of service that has something to do with reverence and compassion for life. This is more meaningful to man than the things he makes or the conveniences he acquires or the ornamental props of his personal kingdoms. For he reaches his full growth only as he believes in the essential beauty of the human soul. It is this that Albert Schweitzer gives him.

Albert Schweitzer is a spiritual immortal. We can be glad that this is so. Each age has need of its saints. A saint becomes a saint when he is claimed by many men as their own, when he awakens in them a desire to know the best that is in them, and the desire to soar morally.

We live at a time when people seem afraid to be themselves, when they seem to prefer a hard, shiny exterior to the genuineness of deep-felt emotion. Sophistication is prized and sentiment is dreaded. It is made to appear that one of the worst blights on a reputation is to be called a do-gooder. The literature of the day is remarkably devoid of themes on the natural goodness or even the potential goodness of man, seeing no dramatic power in the most powerful fact of the human mixture. The values of the time lean to a phony toughness, casual violence, cheap emotion; yet we are shocked when youngsters confess to having tortured and killed because they enjoyed it. Mercy and respect for life are still basic lessons in the taming of the human animal.

It matters not to Schweitzer or to history that he will be dismissed by some as a do-gooder or as a sentimentalist who frittered his life away on Africans who couldn't read or write. "Anyone who proposes to do good," he wrote, "must not expect people to roll stones out of his way, but must accept his lot calmly if they even roll a few more upon it." For the tragedy of life is not in the hurt to a man's name or even in the fact of death itself. The tragedy of life is in what dies inside a man while he lives—the death of genuine feeling, the death of inspired response, the death of the awareness that makes it possible to feel the pain or the glory of other men in oneself.

Schweitzer's aim was not to dazzle an age but to awaken it, to make it comprehend that moral splendor is part of the gift of life, and that each man has unlimited strength to feel human oneness and to act upon it. He has proved that although a man may have no jurisdiction over the fact of his existence, he can hold supreme command over the meaning of existence for him. Thus, no man need fear death; he need fear only that he may die without having known his greatest power—the power of his free will to give his life for others.

If there is a need in America today, it is for Schweitzers among us. We are swollen with meaningless satisfactions and dulled by petty immediacies—but the threat to this nation and its freedoms and to human life in general has never been greater. To the extent that part of this threat is recognized, it is assumed it can be adequately met by a posture of military and material strength. But the crisis is basically moral and demands moral strength.

We can't save the nation by acting as though only the nation is in jeopardy, nor by acting as though the highest value is the nation. The highest value is the human being and the human potential. In order to safeguard this human potential we have to do more than to surround ourselves with high explosives. We have to make the supreme identification with other people, including those who are different from us or who have less than we. If sacrifice is required, we shall have to sacrifice. If we are to lead, what we say and what we do must become more important in our own minds than what we sell or what we use. At a time when men possess the means for demolishing a planet the only business that makes sense is the business of inspired purpose.

We live in eternal dread of hunger; but we shall never escape the hunger inside us if we are starved for inspiration or are empty of vital purpose. And if we see not at all into these things, the things that make for a single body of all those who now live or who have ever lived, then we shall have lived only half a life. It is in this sense that Albert Schweitzer has helped to make men whole.

We can rejoice in this, for Schweitzer has given an infusion of spiritual energy to our age that is real and that will persist.

Returning home, I felt happy that my two specific purposes in going to Lambaréné had been met. But even more important was the fact that the image of Albert Schweitzer I carried away with me was intact—fortified, if anything, by a direct view. For at Lambaréné I learned that a man does not have to be an angel to be a saint.

Departure.

Horizon time.

# APPENDIX

*The first part of Albert Schweitzer's statement, "Peace or Atomic War?" was completed in Lambaréné early in April 1957. It was released to the world by Radio Oslo on April 24, 1957. Four years earlier, Dr. Schweitzer had gone to Oslo to accept the Nobel Peace Prize, the money from which went into the building of the leper hospital at Lambaréné.*

*Again, in April 1958, Radio Oslo issued a second Schweitzer declaration for world broadcast and publication.*

*The text of the complete statement appears on the following pages.*

# Peace or Atomic War?

## by *Albert Schweitzer*

### PART I

SINCE MARCH 1, 1954, hydrogen bombs have been tested by the United States at the Pacific island of Bikini in the Marshall group and by Soviet Russia in Siberia. We know that testing of atomic weapons is something quite different from testing of non-atomic ones. Earlier, when a new type of giant gun had been tested, the matter ended with the detonation. That is not the case after the explosion of a hydrogen bomb. Something remains in the air, namely, an incalculable number of radioactive particles emitting radioactive rays. This was also true of the uranium bombs dropped on Nagasaki and Hiroshima and those which were subsequently tested. However, not much attention was given to this fact because these bombs were smaller and less effective than the hydrogen bombs.

Since radioactive rays of sufficient amount and strength have harmful effects on the human body, it must be considered whether the radiation resulting from the hydrogen explosions that have already taken place represents a danger which would increase with new explosions.

In the course of the three-and-a-half years that have passed since then [the test explosions of the early hydrogen bombs] representatives of the physical and medical sciences have been studying the problem. Observations on the distribution, origin, and nature of radiation have been made. The processes through which the human body is harmfully affected have been analyzed. The material collected, although far from complete, allows us to draw the conclusion that radiation resulting from the explosions which have already taken place represents a danger to the human race—a danger not to be underrated—and that further explosions of atomic bombs will increase this danger to an alarming extent.

Although this conclusion has repeatedly been expressed, especially during the last few months, it has not, strange to say, influenced public opinion to the extent that one might have expected. Individuals and peoples have not been aroused to give to this danger the attention it unfortunately deserves. It must be demonstrated and made clear to them.

I raise my voice, together with those of others who have lately felt it their duty to act, through speaking and writing, in warning of the danger. My age and the generous understanding so many people have shown of my work permit me to hope that my appeal may contribute to preparing the way for the insights so urgently needed.

My thanks go to the radio station in Oslo, the city of the Nobel Peace Prize, for making it possible for what I feel I have to say to reach far-off places.

What is radioactivity?

Radioactivity consists of rays differing from those of light in being invisible and able to pass not only through glass but also through thin metal discs and layers of cell tissue in the human and animal bodies. Rays of this kind were first discovered in 1895 by the physicist Wilhelm Roentgen of Munich, and named after him.

In 1896 the French physicist Henri Becquerel demonstrated that rays of this kind occur in nature. They are emitted from uranium, an element known since 1786.

In 1898 Pierre Curie and his wife discovered in the mineral pitchblende, a uranium ore, the strongly radioactive element radium.

The joy that such rays were at the disposal of humanity was at first unmixed. For the rays appeared to influence the relatively fast growing and decaying cells of malignant tumors and sarcomas. If exposed to these rays repeatedly for a longer period, some of the terrible neoplasms could be destroyed.

After a time it was found, however, that the destruction of cancer cells does not always mean the cure of cancer, and that the normal cells of the body may be seriously damaged if long exposed to radio-activity.

When Mme. Curie, after handling uranium ore for four years, finally held the first gram of radium in her hand, there appeared abrasions in the skin which no treatment could cure. With the years she grew steadily sicker from a disease caused by radioactive rays which damaged her bone marrow and through this her blood. In 1934 death ended her suffering.

Even so, for many years we were not aware of the grave risks in-volved in X-rays to those constantly exposed to them. Through oper-ating X-ray apparatus thousands of doctors and nurses have incurred incurable diseases.

Radioactive rays are material things. Through them the radioactive element constantly and forcefully emits tiny particles of itself. These are of three kinds, named after the three first letters of the Greek alpha-bet: *alpha, beta, gamma.* The gamma rays are the hardest and have the strongest effect.

The reason why elements emit radioactive rays is that they are constantly decaying, and radioactivity is the energy they liberate little by little. There are other elements besides uranium and radium which are radioactive. To the radiation from the elements in the earth is added some radiation from space. Fortunately, the air mass 250 miles high that surrounds our earth protects us against this radiation. Only a very small fraction of it reaches us.

We are, then, constantly being exposed to radioactive radiation coming from the earth and from space. It is so weak, however, that it does not hurt us. Stronger sources of radiation, as for instance X-ray machines and exposed radium, have, as we know, harmful effects if one is exposed to them for some time.

Radioactive rays are, as I said, invisible. How can we tell that they are there and how strong they are?

Thanks to the German physicist Hans Geiger, who died in 1945 as a victim to X-rays, we have an instrument which makes that possible. This instrument, called the Geiger counter, consists of a metal tube containing rarefied air. In it are two metal electrodes between which there is a high potential. Radioactive rays from the outside affect the tube and release a discharge between the two electrodes. The stronger the radiation, the quicker the discharges follow one another. A small device connected to the tube makes the discharge audible. The Geiger counter performs a veritable drum-roll when the discharges are strong.

There are two kinds of atom bomb—uranium bombs and hydrogen bombs. The effect of a uranium bomb is due to a process which liberates energy through the fission of uranium. In the hydrogen bomb the liberation of energy is the result of the transformation of hydrogen into helium.

It is interesting to note that this latter process is similar to that which takes place in the center of the sun, supplying it with the self-renewing energy which it emits in the form of light and heat.

In principle, the effect of both bombs is the same. But according to various estimates the effect of one of the latest hydrogen bombs is 2,000 times stronger than the one dropped on Hiroshima.

To these two bombs has recently been added the cobalt bomb, a kind of super atom bomb. It is a hydrogen bomb surrounded by a layer of cobalt. Its effect is estimated to be many times stronger than that of any hydrogen bomb made so far.

The explosion of an atom bomb creates an inconceivably large number of exceedingly small particles of radioactive elements which decay like uranium or radium. Some of these particles decay very quickly, others more slowly, and some with extraordinary slowness. The strongest of these elements cease to exist only ten seconds after the detonation of the bomb. But in this short time they may have killed a great number of people in a circumference of several miles.

What remains are the less powerful elements. In our time it is with these we have to contend. It is of the danger arising from the radio-active rays emitted by these elements that we must be aware.

Of these elements some exist for hours, some for weeks, months, years, or even millions of years, undergoing continuous decay. They float in the higher strata of air as clouds of radioactive dust. The heavy particles fall down first. The lighter ones will stay in the air for a longer time or come down with rain or snow. How long it will take until everything carried up in the air by past explosions has disappeared no one can say with certainty. According to some estimates, this will not happen sooner than thirty or forty years from now.

As a boy I witnessed how dust hurled into the air from the explosion in 1883 of the island Krakatoa in the Sunda group was so noticeable for two years afterwards that sunsets were given extraordinary splendor by it.

What we can state with certainty, however, is that the radioactive clouds will constantly be carried by the winds around the globe and that some of the dust, by its own weight, or by being brought down little by little, by rain, snow, mist, and dew, will fall down on the hard surface of the earth, and into rivers and oceans.

Of what nature are these radioactive elements, particles of which, carried up in the air by the explosion of atom bombs, are now falling down again?

They are strange variants of the usual non-radioactive elements, having the same chemical properties, but a different atomic weight. Their names are always accompanied by their atomic weights. The same element can occur in several radioactive variants. Besides Iodine 131, which lasts for only sixteen days, we have Iodine 129, which lasts for 200,000,000 years.

Dangerous elements of this kind are Phosphorus 32, Calcium 45, Iodine 131, Iron 55, Bismuth 210, Plutonium 239, Cerium 144, Strontium 89, Caesium 137. If the hydrogen bomb is covered with cobalt, Cobalt 60 must be added to the list.

Particularly dangerous are the elements combining long life with a relatively strong, efficient radiation. Among them Strontium 90 takes the first place. It is present in very large amounts in radioactive dust. Cobalt 60 must also be mentioned as particularly dangerous.

The radioactivity in the air, increased through these elements, will not harm us from the outside, not being strong enough to penetrate the skin. It is another matter with respiration, through which radioactive elements can enter our bodies. But the danger which has to be stressed above all the others is the one which arises from our drinking radioactive water and eating radioactive food as a consequence of the increased radioactivity in the air.

Since the explosions of Bikini and Siberia, rain falling over Japan has, from time to time, been so radioactive that the water from it cannot be drunk. Not only that: reports of radioactive rainfall are coming from all parts of the world where analyses have recently been made. In several places the water has proved to be so radioactive that it was unfit for drinking.

Well-water becomes radioactive to any considerable extent only after longer periods of heavy rainfall.

Wherever radioactive rainwater is found, the soil is also radioactive —and in a higher degree. The soil is made radioactive not only by the downpour, but also from radioactive dust falling on it. And with the soil the vegetation will also have become radioactive, because radioactive elements deposited in the soil pass into the plants, where they are stored. This is of importance, for as a result of this process we may be threatened by a considerable amount of radioactive elements. Radioactive elements in grass, when eaten by animals whose meat is used for food, will be absorbed and stored in our bodies. Or we may absorb them by drinking milk from cows grazing on contaminated soil. In that way small children run an especially dangerous risk of absorbing radioactive elements. When we eat contaminated cheese and fruits the radioactive elements stored in them are transferred to us.

What this storing of radioactive material implies is clearly demonstrated by the observations made when the radioactivity of the Columbia River in North America was analyzed. The radioactivity was caused by the atomic plants at Hanford, which produce plutonium for atomic bombs and which empty their waste water into the river. The radioactivity of the river water was insignificant. But the radioactivity of the river plankton was 2,000 times higher, that of the ducks eating plankton 40,000 times higher, that of the fish 15,000 times higher. In young swallows fed on insects caught by their parents in the river the radioactivity was 500,000 times higher, and in the egg yolks of water birds more than 1,000,000 times higher.

From official and unofficial sources we have been assured, time and time again, that the increase in radioactivity of the air does not exceed the amount the human body can tolerate without any harmful effects. This is just evading the issue. Even if we are not directly affected by the radioactive material in the air, we are indirectly affected through that which has fallen down, is falling down, and will fall down. We are absorbing this through radioactive drinking water and through animal and vegetable foodstuffs, to the same extent as radioactive elements are stored in the vegetation of the region in which we live. Unfortunately for us, nature hoards what is falling down from the air.

None of the radioactivity of the air created by the explosion of atom bombs is so unimportant that it may not, in the long run, become a danger to us through increasing the amount of radioactivity stored in our bodies.

What we absorb of radioactivity is not spread evenly in all cellular tissue. It is deposited in certain parts of our body, particularly in the bone tissue and also in the spleen and in the liver. From those sources the organs which are especially sensitive to it are exposed to radiation. What the radiation lacks in strength is compensated for by time. It works day and night without interruption.

How does radiation affect the cells of an organ? Through being ionized, that is to say, electrically charged. This change means that the chemical processes which make it possible for the cells to do their job in our body no longer function as they should. They are no longer able to perform the tasks which are of vital importance to us. We must also bear in mind that a great number of the cells of an organ may degenerate or die as a result of radiation.

What are the diseases caused by internal radiation? The same diseases that are known to be caused by external radiation.

They are mainly serious blood diseases. The cells of the red bone marrow, where the red and the white blood corpuscles are formed, are very sensitive to radioactive rays. It is these corpuscles, found in great numbers in the blood, which make it possible for it to play such an important part. If the cells in the bone marrow are damaged by radiation they will produce too few or abnormal, degenerating blood corpuscles. Both cases lead to blood diseases and, frequently, to death. These were the diseases that killed the victims of X-rays and radium rays.

It was one of these diseases that attacked the Japanese fishermen who were surprised in their vessel by radioactive ashes falling down 240 miles from Bikini after the explosion of a hydrogen bomb. Being strong and only mildly affected, all but one were saved by continued blood transfusions.

In the cases cited the radiation came from the outside. It is unfortunately very probable that internal radiation affecting the bone marrow and lasting for years will have the same effect, particularly since the radiation goes from the bone tissue to the bone marrow. As I have said, the radioactive elements tend to be stored in the bone tissue.

Internal radiation threatens not only our own health but also that of our descendants. The cells of the reproductive organs are particularly vulnerable to radiation, which attacks the nucleus to such an extent that it can be seen in the microscope. Profound damage to these cells results in corresponding damage to our descendants, such as stillbirths and the births of babies with mental or physical defects.

In this context also, we can point to the effects of radiation coming from the outside. It is a fact—even if the statistical material being published in the press needs checking—that in Nagasaki, during the years following the dropping of the atom bomb, an exceptionally high occurrence of stillbirths and of deformed children was observed.

In order to establish the effect of radioactive radiation on posterity, comparative studies have been made between the descendants of doctors who have been using X-ray apparatus over a period of years and descendants of doctors who have not. The material of this study comprises about 3,000 doctors in each group. A noticeable difference was found. The descendants of radiologists showed a percentage of stillbirths of 14.03, while the percentage among the non-radiologists was 12.22. In the first group 6.01 per cent of the children had congenital defects, while only 4.82 per cent in the second. The number of healthy children in the first group was 80.42 per cent; in the other it was significantly higher, viz. 83.23 per cent.

It must be remembered that even the weakest of internal radiation can have harmful effects on our descendants. The total effect of the damage done to descendants of ancestors who have been exposed to radioactive rays will not, in accordance with the laws of genetics, be apparent in the generations coming immediately after us. The full effects will appear only 100 or 200 years later.

As the matter stands we cannot at present cite cases of serious damage done by internal radiation. To the extent that such radiation exists it is not sufficiently strong and has not lasted long enough to have caused the damage in question. We can only conclude from the harmful effects known to be caused by external radiation what we must expect in the future from internal radiation.

If the effect of the latter is not as strong as that of the former, it may become so through working little by little and without interruption. The final result will be the same in both cases. Their effects add up.

We must also remember that internal radiation, unlike that from outside, does not have to penetrate layers of skin, tissues, and muscles to hit the organs. It works at close range and without any weakening of its force.

When we realize under what conditions internal radiation is working, we cease to underrate it. Even if it is true that, when speaking of the dangers of internal radiation, we can point to no actual case and only express our fear, that fear is so solidly founded on facts that it attains the weight of reality in determining our attitude. We are forced to regard every increase in the existing danger through further creation of radioactive elements by atom bomb explosions as a catastrophe for the human race, a catastrophe that must be prevented.

There can be no question of doing anything else, if only for the reason that we cannot take the responsibility for the consequences it might have for our descendants. They are threatened by the greatest and most terrible danger.

That radioactive elements created by us are found in nature is an astounding event in the history of the earth and of the human race. To fail to consider its importance and its consequences would be a folly for which humanity would have to pay a terrible price. We are committing a folly in thoughtlessness. We must not fail to pull ourselves together before it is too late. We must muster the insight, the seriousness, and the courage to leave folly and to face reality.

This is at bottom what the statesmen of the nations producing atomic bombs are thinking, too. Through the reports they are receiving they are sufficiently informed to arrive at their own judgments, and we must also assume that they are alive to their responsibility.

At any rate, America and Soviet Russia and Britain are telling one another again and again that they want nothing more than to reach an agreement to end the testing of atomic weapons. At the same time, however, they declare that they cannot stop the tests so long as there is no such agreement.

Why do they not come to an agreement? The real reason is that in their own countries there is no public opinion asking for it. Nor is there any such public opinion in other countries, with the exception of Japan. This opinion has been forced upon the Japanese people because, little by little, they will be hit in a most terrible way by the evil consequences of all the tests.

An agreement of this kind presupposes reliability and trust. There must be guarantees preventing the agreement from being signed by anyone intending to win important tactical advantages foreseen only by him. Public opinion in all nations concerned must inspire and accept the agreement.

When public opinion has been created in the countries concerned and among all nations, an opinion informed of the dangers involved in going on with the tests and led by the reason which this information imposes, then the statesmen may reach an agreement to stop the experiments.

A public opinion of this kind stands in no need of plebiscites or committees to express itself. It works through just being there.

The end of further experiments with atom bombs would be like the early sunrays of hope which suffering humanity is longing for.

# PART II

IN APRIL of last year I, together with others, raised my voice to draw attention to the great danger of radioactive poisoning of the air and the earth, following tests with atomic and hydrogen bombs. With others, I appealed to the nuclear powers to come to a workable agreement to stop the tests as soon as possible and declare their genuine desire to renounce the use of nuclear weapons.

At that time there appeared to be reasonable hope that this step would be taken. It was not. The negotiations in London last summer achieved nothing. The conference arranged by the United Nations in the autumn of last year suffered the same fate when the Soviet Union withdrew from the discussions.

The question of nuclear arms control, however, cannot be put aside. Any discussions among the major nations will have to consider this problem.

Cessation of nuclear tests has often been proposed as the first step in any comprehensive and workable plan for arms control.

One might have thought it would be comparatively simple for all those involved to agree on this first step. No nuclear power would have to sacrifice any of the atomic weapons in its possession. The disadvantage of not being able to try out new bombs or nuclear devices would be the same for all.

237

The United States and Great Britain have been reluctant to take the first step. They spoke against it when the matter was discussed in the spring of 1957. Since then many statements have been issued claiming that radioactivity resulting from nuclear tests is not dangerous. For example, in an official statement coming from the United States, we read the following: "The necessary steps should be taken to correct the present confusion of the general public [with respect to the effects of testing]. . . . The present and potential effects on heredity from the gradual increase of radioactivity in the air are kept within tolerable limits. . . . The possibility of harmful effects which people believe to be outside control has a strong emotional impact. . . . The continuation of nuclear tests is necessary and justified in the interests of national security."

Despite these assurances, however, people are becoming increasingly apprehensive concerning the possible dangers resulting from nuclear tests.

The reasoning behind the somewhat obscure statement that "the effects on heredity from the gradual increase of radioactivity in the air are kept within tolerable limits" is that the number of deformed children that will be born as a result of the harm done to the sexual cells supposedly will not be large enough to justify stopping the tests.

During this campaign of reassurance, a prominent American nuclear physicist even declared that the luminous watchdials in the world represent a greater danger than the radioactive fall-out of nuclear tests thus far.

This campaign of reassurance sets up anticipations of glad tidings that science has succeeded in making the prototype of a hydrogen bomb with a considerably less dangerous radioactive fall-out. The new explosive is called a "clean" hydrogen bomb. The old type is being designated as the "dirty" bomb.

The so-called "clean" hydrogen bomb differs from the other in having a jacket made of a material which does not release immense quantities of radioactive elements at the enormous explosion temperature. That is why it is less harmful, as regards radioactivity, than the usual ones.

However, the new highly praised hydrogen bomb is—let it be said in passing—only relatively clean. Its trigger is a uranium bomb made of the fissionable Uranium-235—an atomic bomb as powerful as the one dropped over Hiroshima. This bomb, when detonated, also produces radioactivity, as do the neutrons released in great numbers at the explosion.

Earlier this year, in an American newspaper, Edward Teller, the father of the "dirty" hydrogen bomb, sings a hymn of praise to the idyllic nuclear war to be waged with completely clean hydrogen bombs. He insists on a continuation of the tests in order to perfect this ideal bomb.

Here are two stanzas from Edward Teller's hymn to idyllic nuclear warfare:

"Further tests will put us into a position to fight our opponents' war machine, while sparing the innocent bystanders."

"Clean weapons of this kind will reduce unnecessary casualties in a future war."

The idea of limited nuclear war is a contradiction in terms. Each side will use all the power at its disposal in an attempt to annihilate the enemy. The U.S. Department of Defense has quite recently declared that the irradiation of whole areas has become a new offensive weapon.

The "clean" hydrogen bomb may be intended, I fear, more for display-case purposes than for use. The intention seems to be to convince people that new nuclear tests will be followed by less and less radiation and that there is no real argument for the discontinuation of the tests.

Those who think that the danger created by nuclear tests is small mainly take the air radiation into consideration, and persuade themselves to believe that the danger limit has not yet been reached.

The results of their arithmetic are not so reliable, however, as they would have us believe. Through the years the toleration limit for radiation has had to be lowered several times. In 1934 it was 100 radiation units per year. At present the limit is officially put at 5. In many countries it is even lower. Dr. Lauriston Taylor (U.S.A.), who is regarded as an authority on protection against radiation, holds—like others—that it is an open question whether there is such a thing as a harmless amount of radiation. He thinks that we can speak only of an amount of radiation which we regard as tolerable.

We are constantly being told about a "maximum permissible

amount" of radiation. What does "permissible" mean? And who has the right to "permit" people to be exposed to these dangers?

When speaking about the risk of radiation we must consider not only the radiation coming from the outside, but also the radioactivity that gets into our bodies.

What is the source of this radioactivity?

The radioactive materials put into the air by nuclear tests do not stay there permanently. In the form of radioactive rain—or even radioactive snow—they fall to the earth. They enter the plants through leaves and roots and stay there. We absorb them by drinking milk from cows or by eating the meat of animals which have fed on it. Radioactive rain contaminates our drinking water.

The most powerful radioactive poisoning occurs in the areas between the Northern latitudes 10° and 60°, because of the numerous nuclear tests conducted mainly in these latitudes by the Soviet Union and the United States.

The radioactive elements absorbed over the years by the body are not evenly distributed in the cellular tissue, but deposited and accumulated at certain points. From these points internal radiation takes place, causing injuries to particularly vulnerable organs. What this kind of radiation lacks in strength is made up for by its longevity, working as it does for years, day and night.

It is a well-known fact that one of the most widespread and dangerous elements absorbed by us is Strontium 90. It is stored in the bones and from there emits its rays into cells of red bone marrow, where the red and white corpuscles are made. If the radiation is too great, blood diseases—fatal in most cases—are the result.

The cells of the reproductive organs are particularly sensitive. Even relatively weak radiation may lead to fatal consequences.

The most sinister aspect of internal as well as external radiation is that years may pass before the evil consequences appear. Indeed, they make themselves felt, not in the first or second generation, but in those that follow. Generation after generation, for centuries to come, will witness the birth of an ever-increasing number of children with mental and physical defects.

It is not for the physicist, choosing to take into account only the radiation from the air, to utter the final word on the dangers of nuclear tests. That right belongs to the biologists and physicians who have studied internal as well as external radiation, and to those scientists who pay attention to the facts established by the biologists and physicians.

The declaration signed by 9,235 scientists of all nations, handed to the Secretary General of the U.N. by Dr. Linus Pauling on January 13, 1958, gave the campaign of reassurance a serious blow. The scientists declared that the radioactivity gradually created by nuclear tests represents a grave danger for all parts of the world, particularly serious because its consequences will be an increasing number of deformed children in the future. For this reason they insist on an international agreement putting an end to the nuclear tests.

The declaration signed by the 9,235 scientists did well in stressing the danger of the harmful effects of nuclear tests on future generations resulting, according to biologists and physicians, from the radiation to which we are now being exposed.

We must not disregard our responsibility to guard against the possibility that thousands of childen may be born with the most serious mental and physical defects. It will be no excuse for us to say later that we were unaware of that possibility. Only those who have never been present at the birth of a deformed baby, never witnessed the whimpering cries of its mother, should dare to maintain that the risk of nuclear testing is small. The well-known French biologist and geneticist Jean Rostand calls the continuation of nuclear tests "a crime into the future" (*le crime dans l'avenir*). It is the particular duty of women to prevent this sin against the future. It is for them to raise their voices against it in such a way that they will be heard.

No longer can we take any comfort from the fact that the scientists do not agree on the danger of radiation, nor that we must await the decision of international bodies before making positive statements about radiation. Despite all the claims of safety, the truth about the danger of nuclear explosions marches imperturbably along, influencing an ever-increasing section of public opinion. In the long run, even the most well-organized propaganda can do nothing against the truth.

It is a strange fact that few people have taken into consideration that the question of nuclear testing is not one which concerns the nuclear powers exclusively, a question for them to decide at their pleasure. Who has given these countries the right to experiment, in times of peace, with weapons involving the most serious risks for the whole world? What has international law—enthroned by the United Nations and so highly praised in our time—to say on this matter? Does it no longer look out on the world from its temple? Then let it go out, so that it may face the facts and do its duty accordingly.

International law should consider at once the compelling case of Japan. That country has suffered heavily from the effects of nuclear tests. The radioactive clouds created by the Soviet tests in Northeast Siberia and by the American tests in the Pacific Ocean are carried by the winds over Japan. The resultant radioactive poisoning is considerable. Powerful radioactive rainfalls are quite common. The radioactive poisoning of the soil and the vegetation is so heavy that the inhabitants of some districts ought to abstain from using their harvest for food. People are eating rice contaminated by radioactive strontium, a substance particularly dangerous for children. The ocean surrounding Japan is also at times dangerously radioactive, and thereby the very food supply of the country—in which fish has always played an important part—is being threatened.

As every new nuclear test makes a bad situation worse, the Japanese ministers, having heard of plans for new tests to the north or south of Japan, have presented their country's urgent appeal in Washington or Moscow, beseeching the American or Soviet authorities to give up their plans.

We generally learn about these appeals and the refusals through short newspaper items. Unfortunately, there have been few responsible editorials drawing our attention to the stories behind the news—the misery of human beings who are now in jeopardy. In that respect we and the press are guilty of a lack of compassion. Even guiltier, however, is international law, which has kept silent and indifferent on this question, year after year.

It is high time to recognize that the question of nuclear testing is a matter for world law to consider. Mankind is imperiled by the tests. Mankind insists that they stop, and has every right to do so.

If anything is left of international law in our civilization, then the nations responsible for nuclear tests must renounce them immediately, without making this renunciation dependent on agreements concerning the larger questions of general disarmament. Nuclear tests have nothing to do with disarmament. The nations in question will continue to have the weapons they now have.

There is no time to lose. New tests must not be allowed to increase the already existing danger. It is important to realize that even without new tests the danger will increase during the coming years: a large part of the radioactive elements flung up in the atmosphere and stratosphere at the nuclear experiment is still there. It will come down only after several years—probably about fifteen.

The immediate renunciation of further tests will create a favorable atmosphere for talk on controlling the stockpiles of nuclear weapons and banning their use. When this urgently necessary step has been taken, such negotiations can take place in peace.

That the Soviet Union has announced its willingness to stop its tests is of great importance. The world now looks to the United States and Great Britain for the kind of moral initiative and action that goes along with great leadership.

\*    \*    \*

Today we are faced with the menacing possibility of an outbreak of atomic war between Soviet Russia and the United States. It can only be averted if the two powers decide to renounce atomic arms.

How did this situation arise?

In 1945 America succeeded in producing an atomic bomb with Uranium 235. On August 6, 1945, this bomb was dropped on Hiroshima. Another atomic bomb was dropped on Nagasaki on August 9.

When America came into the possession of such a bomb it held a military advantage over other countries.

In July 1949 the Soviet Union also test-exploded its first nuclear bomb. Its power was approximately equal to the American bomb then existing.

On October 3, 1952, England exploded its first atomic bomb on the Isle of Montebello (situated on the northwest coast of Australia).

In the quest for nuclear supremacy, both the Soviet Union and the United States moved toward the development of a nuclear weapon many times more powerful—the hydrogen bomb. A series of tests was undertaken by the United States in the Marshall Islands beginning in May, 1951, and culminating in a successfully exploded hydrogen bomb in March 1954.

The actual power of the explosion was far stronger than had been originally calculated.

At approximately the same time, the Soviet Union also started its experimentations, exploding its first hydrogen bomb on August 12, 1953.

Today, guided missiles can be launched from their starting points and directed with accuracy at distant targets. The larger explosives are carried by missiles containing the fuel necessary for their propulsion. The gases from this fuel rush with tremendous velocity through a narrow opening. Science is in the process of discovering a fuel which is similar and more efficacious to deal with.

It is said that the Soviet Union already has available rockets with a range up to 600 miles. Soon to come are rockets with a range up to 1,080 miles—if they are not already in use.

It is said that America is attempting to develop rockets with a range of 1,440 miles.

Whether the intercontinental ballistic missile, with its range of 4,800 miles, already exists cannot be ascertained. The Soviet Union has claimed it already has such a missile.

Even without respect to intercontinental ballistic missiles, submarines could launch nuclear attacks on the United States.

The long-range rockets attain unbelievable speed. It is expected that an intercontinental rocket would not take more than twenty minutes to cross the ocean with a payload of nuclear explosive weighing from one to five tons.

How could an atomic war break out today? Not long ago there was talk of local or limited wars that could be contained. But today there is little difference between a local war or a global war. Rocket missiles will be used up to a range of 1,440 miles. The destruction should not be underestimated, even if caused only by a Hiroshima-type bomb.

It can hardly be expected that an enemy will refrain from using atomic bombs or the most devastating hydrogen bombs on large cities at the very outset of a war. One hydrogen bomb now exists that is a thousand times more powerful than the atomic bomb. It will have a destructive radius of many miles. The heat will be 100 million degrees. One can imagine how large would be the number of city-dwellers who would be destroyed by the pressure of the explosion, by flying fragments of glass, by heat and fire and by radio-active waves, even if the attack is only of short duration. The deadly radioactive contamination, as a consequence of the explosion, would have a range of some 45,000 square miles.

An American general has said to some Congressmen: "If at an interval of ten minutes 110 hydrogen bombs are dropped over the U.S.A. there would be a casualty list of about 70 million people; besides, some thousands of square miles would be made useless for a whole generation. Countries like England, West Germany, and France could be finished off with fifteen to twenty hydrogen bombs."

President Eisenhower has pointed out, after watching maneuvers under atomic attack, that defense measures in a future atomic war become useless. In these circumstances all one can do is to pray.

Indeed, not much more can be done in view of an attack by hydrogen bombs than to advise all people living to hide beneath a very strong wall made of stone or cement, and to throw themselves on the ground and to cover the back of their heads, and the body if possible, with cloth. In this way it may be possible to escape anni-hilation and death through radiation. It is very important that the immediate survivors are given non-radioactive food and drink, and that they be removed immediately from the radioactive district.

It is impossible, however, to erect walls and concrete ceilings of adequate thickness to cover an entire city. Where would the material and the means come from? How would a population have time to run to safety in such bunkers?

In an atomic war there would be neither conqueror nor vanquished. During such a bombardment both sides would suffer the same fate. A continuous destruction would take place and no armistice or peace proposals could bring it to an end.

When people deal with atomic weapons, it is not a matter of su-perior arms which will decide the issue between them, but only: "Now we want to commit suicide together, destroying each other mutually . . ."

There is a reason for an English M.P. saying: "He who uses atomic weapons becomes subject to the fate of a bee; namely, when it stings it will perish."

Radioactive clouds resulting from a war between East and West would imperil humanity everywhere. There would be no need to use up the remaining stock of atomic and hydrogen bombs, now running literally into the thousands.

A nuclear war is therefore the most senseless and lunatic act that could ever take place. This must be prevented.

When America had its atomic monopoly, it was not necessary to equip its allies with nuclear weapons. Owing to the end of the monopoly, however, this situation is changing. A whole family of nuclear weapons now exists that can be fitted into the military capability of smaller nations.

As a result, the United States is considering a departure from its stated principle not to put atomic weapons into the hands of other countries. If it does so, this could have the gravest consequences. On the other hand, it is comprehensible that the United States wishes to supply the NATO countries with such new weapons for defense against the Soviet Union. The existence of such arms constitutes a new cause of war between the Soviet Union and the U.S., one that did not exist before. Thus, the ground is laid open for nuclear conflict on European soil. The Soviet Union can be reached with long-range rockets from European soil, as far as Moscow and Kharkov, up to 2,400 miles away. Similarly, London, Paris, and Rome are within easy reach of Soviet rocketry.

Rockets of an average range may be used for defense purposes by Turkey and Iran against the Soviet Union. They could penetrate deeply into its country with arms accepted from America.

The Soviet Union is countering those measures. Both America and the Soviet Union may now seek alliances with the Middle East by offering those countries various kinds of financial support. Therefore, events in the Middle East could endanger the peace of the world.

The danger of an atomic war is being increased by the fact that no warning would be given in starting such a war. Indeed, it could erupt merely on the basis of some incident. Thus, the time factor enters—the side that attacks first would have the initial advantage over the attacked. At the very start, the attacked would find himself sustaining losses which would considerably reduce his fighting capacity.

As a result, one has to be on the alert all the time. This factor constitutes an extreme danger in the event of a sudden outbreak of an atomic war. When one has to act with such speed, he has to reckon with the possibility that an error may occur on what is registered on the radar screen, and that this could result in the outbreak of an atomic war.

Attention was drawn to this danger by the American General Curtis LeMay. Quite recently the world found itself in such a situation. The radar station of the American Air Force and American Coastal Command indicated that an invasion of unidentified bombers was on the way. Upon this warning, the general in command of the strategic bomb force ordered that reprisal bombardment should be made. However, realizing that he was taking a great responsibility, he hesitated. Shortly afterward it was pointed out that the radar stations had committed a technical error. What could have happened if a less balanced general had been in his place!

In the future such dangers are likely to increase. Because small rockets exist which pass through the air with terrific speed and are over the target within a few minutes, defense possibilities become very limited. Only seconds remain to identify the markings on the radar screen, so that the counter-attack can spring into being. The theoretical defense consists in sending out missiles to explode the attacking missiles of the enemy before they complete their job, and also in releasing bombers with a view to destroying the ramps from which they are launched.

Such split-second operations cannot be left to the human brain. It works too slowly. The job has therefore been entrusted to an electronic brain.

Such are the heights of our civilization that a cold electronic brain rather than the moral conscience of man may decide human destiny. Are we so certain that an arithmetical or mechanical decision is really superior? The mechanism of the electronic brain may become faulty. It is dependent on the absolute reliability of its complicated functions. Everything has to click to the minutest detail.

Under the circumstances, the greater the number of countries, large or small, that become part of the nuclear arms terror the greater the terror. Naturally, America must assume that the weapons it entrusts to other nations will not be used irresponsibly. But accidents can happen. *Who* can guarantee that there may not be a "blacksheep" acting on his own, without troubling about the consequences? Who is *able* to keep *all* countries under a situation of rational control? The dam is punctured and it may break down.

That such worries have become very real is shown by the reasoning of the 9,235 scientists who on January 13 petitioned the United Nations to cease atomic tests. The statement says: "As long as atomic weapons remain in the hands of the three great powers, agreement on control is possible. However, if the tests continue and extend to other countries in possession of atomic weapons, the risks and responsibilities in regard to an outbreak of an atomic war become all the greater. From every point of view the danger in a future atomic war becomes all the more intense, so that an urgent renunciation of atomic weapons becomes absolutely imperative."

America has wisely declared that its objective is to outlaw nuclear weapons. Yet at the same time America seems to be moving away from the measures necessary to achieve it. America insists that the missiles it offers to other countries be accepted as quickly as possible. It wishes to hold such a position as to be able to maintain peace by nuclear deterrent. It happens, however, that most of the NATO countries are in no hurry to acquire such weapons because of an increasingly strengthening public opinion.

In recent months public opinion in Europe has been convinced that under no circumstances should Europe be allowed to become a battlefield for an atomic war between the Soviet Union and America. From this position it will not deviate. The time is past when a European power could plan secretly to establish itself as a big power by manufacturing atomic weapons exclusively for its own use. In view of the fact that no public opinion would agree to such an undertaking, it becomes senseless even to prepare secretly for achieving such a plan.

Gone, too, is the time when NATO generals and European governments can decide on the establishment of launching sites and stockpiling of atomic weapons. In view of the fact that the dangers of atomic war and its consequences cannot be avoided, political procedure as employed hitherto can no longer be considered.

Only agreements that are sanctioned by public opinion are now valid.

* * *

What about the negotiations that could lead to the renunciation of nuclear weapons?

One reads and hears that the success of the projected Summit Conference must depend entirely on its every detail being diplomaically prepared beforehand. The best diplomacy is objectivity. One good way of preparing for a conference (if a respectful and well-meaning criticism is permissible) would be for the statesmen and other representatives to make a change from their present undiplomatic way of dealing with each other and to become diplomatic. Many unnecessary, thoughtless, discourteous, foolish, and offensive remarks have been made by both sides, and this has not been advantageous to the political atmosphere.

It would be fitting if those who have the authority to take the responsibility, and not those who have only nominal authority and who cannot move an inch from their instructions, would confer together.

It would be fitting to go ahead with the conference. For more than five months East and West have talked and written to one another, without any conclusions as to the date and the work program being reached. Public opinion everywhere is finding it difficult to accept this state of affairs and is beginning to ask itself whether a conference which comes into being so limpingly has any hope of really achieving anything.

It would be fitting to hold the conference in a town in some neutral European country, for example, Geneva, as was the case in 1955.

It would be fitting that at this conference only questions that have to do directly with the control and renunciation of nuclear weapons should be discussed.

It would be fitting if not too many people were present at the summit meeting. Only the highest personalities of the three nuclear powers together with their experts and advisers should take their seats there.

Attendance could also be opened on a consultative basis to the representatives of those peoples who—like the NATO countries with

America—have connections in nuclear matters; they could then state their opinions on the decisions that hold such grave consequences also for them.

Apart from this, experience teaches us that unnecessarily large attendance brings no advantage to a conference.

The Summit Conference, therefore, is in no way an international or half international one, even though its decisions are of great importance to the whole of mankind.

The three nuclear powers and they alone must decide, in awareness of their repsonsibility to their peoples and to all mankind, whether or not they will renounce the testing and the use of nuclear weapons.

As for planning the conference, impartiality may justify one remark, which is that to date such planning has not been done objectively, and has therefore led nowhere. This leads to the thought that the outcome of a Summit Conference is bound to reflect what went into it.

What is the difference between the partial and the impartial, the fitting and the unfitting in this matter? It lies in the answer to the question on what basis the three nuclear powers decide whether or not to renounce the testing and the use of nuclear weapons.

The unobjective reply would be that the decision will depend on whether an agreement is first reached on comprehensive disarmament or not.

This is a false logic; it presumes that there could be an agreement acceptable to both the East and the West on this issue. But previous negotiations have shown that this is not to be expected; they became stalled right at the start because East and West have been unable to reach agreement even on the conditions under which such discussions should take place.

The anticipated procedure itself is by its very nature not impartial. It is based on false logic. The two vital issues so essential to the very existence of mankind—the cessation of tests and the disposal of nuclear weapons—cannot be made dependent on the Heavens performing the impossible political miracle that alone could insure that none of the three nuclear powers would have any objections to a complete agreement on disarmament.

The fact is that the testing and use of nuclear weapons carry in themselves the absolute reasons for their being renounced. Prior agreement on any other conditions cannot be considered.

Both cause the deepest damage to human rights. The tests do harm to peoples far from the territories of the nuclear powers—endangering their lives and their health—and this in peace time. An atomic war, with its resultant radioactivity, would make the land of peoples not participating in such a war, unlivable. It would be the most unimaginably senseless and cruel way of endangering the existence of mankind. That is why it must not be allowed to happen.

The three nuclear powers owe it to themselves and to mankind to reach agreement on these absolute essentials without first dealing with prior conditions.

The negotiations about disarmament are therefore not the forerunner of such agreement but the outcome of it. They start from the point where agreement on the nuclear issues has been reached, and their goal is to reach the point where the three nuclear powers and the peoples connected with them must agree on guarantees that will seek to avert the danger of a threat of a non-atomic nature taking the place of the previous danger. Everything that the diplomats will have done objectively to prepare the preliminaries to the conference will keep its meaning even if used, not before renunciation, but as the result of it.

Should agreement be reached on the outlawing of nuclear weapons, this by itself will lead to a great improvement in the political situation. As a result of such an agreement, time and distance would again become realities with their own right.

Nuclear weapons, used in conjunction with missiles, change a distant war to a war fought at close range. The Soviet Union and the United States have become next-door neighbors in the modern world but live in constant fear of their lives every minute.

But if nuclear arms should be abolished, the proximity factor would be made less explosive.

Today America has her batteries of nuclear missiles readily available in Europe. Europe has become a connecting land strip between America and Russia, as if the Atlantic had disappeared and the continents had been joined.

But if atomic missiles are outlawed on the basis of effective and enforceable control, this unnatural state of affairs would come to an end. America would again become wholly America; Europe wholly Europe; the Atlantic again wholly the Atlantic Ocean.

The great sacrifices that America brought to Europe during the Second World War and in the years following it will not be forgotten. The many-sided and great help that Europe received from her and the thanks owing for this will not be forgotten.

But the unnatural situation created by the two world wars, which led to a dominating military presence in Europe, cannot continue . indefinitely. It must gradually cease to exist—both for the sake of Europe and for the sake of America.

Now there will be shocked voices from all sides: What will become of poor Europe if American atomic weapons no longer defend it from within and from without? Will Europe be delivered to the Soviet? Must it be prepared to languish in a Communist-Babylonian imprisonment for long years?

What Europe and the Europeans have to agree about is that they belong together for better or for worse. This is a new historical fact that can no longer be by-passed politically.

Another factor that must be recognized politically is that it is no longer a question of subjugating peoples, but learning to get along with them intellectually, culturally, spiritually.

A Europe standing on its own has no reason to despair.

Disarmament discussions between the three nuclear powers must seek the guarantees that can bring about actual, total, and durable disposal of nuclear weapons. The question of control and safeguards is a vital one. Reciprocal agreement will have to be reached about allowing international commissions to inspect and investigate on national soil.

One talks of giving aircraft belonging to a world police the right to fly at medium and high altitudes for purposes of aerial inspection.

One asks to what extent a state would be willing to subject itself to such control? It may be said that unfortunate incidents could easily occur as a result. And what about the power that should be entrusted to such a world control? Even the widest form of such control could never insure that everywhere and all the time war could be avoided. But it represents a reasonable basis on which, given time and some relaxation of tension, a workable world system of security might be built.

The same applies also in another matter. As a result of renouncing nuclear arms, the Soviet Union's military might insofar as Europe is concerned would be less affected than that of America. There would remain to the Soviet the many armed divisions with conventional weapons; with those divisions it could easily over-run the NATO states in western Europe—particularly Western Germany—without its being possible for anyone to come to their aid. With this in mind, the Soviet Union should agree in the course of disarmament negotiations to reduce her army, and to commit herself not to undertake steps against Germany. But here, too, no manner of detailed agreements and internationally guaranteed disarmament agreements would be enough. Therefore, we must strive continually to improve the situation, building brick by brick.

We live at a time when the good faith of peoples is doubted more than ever before. Expressions casting doubt on the trustworthiness of the next nation are bandied back and forth. They are based on what happened in the two world wars when the nations experienced dishonesty, injustice, and inhumanity from one another. How can a new trust come about?

We cannot continue in a situation of paralyzing mistrust. If we want to work our way out of the desperate situation in which we find ourselves another spirit must enter into the people. It can come only if the awareness of its necessity suffices to give us strength to believe in its coming. We must presuppose the awareness of this need in all the peoples who have suffered along with us. We must approach them in the spirit that we are human beings, all of us, and that we feel ourselves fitted to feel with each other; to think and to will together in the same way.

The awareness that we are all human beings together has become lost in war and politics. We have reached the point of regarding each other as only members of a people allied with us or against us, and our attitudes, prejudices, sympathies, or antipathies are all conditioned by that fact. Now we must rediscover the fact that we—all together—are human beings, and that we must strive to concede to each other what moral capacity we have.

In that way we can begin to believe that in other peoples too there will arise the need for a new spirit; and that can be the beginning of a feeling of mutual trustworthiness toward each other. The spirit is a mighty force for transforming things. Let us have hope that the spirit can bring people and lands back to an awareness of enlightenment.

At this stage we have the choice of two risks. The one consists in continuing the mad atomic arms race with its danger of unavoidable atomic war in the near future. The other is in the renunciation of nuclear weapons, and the hope that America and the Soviet Union, and the peoples associated with them, will manage to live in peace. The first holds no hope of a prosperous future; the second does. We must risk the second.

In President Eisenhower's speech of November 7, 1957, we find the following: "What the world needs more than a gigantic leap into space is a gigantic leap into peace."

This gigantic leap consists in finding the courage to hope that the spirit of good sense will arise in all peoples and in all lands, a spirit sufficiently strong to overcome the insanity and the inhumanity.

Once agreement on renunciation of nuclear arms has been reached, it would be the responsibility of the United Nations to undertake to see that now, as in the future, they would neither be made nor used. The danger that one or another people might attempt to manufacture nuclear weapons will have to be kept in mind for a long time.

The future holds many difficult problems. The most difficult of these will be the rights of access of over-populated countries to neighboring lands.

But if in our time we renounce nuclear arms, we will have taken the first step on the way to the distant goal of the end to war itself. If we do not do this we remain on the road that leads to atomic war and misery in the near future.

Those who are to meet at the summit must be aware of this, so that they can negotiate with propriety, with the right degree of seriousness, and with a full sense of responsibility.

The Summit Conference must not fail. The will of mankind will not permit it.